Springer Series in Reliability Engineering

Series Editor

Professor Hoang Pham
Department of Industrial Engineering
Rutgers
The State University of New Jersey
96 Frelinghuysen Road
Piscataway, NJ 08854-8018
USA

Other titles in this series

Universal Generating Function in Reliability Analysis and Optimization
Gregory Levitin

Warranty Management and Product Manufacture
D.N.P. Murthy and Wallace R. Blischke

System Software Reliability
H. Pham

Toshio Nakagawa

Maintenance Theory of Reliability

With 27 Figures

 Springer

Professor Toshio Nakagawa
Aichi Institute of Technology, 1247 Yachigusa, Yaguasa-cho,
Toyota 470-0392, Japan

British Library Cataloguing in Publication Data
Nakagawa, Toshio
 Maintenance theory of reliability. — (Springer series in
 reliability engineering)
 1. Maintainability (Engineering)   2. Reliability (Engineering)
 3. Maintenance
 I. Title
 620'.0045

Library of Congress Cataloging-in-Publication Data
Nakagawa, Toshio, 1942–
 Maintenance theory of reliability/Toshio Nakagawa.
     p. cm.
 Includes bibliographical references and index.

    1. Reliability (Engineering)   I. Title.
 TA169.N354 2005
 620'.00452—dc22   2005042766

Apart from any fair dealing for the purposes of research or private study, or criticism or review, as permitted under the Copyright, Designs and Patents Act 1988, this publication may only be reproduced, stored or transmitted, in any form or by any means, with the prior permission in writing of the publishers, or in the case of reprographic reproduction in accordance with the terms of licences issued by the Copyright Licensing Agency. Enquiries concerning reproduction outside those terms should be sent to the publishers.

Springer Series in Reliability Engineering series ISSN 1614-7839

ISBN: 978-1-84996-966-6
Springer Science+Business Media
springeronline.com

© Springer-Verlag London Limited 2005
Softcover reprint of the hardcover 1st edition 2005

The use of registered names, trademarks, etc., in this publication does not imply, even in the absence of a specific statement, that such names are exempt from the relevant laws and regulations and therefore free for general use.

The publisher makes no representation, express or implied, with regard to the accuracy of the information contained in this book and cannot accept any legal responsibility or liability for any errors or omissions that may be made.

Typesetting: Output-ready by the author

9 8 7 6 5 4 3 2 1   Printed on acid-free paper

Preface

Many serious accidents have happened in the world where systems have been large-scale and complex, and have caused heavy damage and a social sense of instability. Furthermore, advanced nations have almost finished public infrastructure and rushed into a maintenance period. Maintenance will be more important than production, manufacture, and construction, that is, *more maintenance* for environmental considerations and for the protection of natural resources. From now on, the importance of maintenance will increase more and more. In the past four decades, valuable contributions to maintenance policies in reliability theory have been made. This book is intended to summarize the research results studied mainly by the author in the past three decades.

The book deals primarily with standard to advanced problems of maintenance policies for system reliability models. System reliability can be mainly improved by repair and preventive maintenance, and replacement, and reliability properties can be investigated by using stochastic process techniques. The optimum maintenance policies for systems that minimize or maximize appropriate objective functions under suitable conditions are discussed both analytically and practically.

The book is composed of nine chapters. Chapter 1 is devoted to an introduction to reliability theory, and briefly reviews stochastic processes needed for reliability and maintenance theory. Chapter 2 summarizes the results of repair maintenance, which is the most basic maintenance in reliability. The repair maintenance of systems such as the one-unit system and multiple-unit redundant systems is treated. Chapters 3 through 5 summarize the results of three typical maintenance policies of age, periodic, and block replacements. Optimum policies of three replacements are discussed, and their several modified and extended models are proposed. Chapter 6 is devoted to optimum preventive maintenance policies for one-unit and two-unit systems, and the useful modified preventive policy is also proposed. Chapter 7 summarizes the results of imperfect maintenance models. Chapter 8 is devoted to optimum inspection policies. Several variant inspection models with approximate inspection policies, inspection policies for a standby unit, a storage system and intermittent faults, and finite inspection models are proposed. Chapter 9 presents five maintenance models such as discrete replacement and inspection models, finite replacement models, random maintenance models, and replacement models with spares at continuous and discrete times.

This book gives a detailed introduction to maintenance policies and provides the current status and further studies of these fields, emphasizing mathematical formulation and optimization techniques. It will be helpful for reliability engineers and managers engaged in maintenance work. Furthermore, sufficient references leading to further studies are cited at the end of each chapter. This book will serve as a textbook and reference book for graduate students and researchers in reliability and maintenance.

I wish to thank Professor Shunji Osaki, Professor Kazumi Yasui and all members of the Nagoya Computer and Reliability Research Group for their cooperation and valuable discussions. I wish to express my special thanks to Professor Fumio Ohi and Dr. Bibhas Chandra Giri for their careful reviews of this book, and Dr. Satoshi Mizutani for his support in writing this book. Finally, I would like to express my sincere appreciation to Professor Hoang Pham, Rutgers University, and editor Anthony Doyle, Springer-Verlag, London, for providing the opportunity for me to write this book.

Toyota, Japan *Toshio Nakagawa*
June 2005

Contents

1 Introduction 1
   1.1 Reliability Measures 4
   1.2 Typical Failure Distributions 13
   1.3 Stochastic Processes 19
      1.3.1 Renewal Process 20
      1.3.2 Alternating Renewal Process 24
      1.3.3 Markov Processes 26
      1.3.4 Markov Renewal Process with Nonregeneration Points . 30
   References 35

2 Repair Maintenance 39
   2.1 One-Unit System 40
      2.1.1 Reliability Quantities 40
      2.1.2 Repair Limit Policy 51
   2.2 Standby System with Spare Units 55
      2.2.1 Reliability Quantities 56
      2.2.2 Optimization Problems 59
   2.3 Other Redundant Systems 62
      2.3.1 Standby Redundant System 63
      2.3.2 Parallel Redundant System 65
   References 66

3 Age Replacement 69
   3.1 Replacement Policy 70
   3.2 Other Age Replacement Models 76
   3.3 Continuous and Discrete Replacement 83
   References 92

4 Periodic Replacement 95
   4.1 Definition of Minimal Repair 96
   4.2 Periodic Replacement with Minimal Repair 101

	4.3	Periodic Replacement with Nth Failure 104
	4.4	Modified Replacement Models 107
	4.5	Replacements with Two Different Types 110
		References 114

5 Block Replacement 117
    5.1 Replacement Policy 117
    5.2 No Replacement at Failure 120
    5.3 Replacement with Two Variables 121
    5.4 Combined Replacement Models 125
        5.4.1 Summary of Periodic Replacement 125
        5.4.2 Combined Replacement 126
    References 132

6 Preventive Maintenance 135
    6.1 One-Unit System with Repair 136
        6.1.1 Reliability Quantities 136
        6.1.2 Optimum Policies 139
        6.1.3 Interval Reliability 140
    6.2 Two-Unit System with Repair 144
        6.2.1 Reliability Quantities 145
        6.2.2 Optimum Policies 150
    6.3 Modified Discrete Preventive Maintenance Policies 154
        6.3.1 Number of Failures 155
        6.3.2 Number of Faults 160
        6.3.3 Other PM Models 165
    References 167

7 Imperfect Preventive Maintenance 171
    7.1 Imperfect Maintenance Policy 173
    7.2 Preventive Maintenance with Minimal Repair 175
    7.3 Inspection with Preventive Maintenance 182
        7.3.1 Imperfect Inspection 183
        7.3.2 Other Inspection Models 185
        7.3.3 Imperfect Inspection with Human Error 187
    7.4 Computer System with Imperfect Maintenance 188
    7.5 Sequential Imperfect Preventive Maintenance 191
    References 197

8 Inspection Policies 201
    8.1 Standard Inspection Policy 202
    8.2 Asymptotic Inspection Schedules 207
    8.3 Inspection for a Standby Unit 212
    8.4 Inspection for a Storage System 216
    8.5 Intermittent Faults 220

	8.6 Inspection for a Finite Interval 224
	References 229
9	**Modified Maintenance Models** 235
	9.1 Modified Discrete Models 236
	9.2 Maintenance Policies for a Finite Interval 241
	9.3 Random Maintenance Policies 245
	9.3.1 Random Replacement 246
	9.3.2 Random Inspection 253
	9.4 Replacement Maximizing MTTF 258
	9.5 Discrete Replacement Maximizing MTTF 261
	9.6 Other Maintenance Policies 263
	References 264
Index	 267

1
Introduction

Reliability theory has grown out of the valuable experiences from many defects of military systems in World War II and with the development of modern technology. For the purpose of making good products with high quality and designing highly reliable systems, the importance of reliability has been increasing greatly with the innovation of recent technology. The theory has been actually applied to not only industrial, mechanical, and electronic engineering but also to computer, information, and communication engineering. Many researchers have investigated statistically and stochastically complex phenomena of real systems to improve their reliability.

Recently, many serious accidents have happened in the world where systems have been large-scale and complex, and they not only caused heavy damage and a social sense of instability, but also brought an unrecoverable bad influence on the living environment. These are said to have occurred from various sources of equipment deterioration and maintenance reduction due to a policy of industrial rationalization and personnel cuts.

Anyone may worry that big earthquakes in the near future might happen in Japan and might destroy large old plants such as chemical and power plants, and as a result, inflict serious damage to large areas.

Most industries at present restrain themselves from making investments in new plants and try to run current plants safely and efficiently as long as possible. Furthermore, advanced nations have almost finished public infrastructure and will now rush into a maintenance period [1]. From now on, *maintenance* will be more important than redundancy, production, and construction in reliability theory, *i.e.*, *more maintenance than redundancy* and *more maintenance than production*. Maintenance policies for industrial systems and public infrastructure should be properly and quickly established according to their occasions. From these viewpoints, reliability researchers, engineers, and managers have to learn maintenance theory simply and throughly, and apply them to real systems to carry out more timely maintenance.

The book considers systems that perform some mission and consist of several units, where unit means item, component, part, device, subsystem, equipment, circuit, material, structure, or machine. Such systems cover a very wide class from simple parts to large-scale space systems. System reliability can be evaluated by unit reliability and system configuration, and can be improved by adopting some appropriate maintenance policies. In particular, the following three policies are generally used.

(1) Repair of failed units
(2) Provision of redundant units
(3) Maintenance of units before failure

The first policy is called *corrective maintenance* and adopted in the case where units can be repaired and their failures do not adversely affect a whole system. If units fail then they may begin to be repaired immediately or may be scrapped. After the repair completion, units can operate again.

The second policy is adopted in the case where system reliability can be improved by providing redundant and spare units. In particular, standby and parallel systems are well known and used in practice.

Maintenance of units after failure may be costly, and sometimes requires a long time to effect corrective maintenance of the failed units. The most important problem is to determine when and how to maintain preventively units before failure. However, it is not wise to maintain units with unnecessary frequency. From this viewpoint, the commonly considered maintenance policies are *preventive replacement* for units without repair and *preventive maintenance* for units with repair on a specific schedule. Consequently, the object of maintenance optimization problems is to determine the frequency and timing of corrective maintenance, preventive replacement, and/or preventive maintenance according to costs and effects.

Units under age replacement and preventive maintenance are replaced or repaired at failure, or at a planned time after installation, whichever occurs first. Units under periodic and block replacements are replaced at periodic times, and undergo repair or replacement of failure between planned replacements. It is assumed throughout Chapters 3 to 6 that units after any maintenance become as good as new; *i.e., maintenance is perfect*, unless otherwise stated. But, units after maintenance in Chapter 7 might be younger, however, they do not become new; *i.e., maintenance is imperfect*. In either case, it may be wise to carry out some maintenance of operating units to prevent failures when the failure rate increases with age.

In the above discussions, we have concentrated on the behavior of operating units. Another point of interests is that of failed units undergoing repair. We obtain in Chapter 2 reliability quantities of repairable units such as mean time to failure, availability, and expected number of failures. If the repair of a failed unit takes a long time, it may be better to replace it than to repair it. This policy is achieved by stopping the repair if it is not completed within a specified time, and by replacing a failed unit with a new one. This policy is called a *repair limit policy*, and is a striking contrast to the preventive maintenance policy.

We need to check units such as standby and storage units whose failures can be detected only through inspection, which is called *inspection policy*. For example, consider the case where a standby unit may fail. It may be catastrophic and dangerous that a standby unit has failed when an original unit fails. To avoid such a situation, we should check a standby unit to see whether it is good. If the failure is detected then the maintenance suitable for the unit should be done immediately.

Most systems in offices and industry are successively executing jobs and computer processes. For such systems, it would be impractical to do maintenance on them at planned times. Random replacement and inspection policies, in which units are replaced and checked, respectively, at random times, are proposed in Chapter 9.

For systems with redundant or spare units, we have to determine how many units should be provided initially. It would not be advantageous to hold too many units in order to improve reliability, or to hold too few units in order to reduce costs. As one technique of determining the number of units, we may compute an optimum number of units that minimize the expected cost, or the minimum number such that the probability of failure is less than a specified value. If the total cost is given, we may compute the maximum number of units within a limited cost. Furthermore, we are interested in an optimization problem: when to replace units with spare ones in order to lengthen the time to failure.

Failures occur in several different types of failure modes such as wear, fatigue, fracture, crack, breaking, corrosion, erosion, instability, and so on. *Failure* is classified into *intermittent failure* and *extended failure* [2, 3]. Furthermore, extended failure is divided into *complete failure* and *partial failure*, both of which are classified into *sudden failure* and *gradual failure*. Extended failure is also divided into *catastrophic failure* which is both sudden and complete, and *degraded failure* which is both partial and gradual.

In such failure studies, the time to failure is mostly observed on *operating time* or *calendar time*, however, it is often measured by the number of cycles to failure and combined scales. A good time scale of failure maintenance models was discussed in [4, 5]. Furthermore, alternative time scales for cars with random usage were defined and investigated in [6]. In other cases, the lifetimes are sometimes not recorded at the exact instant of failure and are collected statistically at discrete times. Rather some units may be maintained preventively in their idle times, and intermittently used systems maintained after a certain number of uses. In any case, it would be interesting and possibly useful to solve optimization problems with discrete times.

It is supposed that the planning time horizon for most units is infinite. In this case, as the measures of reliability, we adopt the mean time to failure, the availability, and the expected cost per unit of time. It is appropriate to adopt as objective functions the expected cost from the viewpoint of economics, the availability from overall efficiency, and the mean time to failure from reliability. Practically, the working time of units may be finite. The total expected cost until maintenance is adopted for a finite time interval as an objective function, and optimum policy that minimizes it is discussed by using the partition method derived in Chapter 8.

The known results of maintenance and associated optimization problems were summarized in [7–11]. Since then, many papers have been published and reviewed in [12–19]. The recently published books [20–25] collected many reliability and maintenance models, discussed their optimum policies, and applied them to actual systems.

Most of the contents of this book are our original work based on the book of Barlow and Proschan: reliability measures, failure distributions, and stochastic processes needed for learning reliability theory are summarized briefly in Chapter 1. These results are introduced without detailed explanations and proofs. However, several examples are given to help us to understand them easily.

Some fundamental repair models in reliability theory are analyzed in Chapter 2, and useful reliability quantities of such repairable systems are analytically obtained, using the techniques in Chapter 1. Several replacement policies are contained systematically from elementary knowledge to advanced studies in Chapters 3 through 5. Several preventive maintenance and imperfect policies are introduced and analyzed in Chapters 6 and 7. The results and methods presented in Chapters 3 through 7 can be applied practically to real systems by modifying and extending them according to circumstances. Moreover, they might include scholarly research materials for further studies. The most important thing in reliability engineering is when to check units suitably and how to seek fitting maintenance for them. Many inspection models based on the results of Barlow and Proschan are summarized in Chapter 8, and would be useful for us to plan maintenance schemes and to carry them into execution. Finally, several modified maintenance models are surveyed in Chapter 9, and give further topics of research.

1.1 Reliability Measures

We are interested in certain quantities for analyzing reliability and maintenance models. The first problem is that of how long a unit can operate without failure, *i.e.*, *reliability*, which is defined as the probability that it will perform a required function under stated conditions for a stated period of time [26]. *Failure* might be defined in many ways, and usually means mechanical breakdown, deterioration beyond a threshold level, appearance of certain defects in system performance, or decrease in system performance below a critical level [4]. *Failure rate* is a good measure for representing the operating characteristics of a unit that tends to frequency as it ages. When units are replaced upon failure or are preventively maintained, we are greatly concerned with the ratio at which units can operate, *i.e.*, *availability*, which is defined as the probability that it will be able to operate within the tolerances at a given instant of time [27].

This section defines reliability function, failure rate, and availability, and obtains their properties needed for solving future optimization problems of maintenance policies, which are treated in the sequel chapters.

(1) Reliability

Suppose that a nonnegative random variable X ($X \geq 0$) which denotes the failure time of a unit, has a cumulative probability distribution $F(t) \equiv \Pr\{X \leq t\}$ with right continuous, and a probability density function $f(t)$ ($0 \leq t < \infty$); i.e., $f(t) = \mathrm{d}F(t)/\mathrm{d}t$ and $F(t) = \int_0^t f(u)\mathrm{d}u$. They are called *failure time distribution* and *failure density function* in reliability theory, and are sometimes called simply a failure distribution $F(t)$ and a density function $f(t)$.

The survival distribution of X is

$$R(t) \equiv \Pr\{X > t\} = 1 - F(t) = \int_t^\infty f(u)\,\mathrm{d}u \equiv \overline{F}(t) \quad (1.1)$$

which is called the *reliability function*, and its mean is

$$\mu \equiv E\{X\} = \int_0^\infty t f(t)\,\mathrm{d}t = \int_0^\infty R(t)\,\mathrm{d}t \quad (1.2)$$

if it exists, which is called *MTTF (mean time to failure)* or *mean lifetime*. It is usually assumed throughout this book that $0 < \mu < \infty$, $F(0-) = F(0+) = 0$, and $F(\infty) = \lim_{t \to \infty} F(t) = 1$; i.e., $R(0) = 1$ and $R(\infty) = 0$, unless otherwise stated. Note that $F(t)$ is nondecreasing from 0 to 1 and $R(t)$ is nonincreasing from 1 to 0.

(2) Failure Rate

The notion of aging, which describes how a unit improves or deteriorates with its age, plays a role in reliability theory [28]. Aging is usually measured based on the term of a failure rate function. That is, *failure rate* is the most important quantity in maintenance theory, and important in many different fields, *e.g.*, statistics, social sciences, biomedical sciences, and finance [29–31]. It is known by different names such as *hazard rate*, *risk rate*, *force of mortality*, and so on [32]. In particular, Cox's proportional hazard model is well known in the fields of biomedical statistics and default risk [33, 34]. The existing literature on this model was reviewed in [35].

We define *instant failure rate function* $h(t)$ as

$$h(t) \equiv \frac{f(t)}{\overline{F}(t)} = -\frac{1}{\overline{F}(t)}\frac{\mathrm{d}\overline{F}(t)}{\mathrm{d}t} \quad \text{for } F(t) < 1 \quad (1.3)$$

which is called simply the *failure rate* or *hazard rate*. This means physically that $h(t)\Delta t \approx \Pr\{t < X \le t+\Delta t | X > t\}$ represents the probability that a unit with age t will fail in an interval $(t, t+\Delta t]$ for small $\Delta t > 0$. This is generally drawn as a bathtub curve. Recently, the reversed failure rate is defined by $f(t)/F(t)$ for $F(t) > 0$, where $f(t)\Delta t/F(t)$ represents the probability of a failure in an interval $(t-\Delta t, t]$ given that it has occurred in $(0, t]$ [36, 37].

Furthermore, $H(t) \equiv \int_0^t h(u)du$ is a *cumulative hazard function*, and has the relation

$$R(t) = \exp\left[-\int_0^t h(u)\,du\right] = e^{-H(t)}; \quad i.e., \quad H(t) = -\log R(t). \quad (1.4)$$

Thus, $F(t)$, $R(t)$, $f(t)$, $h(t)$, and $H(t)$ determine one another. In addition, because $e^a \ge 1+a$, we have the inequalities

$$\frac{H(t)}{1+H(t)} \le F(t) \le H(t) \le \frac{F(t)}{1-F(t)}$$

which would give good inequalities for small $t > 0$.

In particular, a random variable $Y \equiv H(X)$ has the following distribution

$$\Pr\{Y \le t\} = \Pr\{H(X) \le t\} = \Pr\{X \le H^{-1}(t)\} = 1 - e^{-t},$$

where H^{-1} is the inverse function of H. Thus, Y has an exponential distribution with mean 1, and $E\{H(X)\} = 1$. Moreover, x_1 which satisfies $H(x_1) = 1$ is called *characteristic life* in the probability paper of a Weibull distribution. This represents the mean lifetime that about 63.2% of units have failed until time x_1. Moreover, $H(t)$ is called the *mean value function* and has a close relation to nonhomogeneous Poisson processes in Section 1.3. In this process, x_k which satisfies $H(x_k) = k$ $(k = 1, 2, \dots)$ represents the time that the expected number of failures is k when failures occur at a nonhomogeneous Poisson process. The property of $H(t)/t$, which represents the expected number of failures per unit of time, was investigated in [38].

We denote the following failure rates of a continuous failure distribution $F(t)$ and compare them [39, 40].

(1) Instant failure rate $h(t) \equiv f(t)/\overline{F}(t)$.
(2) Interval failure rate $h(t;x) \equiv \int_t^{t+x} h(u)\,du/x = \log[\overline{F}(t)/\overline{F}(t+x)]/x$ for $x > 0$.
(3) Failure rate $\lambda(t;x) \equiv [F(t+x) - F(t)]/\overline{F}(t)$ for $x > 0$.
(4) Average failure rate $\Lambda(t;x) \equiv [F(t+x) - F(t)]/\int_t^{t+x} \overline{F}(u)du$ for $x > 0$.

Definition 1.1. A distribution F is IFR (DFR) if and only if $\lambda(t;x)$ is increasing (decreasing) in t for any given $x > 0$ [7], where IFR (DFR) means *Increasing Failure Rate (Decreasing Failure Rate)*.

By this definition, we investigate the properties of failure rates.

Theorem 1.1.

(i) If one of the failure rates is increasing (decreasing) in t then the others are increasing (decreasing), and if F is exponential, i.e., $F(t) = 1 - e^{-\lambda t}$, then all failure rates are constant in t, and $h(t) = h(t;x) = \Lambda(t;x) = \lambda$.
(ii) If F is IFR then $\Lambda(t-x;x) \leq h(t) \leq \Lambda(t;x) \leq h(t+x)$, where $\Lambda(t-x;x) = \Lambda(0;t)$ for $x > t$.
(iii) If F is IFR then $\Lambda(t;x) \leq h(t;x)$.
(iv) $h(t;x) \geq \lambda(t;x)/x$ and $\Lambda(t;x) \geq \lambda(t;x)/x$.
(v) $h(t) = \lim_{x \to 0} h(t;x) = \lim_{x \to 0} \lambda(t;x)/x = \lim_{x \to 0} \Lambda(t;x)$.

Proof. The property (v) easily follows from the definition of $h(t)$. Hence, we can prove property (i) if we show that $h(t)$ is increasing (decreasing) in t implies $h(t;x)$, $\lambda(t;x)$, and $\Lambda(t;x)$ all are increasing (decreasing) in t. For example, for $t_1 \leq t_2$,

$$\frac{\overline{F}(t_1+x)}{\overline{F}(t_1)} = \exp\left[-\int_{t_1}^{t_1+x} h(u)\,du\right] \geq (\leq) \exp\left[-\int_{t_2}^{t_2+x} h(u)\,du\right] = \frac{\overline{F}(t_2+x)}{\overline{F}(t_2)}$$

implies that $\lambda(t;x)$ is increasing (decreasing) if $h(t)$ is increasing (decreasing). Similarly, we can prove the other properties.

Suppose that F is IFR. Because

$$\frac{f(v)}{\overline{F}(v)} \leq h(t) \leq \frac{f(u)}{\overline{F}(u)} \leq h(t+x) \qquad \text{for } v \leq t \leq u \leq t+x$$

we easily have property (ii).

Furthermore, letting

$$Q(x) \equiv \int_t^{t+x} h(u)\,du \int_t^{t+x} \overline{F}(u)\,du - x[F(t+x) - F(t)]$$

we have $Q(0) = 0$, and

$$\frac{dQ(x)}{dx} = \int_t^{t+x} [h(t+x) - h(u)][F(t+x) - F(u)]\,du \geq 0$$

because both $h(t)$ and $F(t)$ are increasing in t. This proves property (iii).

Finally, from the property that $\overline{F}(t)$ is decreasing in t, we have

$$\int_t^{t+x} \overline{F}(u)\,du \leq x\overline{F}(t), \qquad \int_t^{t+x} \frac{f(u)}{\overline{F}(u)}\,du \geq \frac{1}{\overline{F}(t)}\int_t^{t+x} f(u)\,du$$

which imply property (iv). All inequalities in results (ii) and (iii) are reversed when F is DFR. ∎

Hereafter, we may call the four failure rates simply the *failure rate* or *hazard rate*. Furthermore, properties of failure rates have been investigated in [8, 28, 41].

Example 1.1. Consider a unit such as a scale and production system that is maintained preventively only at time T ($0 \leq T \leq \infty$). It is supposed that an operating unit has some earnings per unit of time and does not have any earnings during the time interval if it fails before time T. The average time during $(0, T]$ in which we have some earnings is

$$l_0(T) = 0 \times F(T) + T\overline{F}(T) = T\overline{F}(T)$$

and $l_0(0) = l_0(\infty) = 0$. Differentiating $l_0(T)$ with respect to T and setting it equal to zero, we have

$$\overline{F}(T) - Tf(T) = 0; \quad i.e., \quad h(T) = \frac{1}{T}.$$

Thus, an optimum time T_0 that maximizes $l_0(T)$ is given by a unique solution of equation $h(T) = 1/T$ when F is IFR. For example, when $F(t) = 1 - e^{-\lambda t}$, $T_0 = 1/\lambda$; i.e., we should do the preventive maintenance at the interval of mean failure time.

Next, consider a unit with one spare where the first operating unit is replaced before failure at time T ($0 \leq T \leq \infty$) with the spare one which will be operating to failure. Suppose that both units have the identical failure distribution $F(t)$ with finite mean μ. Then, the mean time to either failure of the first or spare unit is

$$l_1(T) = \int_0^T t\,dF(t) + \overline{F}(T)(T + \mu) = \int_0^T \overline{F}(t)\,dt + \overline{F}(T)\mu$$

and $l_1(0) = l_1(\infty) = \mu$, and

$$\frac{dl_1(T)}{dT} = \overline{F}(T)[1 - \mu h(T)].$$

Thus, an optimum time T_1 that maximizes $l_1(T)$ when $h(t)$ is strictly increasing is given uniquely by a solution of equation $h(T) = 1/\mu$. When the failure rate of parts and machines is statistically estimated, T_0 and T_1 would be a simple barometer for doing their maintenance. ∎

A generalized model with n spare units is discussed in Section 9.4. A probability method of provisioning spare parts and several models for forecasting spare requirements and integrating logistics support were provided and discussed in [42, 43].

Example 1.2. Suppose that X denotes the failure time of a unit. Then, the failure distribution of a unit with age T ($0 \leq T < \infty$) is $F(t; T) \equiv \Pr\{T < X \leq t + T | X > T\} = \lambda(T;t) = [F(t+T) - F(T)]/\overline{F}(T)$, and its MTTF is

$$\int_0^\infty \overline{F}(t; T)\,dt = \frac{1}{\overline{F}(T)} \int_T^\infty \overline{F}(t)\,dt \qquad (1.5)$$

which is decreasing (increasing) from μ to $1/h(\infty)$ when F is IFR (DFR), and is called *mean residual life*.

Furthermore, suppose that a unit with age T has been operating without failure. Then, the relative increment of the mean time μ when the unit is replaced with a new spare one and $\int_T^\infty \overline{F}(t)\mathrm{d}t/\overline{F}(T)$ when it keeps on operating [44] is

$$L(T) \equiv \overline{F}(T)\left[\mu - \frac{1}{\overline{F}(T)}\int_T^\infty \overline{F}(t)\,\mathrm{d}t\right] = \mu\overline{F}(T) - \int_T^\infty \overline{F}(t)\,\mathrm{d}t$$

$$\frac{\mathrm{d}L(T)}{\mathrm{d}T} = \overline{F}(T)[1 - \mu h(T)].$$

Thus, an optimum time that maximizes $L(T)$ is given by the same solution of equation $h(T) = 1/\mu$ in Example 1.1.

Next, consider a unit with unlimited spare units in Example 1.1, where each unit has the identical failure distribution $F(t)$ and is replaced before failure at time T ($0 < T \le \infty$). Then, from the renewal-theoretic argument (see Section 1.3.1), its MTTF is

$$l(T) = \int_0^T t\,\mathrm{d}F(t) + \overline{F}(T)[T + l(T)]; \quad i.e., \quad l(T) = \frac{1}{F(T)}\int_0^T \overline{F}(t)\,\mathrm{d}t \tag{1.6}$$

which is decreasing (increasing) from $1/h(0)$ to μ when F is IFR (DFR). When F is IFR, we have from property (ii),

$$\frac{\overline{F}(T)}{\int_T^\infty \overline{F}(t)\,\mathrm{d}t} \ge h(T) \ge \frac{F(T)}{\int_0^T \overline{F}(t)\,\mathrm{d}t}. \tag{1.7}$$

From these inequalities, it is easy to see that $h(0) \le 1/\mu \le h(\infty)$. ∎

Similar properties of the failure rate for a discrete distribution $\{p_j\}_{j=0}^\infty$ can be shown. In this case, the instant failure rate is defined as $h_n \equiv p_n/[1 - P_n]$ ($n = 0, 1, 2, \dots$) and $h_n \le 1$, where $1 - P_n \equiv \overline{P}_n \equiv \sum_{j=n}^\infty p_j$. A modified failure rate is defined as $\lambda_n \equiv -\log(\overline{P}_{n+1}/\overline{P}_n) = -\log(1 - h_n)$, and it is shown that this failure rate is additive for a series system [45].

(3) Availability

Availability is one of the most important measures in reliability theory. Some authors have defined various kinds of availabilities. Earlier literature on availabilities was summarized in [7, 46]. Later, a system availability for a given length of time [47], and a single-cycle availability incorporating a probabilistic guarantee that its value will be reached in practice [48] were defined. By modifying Martz's definition, the availability for a finite interval was defined in [49]. A good survey and a systematic classification of availabilities were given in [50].

We present the definition of availabilities [7]. Let

$$Z(t) \equiv \begin{cases} 1 & \text{if the system is up at time } t \\ 0 & \text{if the system is down at time } t. \end{cases}$$

(a) *Pointwise availability* is the probability that the system will be up at a given instant of time [27]. This availability is given by

$$A(t) \equiv \Pr\{Z(t) = 1\} = E\{Z(t)\}. \tag{1.8}$$

(b) *Interval availability* is the expected fraction of a given interval that the system will be able to operate, which is given by

$$\frac{1}{t} \int_0^t A(u)\, du. \tag{1.9}$$

(c) *Limiting interval availability* is the expected fraction of time in the long run that the system will be able to operate, which is given by

$$A \equiv \lim_{t \to \infty} \frac{1}{t} \int_0^t A(u)\, du. \tag{1.10}$$

In general, the interval availability is defined as

$$A(x, t+x) \equiv \frac{1}{t} \int_x^{t+x} A(u)\, du$$

and its limiting interval availability is

$$A(x) \equiv \lim_{t \to \infty} \frac{1}{t} \int_x^{t+x} A(u)\, du \qquad \text{for any } x \geq 0.$$

The above three availabilities (a), (b), and (c) were expressed as instantaneous, average uptime, and steady-state availability, respectively [46].

Next, consider n cycles, where each cycle consists of the beginning of up state to the terminating of down state.

(d) *Multiple cycle availability* is the expected fraction of a given cycle that the system will be able to operate [47], which is given by

$$A(n) \equiv E\left\{\frac{\sum_{i=1}^n X_i}{\sum_{i=1}^n (X_i + Y_i)}\right\}, \tag{1.11}$$

where X_i (Y_i) represents the uptime (downtime) (see Section 1.3.2).

(e) *Multiple cycle availability with probability* is the value $A_\nu(n)$ that satisfies the following equation [48]

$$\Pr\left\{\frac{\sum_{i=1}^n X_i}{\sum_{i=1}^n (X_i + Y_i)} \geq A_\nu(n)\right\} = \nu \qquad \text{for } 0 \leq \nu < 1. \tag{1.12}$$

Let $U(t)$ ($D(t)$) be the total uptime (downtime) in an interval $(0, t]$; i.e., $U(t) = t - D(t)$.

(f) *Limiting interval availability with probability* is the value $A_\nu(t)$ that satisfies

$$\Pr\left\{\frac{U(t)}{t} \geq A_\nu(t)\right\} = \nu \quad \text{for } 0 \leq \nu < 1. \tag{1.13}$$

The above availabilities of a one-unit system with repair maintenance and their concrete expressions are given in Section 2.1.1. The availabilities of multicomponent systems were given in [51].

A multiple availability which presents the probability that a unit should be available at each instant of demand was defined in [52, 53]. Several other kinds of availabilities such as *random-request availability, mission availability, computation availability,* and *equivalent availability* for specific application systems were proposed in [54].

Furthermore, *interval reliability* is the probability that at a specified time, a unit is operating and will continue to operate for an interval of duration [55]. Repair and replacement are permitted. Then, the interval reliability $R(x; t)$ for an interval of duration x starting at time t is

$$R(x; t) \equiv \Pr\{Z(u) = 1, t \leq u \leq t + x\} \tag{1.14}$$

and its limit of $R(x; t)$ as $t \to \infty$ is called the *limiting interval reliability*. This becomes simply reliability when $t = 0$ and pointwise availability at time t as $x \to 0$. The interval reliability of a one-unit system with repair maintenance is derived in Section 2.1, and an optimum preventive maintenance policy that maximizes it is discussed in Section 6.1.3.

(4) Reliability Scheduling

Most systems usually perform their functions for a job by scheduling time. A job in the real world is done in random environments due to many sources of uncertainty [56]. So, it would be reasonable to assume that a scheduling time is a random variable, and define the reliability as the probability that the job is accomplished successfully by a system.

Suppose that a random variable S ($S > 0$) is the scheduling time of a job, and X is the failure time of a unit. Furthermore, S and X are independent of each other, and have their respective distributions $W(t)$ and $F(t)$ with finite means; i.e., $W(t) \equiv \Pr\{S \leq t\}$ and $F(t) \equiv \Pr\{X \leq t\}$.

We define the reliability of the unit with scheduling time S as

$$R(W) \equiv \Pr\{S \leq X\} = \int_0^\infty \overline{W}(t) \, dF(t) = \int_0^\infty R(t) \, dW(t) \tag{1.15}$$

which is also called expected gain with some weight function $W(t)$ [7].

We have the following results on $R(W)$.

(1) When $W(t)$ is the degenerate distribution placing unit mass at time t, we have $R(W) = R(t)$ which is the reliability function. Furthermore, when $W(t)$ is a discrete distribution

$$W(t) \equiv \begin{cases} 0 & \text{for } 0 \leq t < T_1 \\ \sum_{i=1}^{j} p_i & \text{for } T_j \leq t < T_{j+1} \ (j = 1, 2, \ldots, N-1) \\ 1 & \text{for } t \geq T_N \end{cases}$$

we have

$$R(W) = \sum_{j=1}^{N} p_j R(T_j).$$

(2) When $W(t) = F(t)$ for all $t \geq 0$, $R(W) = 1/2$.
(3) When $W(t) = 1 - e^{-\omega t}$, $R(W) = 1 - F^*(\omega)$, and inversely, when $F(t) = 1 - e^{-\lambda t}$, $R(W) = W^*(\lambda)$, where $G^*(s)$ is the Laplace–Stieltjes transform of any function $G(t)$; i.e., $G^*(s) \equiv \int_0^\infty e^{-st} dG(t)$ for $s > 0$.
(4) When both S and X are normally distributed with mean μ_1 and μ_2, and variance σ_1^2 and σ_2^2, respectively, $R(W) = \Phi[(\mu_2 - \mu_1)/\sqrt{\sigma_2^2 + \sigma_1^2}]$, where $\Phi(u)$ is a standard normal distribution with mean 0 and variance 1.
(5) When S is uniformly distributed on $(0, T]$, $R(W) = \int_0^T R(t) dt / T$, which represents the interval availability during $(0, T]$ and is decreasing from 1 to 0.

Example 1.3. Some work needs to have a job scheduling time set up. If the work is not accomplished until the scheduled time, its time is prolonged, and this causes some losses to the scheduling.

Suppose that the job scheduling time is L $(0 \leq L < \infty)$ whose cost is sL. If the work is accomplished up to time L, it needs cost c_1, and if it is not accomplished until time L and is done during (L, ∞), it needs cost c_f, where $c_f > c_1$. Then, the expected cost until the completion of work is

$$\begin{aligned} C(L) &\equiv c_1 \Pr\{S \leq L\} + c_f \Pr\{S > L\} + sL \\ &= c_1 W(L) + c_f [1 - W(L)] + sL. \end{aligned} \tag{1.16}$$

Because $\lim_{L \to 0} C(L) = c_f$ and $\lim_{L \to \infty} C(L) = \infty$, there exists a finite job scheduling time L^* $(0 \leq L^* < \infty)$ that minimizes $C(L)$.

We seek an optimum time L^* that minimizes $C(L)$. Differentiating $C(L)$ with respect to L and setting it equal to zero, we have $w(L) = s/(c_f - c_1)$, where $w(t)$ is a density function of $W(t)$. In particular, when $W(t) = 1 - e^{-\omega t}$,

$$\omega e^{-\omega L} = \frac{s}{c_f - c_1}. \tag{1.17}$$

Therefore, we have the following results.

(i) If $\omega > s/(c_f - c_1)$ then there exists a finite and unique L^* $(0 < L^* < \infty)$ that satisfies (1.17).

(ii) If $\omega \leq s/(c_f - c_1)$ then $L^* = 0$; i.e., we should not make a schedule for the job. ∎

1.2 Typical Failure Distributions

It is very important to know properties of distributions typically used in reliability theory, and to identify what type of distribution fits the observed data. It helps us in analyzing reliability models to know what properties the failure and maintenance time distributions have. In general, it is well known that failure distributions have the IFR property and maintenance time distributions have the DFR property. Some books of [57, 58] extensively summarized and studied this problem deeply.

This section briefly summarizes discrete and continuous distributions related to the analysis of reliability systems. The failure rate with the IFR property plays an important role in maintenance theory. At the end, we give a diagram of the relationship among the extreme distributions, and define their discrete extreme distributions, including the Weibull distribution. Note that geometric, negative binomial, and discrete Weibull distributions at discrete times correspond to exponential, gamma and Weibull ones at continuous times, respectively.

(1) Discrete Time Distributions

Let X be a random variable that denotes the failure time of units which operate at discrete times. Let the probability function be p_k $(k = 0, 1, 2, \dots)$ and the moment-generating function be $P^*(\theta)$; i.e., $p_k \equiv \Pr\{X = k\}$ and $P^*(\theta) \equiv \sum_{k=0}^{\infty} e^{\theta k} p_k$ for $\theta > 0$ if it exists.

(i) Binomial distribution

$$p_k = \binom{n}{k} p^k q^{n-k} \quad \text{for } 0 < p < 1, \, q \equiv 1 - p$$

$$E\{X\} = np, \quad V\{X\} = npq, \quad P^*(\theta) = (pe^\theta + q)^n$$

$$\sum_{i=k+1}^{n} \binom{n}{i} p^i q^{n-i} = \frac{n!}{(n-k-1)!k!} \int_0^p x^k (1-x)^{n-k-1} \, dx,$$

where the right-hand side function is called the incomplete beta function [7, p. 39].

(ii) Poisson distribution

$$p_k = \frac{\lambda^k}{k!} e^{-\lambda} \quad \text{for } \lambda > 0$$

$$E\{X\} = V\{X\} = \lambda, \quad P^*(\theta) = \exp[-\lambda(1 - e^\theta)].$$

Units are statistically independent and their failure distribution is $F(t) = 1 - e^{-\lambda t}$. Let $N(t)$ be a random variable that denotes the number of failures during $(0, t]$. Then, $N(t)$ has a Poisson distribution $\Pr\{N(t) = k\} = [(\lambda t)^k/k!]e^{-\lambda t}$ in Section 1.3.1.

(iii) Geometric distribution

$$p_k = pq^k \quad \text{for } 0 < q < 1$$

$$E\{X\} = \frac{q}{p}, \quad V\{X\} = \frac{q}{p^2}, \quad P^*(\theta) = \frac{p}{1 - qe^\theta}$$

$$h_k = p.$$

The failure rate is constant, and it has a memoryless property, *i.e.*, the Markov property in Section 1.3.

(iv) Negative binomial distribution

$$p_k = \binom{-\alpha}{k} p^\alpha (-q)^k \quad \text{for } q \equiv 1 - p > 0, \, \alpha > 0$$

$$E\{X\} = \frac{\alpha q}{p}, \quad V\{X\} = \frac{\alpha q}{p^2}, \quad P^*(\theta) = \left(\frac{p}{1 - qe^\theta}\right)^\alpha.$$

The failure rate is increasing (decreasing) for $\alpha > 1$ ($\alpha < 1$) and coincides with the geometric distribution for $\alpha = 1$.

(2) Continuous Time Distributions

Let $F(t)$ be the failure distribution with a density function $f(t)$. Then, its LS transform is given by $F^*(s) \equiv \int_0^\infty e^{-st} \, dF(t) = \int_0^\infty e^{-st} f(t) \, dt$ for $s > 0$.

(i) Normal distribution

$$f(t) = \frac{1}{\sqrt{2\pi}\sigma} \exp\left[-\frac{(t - \mu)^2}{2\sigma^2}\right] \quad \text{for } -\infty < \mu < \infty, \, \sigma > 0$$

$$E\{X\} = \mu, \quad V\{X\} = \sigma^2.$$

(ii) Log normal distribution

$$f(t) = \frac{1}{\sqrt{2\pi}\sigma t} \exp\left[-\frac{1}{2\sigma^2}(\log t - \mu)^2\right] \quad \text{for } -\infty < \mu < \infty, \, \sigma > 0$$

$$E\{X\} = \exp\left(\mu + \frac{1}{2}\sigma^2\right), \quad V\{X\} = \exp[2(\mu + \sigma^2)] - \exp(2\mu + \sigma^2).$$

The failure rate is decreasing in a long time interval, and hence, it is fitted for most maintenance times, and search times for failures.

(iii) **Exponential distribution**

$$f(t) = \lambda e^{-\lambda t}, \quad F(t) = 1 - e^{-\lambda t} \quad \text{for } \lambda > 0$$

$$E\{X\} = \frac{1}{\lambda}, \quad V\{X\} = \frac{1}{\lambda^2}, \quad F^*(s) = \frac{\lambda}{s+\lambda}$$

$$h(t) = \lambda.$$

When a unit has a memoryless property, the failure rate is constant [59, p. 74]. Thus, a unit with some age x has the same exponential distribution $(1 - e^{-\lambda t})$, irrespective of its age; *i.e.*, the previous operating time does not affect its future lifetime.

(iv) **Gamma distribution**

$$f(t) = \frac{\lambda(\lambda t)^{\alpha-1}}{\Gamma(\alpha)} e^{-\lambda t} \quad \text{for } \lambda, \alpha > 0$$

$$E\{X\} = \frac{\alpha}{\lambda}, \quad V\{X\} = \frac{\alpha}{\lambda^2}, \quad F^*(s) = \left(\frac{\lambda}{s+\lambda}\right)^{\alpha}$$

where $\Gamma(\alpha) \equiv \int_0^\infty x^{\alpha-1} e^{-x} dx$ for $\alpha > 0$. The failure rate is increasing (decreasing) for $\alpha > 1$ ($\alpha < 1$) and this coincides with the exponential distribution for $\alpha = 1$. If failures of each unit occur at a Poisson process with rate λ, *i.e.*, each unit fails according to an exponential distribution and is replaced instantly upon failure, the total time until the nth failure has $f(t) = [\lambda(\lambda t)^{n-1}/(n-1)!]e^{-\lambda t}$ ($n = 1, 2, \dots$) which is the n-fold convolution of exponential distribution, and is called the Erlang distribution.

(v) **Weibull distribution**

$$f(t) = \lambda \alpha t^{\alpha-1} \exp(-\lambda t^\alpha), \quad F(t) = 1 - \exp(-\lambda t^\alpha) \quad \text{for } \lambda, \alpha > 0$$

$$E\{X\} = \lambda^{-1/\alpha} \Gamma\left(1 + \frac{1}{\alpha}\right),$$

$$V\{X\} = \lambda^{-2/\alpha} \left\{ \Gamma\left(1 + \frac{2}{\alpha}\right) - \left[\Gamma\left(1 + \frac{1}{\alpha}\right)\right]^2 \right\}$$

$$h(t) = \lambda \alpha t^{\alpha-1}.$$

The failure rate is increasing (decreasing) for $\alpha > 1$ ($\alpha < 1$) and this coincides with the exponential distribution for $\alpha = 1$.

(3) Extreme Distributions

The Weibull distribution is the most popular distribution of failure times for various phenomena [45, 60], and also is applied in many different fields. The literature on Weibull distributions was integrated, reviewed, and discussed,

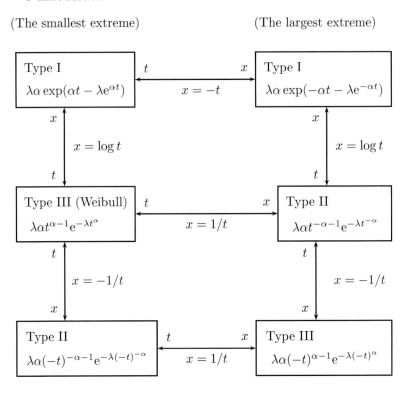

Fig. 1.1. Flow diagram among extreme distributions

and how to formulate Weibull models was shown in [61]. It is also called the Type III asymptotic distribution of extreme values [29], and hence, it is important to investigate the properties of their distributions.

Figure 1.1 shows the flow diagram among extreme density functions [62]. For example, transforming $x = \log t$, i.e., $t = e^x$, in a Type I distribution of the smallest extreme value, we have the Weibull distribution:

$$\lambda\alpha \exp(\alpha x - \lambda e^{\alpha x})\,dx = \lambda\alpha t^{\alpha-1}\exp(-\lambda t^{\alpha})\,dt.$$

The failure rate of the Weibull distribution is $\lambda\alpha t^{\alpha-1}$, which increases with t for $\alpha > 1$. Let us find the distribution for which the failure rate increases exponentially. Substituting $h(t) = \lambda\alpha e^{\alpha t}$ in (1.3) and (1.4), we have

$$f(t) = h(t)\exp\left[-\int_0^t h(u)\,du\right] = \lambda\alpha e^{\alpha t}\exp[-\lambda(e^{\alpha t} - 1)]$$

which is obtained by considering the positive part of Type I of the smallest extreme distribution and by normalizing it.

In failure studies, the time to failure is often measured in the number of cycles to failure, and therefore becomes a discrete random variable. It has

1.2 Typical Failure Distributions

already been shown that geometric and negative binomial distributions at discrete times correspond to exponential and gamma distributions at continuous times, respectively. We are interested in the following question: what discrete distribution corresponds to the Weibull distribution?

Consider the continuous exponential survival function $\overline{F}(t) = e^{-\lambda t}$. Suppose that t takes only the discrete values $0, 1, \ldots$. Then, replacing $e^{-\lambda}$ by q, and t by k formally, we have the geometric survival distribution q^k for $k = 0, 1, 2, \ldots$. This could happen when failures of a unit with an exponential distribution are not revealed unless a specified test has been carried out to determine the condition of the unit and the probability that its failures are detected at the kth test is geometric.

In a similar way, from the survival function $\overline{F}(t) = \exp[-(\lambda t)^\alpha]$ of a Weibull distribution, we define the following discrete Weibull survival function [63].

$$\sum_{j=k}^{\infty} p_j = (q)^{k^\alpha} \quad \text{for } \alpha > 0, \, 0 < q < 1 \quad (k = 0, 1, 2, \ldots).$$

The probability function, the failure rate, and the mean are

$$p_k = (q)^{k^\alpha} - (q)^{(k+1)^\alpha}, \quad h_k = 1 - (q)^{(k+1)^\alpha - k^\alpha}$$

$$E\{X\} = \sum_{k=1}^{\infty} (q)^{k^\alpha}.$$

The failure rate is increasing (decreasing) for $\alpha > 1$ ($\alpha < 1$) and coincides with the geometric distribution for $\alpha = 1$.

When a random variable X has a geometric distribution, i.e., $\Pr\{X \geq k\} = q^k$, the survival function distribution of a random variable $Y \equiv X^{1/\alpha}$ for $\alpha > 0$ is

$$\Pr\{Y \geq k\} = \Pr\{X \geq k^\alpha\} = (q)^{k^\alpha}$$

which is the discrete Weibull distribution. The parameters of a discrete Weibull distribution were estimated in [64]. Furthermore, modified discrete Weibull distributions were proposed in [65].

Failures of some units often depend more on the total number of cycles than on the total time that they have been used. Such examples are switching devices, railroad tracks, and airplane tires. In this case, we believe that a discrete Weibull distribution will be a good approximation for such devices, materials, or structures. A comprehensive survey of discrete distributions used in reliability models was presented in [66].

Figure 1.2 shows the graph of the probability function p_k for $q = 0.6$ and $\alpha = 0.5, 1.0, 1.5$, and 2.0, and Figure 1.3 gives the survival functions of discrete extreme distributions as those in Figure 1.1.

Example 1.4. Consider an n-unit parallel redundant system (see Example 1.6) in a random environment that generates shocks at mean interval

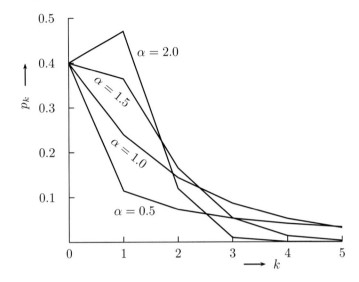

Fig. 1.2. Discrete Weibull probability function $p_k = q^{k^\alpha}$ for $q = 0.6$

Type	The smallest extreme	The largest extreme
I	q^{α^k} ($\alpha > 1$, $-\infty < k < \infty$)	$1 - q^{\alpha^{-k}}$ ($\alpha > 1$, $-\infty < k < \infty$)
II	$q^{(-k)^{-\alpha}}$ ($\alpha > 0$, $-\infty < k \le 0$)	$1 - q^{k^{-\alpha}}$ ($\alpha > 0$, $0 \le k < \infty$)
III	q^{k^α} ($\alpha > 0$, $0 \le k < \infty$)	$1 - q^{(-k)^\alpha}$ ($\alpha > 0$, $-\infty \le k \le 0$)

Fig. 1.3. Survival functions of discrete extreme for $0 < q < 1$

θ [67]. Each unit fails with probability p_k at the kth shock ($k = 1, 2, \ldots$), independently of other units. Then, the mean time to system failure is

$$\mu_n = \theta \sum_{k=1}^{\infty} k \left\{ \left[\sum_{j=1}^{k} p_j \right]^n - \left[\sum_{j=1}^{k-1} p_j \right]^n \right\}$$

$$= \theta \sum_{k=0}^{\infty} \left\{ 1 - \left[\sum_{j=1}^{k} p_j \right]^n \right\} = \theta \sum_{i=1}^{n} \binom{n}{i} (-1)^{i+1} \sum_{k=0}^{\infty} \left[\sum_{j=k+1}^{\infty} p_j \right]^i,$$

where $\sum_{j=1}^{0} \equiv 0$. For example, when shocks occur according to a discrete Weibull distribution $\sum_{j=k}^{\infty} p_j = (q)^{(k-1)^\alpha}$ ($k = 1, 2, \ldots$),

$$\mu_n = \theta \sum_{i=1}^{n} \binom{n}{i} (-1)^{i+1} \sum_{k=0}^{\infty} q^{ik^\alpha}.$$

In particular, when $\alpha = 1$,

$$\mu_n = \theta \sum_{i=1}^{n} \binom{n}{i}(-1)^{i+1}\frac{1}{1-q^i}. \quad \blacksquare$$

1.3 Stochastic Processes

In this section, we briefly present some kinds of stochastic processes for systems with maintenance. Let us sketch the simplest system as an example. It is a one-unit system with repair or replacement whose time is negligible; *i.e.*, a unit is operating and is repaired or replaced when it fails, where the time required for repair or replacement is negligible. When the repair or replacement is completed, the unit becomes as good as new and begins to operate. The system forms a renewal process, *i.e.*, a renewal theory arises from the study of *self-renewing aggregates*, and plays an important role in the analysis of probability models with sums of independent nonnegative random variables. We summarize the main results of a renewal theory for future studies of maintenance models in this book.

Next, consider a one-unit system where the repair or replacement time needs a nonnegligible time; *i.e.*, the system repeats up and down alternately. The system forms an alternating renewal process that repeats two different renewal processes alternately. Furthermore, if the duration times of up and down are multiples of a period of time, then the system can be described by a *discrete time parameter Markov chain*. If the duration times of up and down are distributed exponentially, then the system can be described by a *continuous time parameter Markov process*. In general, Markov chains or processes have the Markovian property: the future behavior depends only on the present state and not on its past history. If the duration times of up and down are distributed arbitrarily, then the system can be described by a *semi-Markov process* or *Markov renewal process*.

Because the mechanism of failure occurrences may be uncertain in complex systems, we have to observe the behavior of such systems statistically and stochastically. It would be very effective in the reliability analysis to deal with maintenance problems underlying stochastic processes, which justly describe a physical phenomenon of random events. Therefore, this section summarizes the theory of renewal processes, Markov chains, semi-Markov processes, and Markov renewal processes for future studies of maintenance models. More general theory and applications of renewal processes are found in [68, 69].

Markov chains are essential and fundamental in the theory of stochastic processes. On the other hand, semi-Markov processes or Markov renewal processes are based on a marriage of renewal processes and Markov chains, which were first studied by [70]. Pyke gave a careful definition and discussions of Markov renewal processes in detail [71, 72]. In reliability, these processes are one of the most powerful mathematical techniques for analyzing maintenance and random models. A table of applicable stochastic processes associated with repairman problems was shown in [7].

State space is usually defined by the number of units that is functioning satisfactorily. As far as the applications are concerned, we consider only a finite number of states. We mention only the theory of stationary Markov chains with finite-state space for analysis of maintenance models. It is shown that transition probabilities, first-passage time distributions, and renewal functions are given in terms of one-step transition probabilities. Furthermore, some limiting properties are summarized when all states communicate.

We omit the proofs of results and derivations. For more detailed discussions and applications of Markov processes, we refer readers to the books [59, 73–75].

1.3.1 Renewal Process

Consider a sequence of independent and nonnegative random variables $\{X_1, X_2, \ldots\}$, in which $\Pr\{X_i = 0\} < 1$ for all i because of avoiding the triviality. Suppose that $X_2, X_3, \ldots$ have an identical distribution $F(t)$ with finite mean μ, however, X_1 possibly has a different distribution $F_1(t)$ with finite mean μ_1, in which both $F_1(t)$ and $F(t)$ are not degenerate at time $t = 0$, and $F_1(0) = F(0) = 0$.

We have three cases according to the following types of $F_1(t)$.

(1) If $F_1(t) = F(t)$, i.e., all random variables are identically distributed, the process is called an *ordinary renewal process* or *renewal process* for short.
(2) If $F_1(t)$ and $F(t)$ are not the same, the process is called a *modified* or *delayed renewal process*.
(3) If $F_1(t)$ is expressed as $F_1(t) = \int_0^t [1 - F(u)] du/\mu$ which is given in (1.30), the process is called an *equilibrium* or *stationary renewal process*.

Example 1.5. Consider a unit that is replaced with a new one upon failure. A unit begins to operate immediately after the replacement whose time is negligible. Suppose that the failure distribution of each new unit is $F(t)$. If a new unit is installed at time $t = 0$ then all failure times have the same distribution, and hence, we have an ordinary renewal process. On the other hand, if a unit is in use at time $t = 0$ then X_1 represents its residual lifetime and could be different from the failure time of a new unit, and hence, we have a modified renewal process. In particular, if the observed time origin is sufficiently large after the installation of a unit and X_1 has a failure distribution $\int_0^t [1 - F(u)] du/\mu$, we have an equilibrium renewal process. ∎

Letting $S_n \equiv \sum_{i=1}^n X_i$ ($n = 1, 2, \ldots$) and $S_0 \equiv 0$, we define $N(t) \equiv \max_n \{S_n \leq t\}$ which represents the number of renewals during $(0, t]$. Renewal theory is mainly devoted to the investigation into the probabilistic properties of $N(t)$.

Denoting

1.3 Stochastic Processes

$$F^{(0)}(t) \equiv \begin{cases} 1 & \text{for } t \geq 0 \\ 0 & \text{for } t < 0 \end{cases} \quad F^{(n)}(t) \equiv \int_0^t F^{(n-1)}(t-u)\,dF(u) \quad (n = 1, 2, \dots);$$

i.e., letting $F^{(n)}$ be the n-fold Stieltjes convolution of F with itself, represents the distribution of the sum $X_2 + X_3 + \cdots + X_{n+1}$. Evidently,

$$\Pr\{N(t) = 0\} = \Pr\{X_1 > t\} = 1 - F_1(t)$$
$$\Pr\{N(t) = n\} = \Pr\{S_n \leq t \text{ and } S_{n+1} > t\}$$
$$= F_1(t) * F^{(n-1)}(t) - F_1(t) * F^{(n)}(t) \quad (n = 1, 2, \dots), \quad (1.18)$$

where the asterisk denotes the pairwise Stieltjes convolution; i.e., $a(t) * b(t) \equiv \int_0^t b(t-u)\,da(u)$.

We define the expected number of renewals in $(0,t]$ as $M(t) \equiv E\{N(t)\}$, which is called the *renewal function*, and $m(t) \equiv dM(t)/dt$, which is called the *renewal density*. From (1.18), we have

$$M(t) = \sum_{k=1}^{\infty} k \Pr\{N(t) = k\} = \sum_{k=1}^{\infty} F_1(t) * F^{(k-1)}(t). \quad (1.19)$$

It is fairly easy to show that $M(t)$ is finite for all $t \geq 0$ because $\Pr\{X_i = 0\} < 1$. Furthermore, from the notation of convolution,

$$M(t) = F_1(t) + \sum_{k=1}^{\infty} \int_0^t F^{(k)}(t-u)\,dF_1(u) = \int_0^t [1 + M_0(t-u)]\,dF_1(u) \quad (1.20)$$

$$m(t) = f_1(t) + \int_0^t m_0(t-u) f_1(u)\,du,$$

where $M_0(t)$ is the renewal function of an ordinary renewal process with distribution F; i.e., $M_0(t) \equiv \sum_{k=1}^{\infty} F^{(k)}(t)$, $m_0(t) \equiv dM_0(t)/dt = \sum_{k=1}^{\infty} f^{(k)}(t)$, and f and f_1 are the respective density functions of F and F_1. The LS transform of $M(t)$ is given by

$$M^*(s) \equiv \int_0^{\infty} e^{-st}\,dM(t) = \frac{F_1^*(s)}{1 - F^*(s)}, \quad (1.21)$$

where, in general, $\Phi^*(s)$ is the LS transform of $\Phi(t)$; i.e., $\Phi^*(s) \equiv \int_0^{\infty} e^{-st}\,d\Phi(s)$ for $s > 0$. Thus, $M(t)$ is determined by $F_1(t)$ and $F(t)$. When $F_1(t) = F(t)$, $M_0(t) = M(t)$, and Equation (1.21) implies $F^*(s) = M^*(s)/[1 + M^*(s)]$, and hence, $F(t)$ is also determined by $M(t)$ because the LS transform determines the distribution uniquely. The Laplace inversion method is referred to in [76, 77].

We summarize some important limiting theorems of renewal theory for future references.

Theorem 1.2.

(i) With probability 1, $\dfrac{N(t)}{t} \to \dfrac{1}{\mu}$ as $t \to \infty$.

(ii) $\dfrac{M(t)}{t} \to \dfrac{1}{\mu}$ as $t \to \infty$. \hfill (1.22)

It is well known that when $F_1(t) = F(t) = 1 - e^{-t/\mu}$, $M(t) = t/\mu$ for all $t \geq 0$, and hence, $M(t+h) - M(t) = h/\mu$. Furthermore, when the process is an equilibrium renewal process, we also have that $M(t) = t/\mu$.

Before mentioning the following theorems, we define that a nonnegative random X is called a *lattice* if there exists $d > 0$ such that $\sum_{n=0}^{\infty} \Pr\{X = nd\} = 1$. The largest d having this property is called the *period* of X. When X is a lattice, the distribution $F(t)$ of X is called a *lattice distribution*. On the other hand, when X is not a lattice, F is called a *nonlattice distribution*.

Theorem 1.3.

(i) If F is a nonlattice distribution,
$$M(t+h) - M(t) \to \frac{h}{\mu} \quad \text{as } t \to \infty. \tag{1.23}$$

(ii) If $F(t)$ is a lattice distribution with period d,
$$\Pr\{\text{Renewal at } nd\} \to \frac{d}{\mu} \quad \text{as } t \to \infty. \tag{1.24}$$

Theorem 1.4. If $\mu_2 \equiv \int_0^\infty t^2 dF(t) < \infty$ and F is nonlattice,
$$M(t) = \frac{t}{\mu} + \frac{\mu_2}{2\mu^2} - 1 + o(1) \quad \text{as } t \to \infty. \tag{1.25}$$

From this theorem, $M(t)$ and $m(t)$ are approximately given by
$$M(t) \approx \frac{t}{\mu} + \frac{\mu_2}{2\mu^2} - 1, \quad m(t) \approx \frac{1}{\mu} \tag{1.26}$$

for large t. Furthermore, the following inequalities of $M(t)$ when F is IFR are given [7],
$$\frac{t}{\mu} - 1 \leq \frac{t}{\int_0^t \overline{F}(u)\,du} - 1 \leq M(t) \leq \frac{tF(t)}{\int_0^t \overline{F}(u)\,du} \leq \frac{t}{\mu}. \tag{1.27}$$

Next, let $\delta(t) \equiv t - S_{N(t)}$ and $\gamma(t) \equiv S_{N(t)+1} - t$, which represent the current age and the residual life, respectively. In an ordinary renewal process, we have the following distributions of $\delta(t)$ and $\gamma(t)$ when F is not a lattice.

Theorem 1.5.

$$\Pr\{\delta(t) \leq x\} = \begin{cases} F(t) - \int_0^{t-x}[1 - F(t-u)]\,\mathrm{d}M(u) & \text{for } x \leq t \\ 1 & \text{for } x > t \end{cases} \quad (1.28)$$

$$\Pr\{\gamma(t) \leq x\} = F(t+x) - \int_0^t [1 - F(t+x-u)]\,\mathrm{d}M(u) \quad (1.29)$$

and their limiting distribution is

$$\lim_{t\to\infty} \Pr\{\delta(t) \leq x\} = \lim_{t\to\infty} \Pr\{\gamma(t) \leq x\} = \frac{1}{\mu}\int_0^x [1 - F(u)]\,\mathrm{d}u. \quad (1.30)$$

It is of interest that the mean of the above limiting distribution is

$$\frac{1}{\mu}\int_0^\infty x[1 - F(x)]\,\mathrm{d}x = \frac{\mu}{2} + \frac{\mu_2 - \mu^2}{2\mu} \quad (1.31)$$

which is greater than half of the mean interval time μ [68]. Moreover, the stochastic properties of $\gamma(t)$ were investigated in [78, 79].

If the number $N(t)$ of some event during $(0, t]$ has the following distribution

$$\Pr\{N(t) = n\} = \frac{[H(t)]^n}{n!}e^{-H(t)} \quad (n = 0, 1, 2, \dots) \quad (1.32)$$

and has the property of independent increments, then the process $\{N(t), t \geq 0\}$ is called a *nonhomogeneous Poisson process* with *mean value function* $H(t)$. Clearly, $E\{N(t)\} = H(t)$ and $h(t) \equiv \mathrm{d}H(t)/\mathrm{d}t$, i.e., $H(t) = \int_0^t h(u)\mathrm{d}u$, is called an *intensity function*.

Suppose that a unit fails and undergoes minimal repair; i.e., its failure rate remains undisturbed by any minimal repair (see Section 4.1). Then, the number $N(t)$ of failures during $(0, t]$ has a Poisson distribution in (1.32). In this case, we say that failures of a unit occur at a nonhomogeneous Poisson process, and $H(t)$ and $h(t)$ correspond to the cumulative hazard function and failure rate of a unit with itself, respectively.

Finally, we introduce a renewal reward process [73] or cumulative process [69]. For instance, if we consider the total reward produced by the successive production of a machine, then the process forms a renewal reward process, where the successive production can be described by a renewal process and the total rewards caused by production may be additive.

Define that a reward Y_n is earned at the nth renewal time $(n = 1, 2, \dots)$. When a sequence of pairs $\{X_n, Y_n\}$ is independent and identically distributed, $Y(t) \equiv \sum_{n=1}^{N(t)} Y_n$ is denoted by the total reward earned during $(0, t]$. When successive shocks of a unit occur at time interval X_n and each shock causes an amount of damage Y_n to a unit, the total amount of damage is given by $Y(t)$ [69, 80].

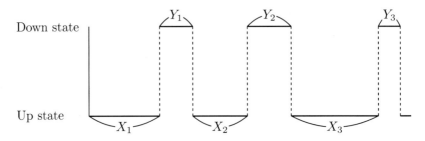

Fig. 1.4. Realization of alternating renewal process

Theorem 1.6. Suppose that $E\{Y\} \equiv E\{Y_n\}$ are finite.

(i) With probability 1, $\dfrac{Y(t)}{t} \to \dfrac{E\{Y\}}{\mu}$ as $t \to \infty$. (1.33)

(ii) $\dfrac{E\{Y(t)\}}{t} \to \dfrac{E\{Y\}}{\mu}$ as $t \to \infty$. (1.34)

In the above theorems, we interpret $a/\mu = 0$ whenever $\mu = \infty$ and $|a| < \infty$. Theorem 1.6 can be easily proved from Theorem 1.2 and the detailed proof can be found in [73]. This theorem shows that if one cycle is denoted by the time interval between renewals, the expected reward per unit of time for an infinite time span is equal to the expected reward per one cycle, divided by the mean time of one cycle. This is applied throughout this book to many optimization problems that minimize cost functions.

1.3.2 Alternating Renewal Process

Alternating renewal processes are the processes that repeat *on* and *off* or *up* and *down* states alternately [69]. Many redundant systems generate alternating renewal processes. For example, we consider a one-unit system with repair maintenance in Section 2.1. The unit begins to operate at time 0, and is repaired upon failure and returns to operation. We could consider the time required for repair as the replacement time. It is assumed in any event that the unit becomes as good as new after the repair or maintenance completion. It is said that the system forms an *ordinary alternating renewal process* or simply an *alternating renewal process*. If we take the time origin a long way from the beginning of an operating unit, the system forms an *equilibrium alternating renewal process*.

Furthermore, consider an n-unit standby redundant system with r repair-persons ($1 \leq r \leq n$) and one operating unit supported by $n-1$ identical spare units [7, p. 150; 81]. When each unit fails randomly and the repair times are exponential, the system forms a *modified alternating renewal process*.

1.3 Stochastic Processes

We are concerned with only the off time properties and apply them to reliability systems. Consider an alternating renewal process $\{X_1, Y_1, X_2, Y_2, \dots\}$, where X_i and Y_i ($i = 1, 2, \dots$) are independent random variables with distributions F and G, respectively (see Figure 1.4). The alternating renewal process assumes on and off states alternately with distributions F and G.

Let $N(t)$ and $D(t)$ be the number of up states and the total amount of time spent in down states during $(0, t]$, respectively. Then, from a well-known formula of the sum of independent random variables,

$$\Pr\{Y_1 + Y_2 + \cdots + Y_n \leq x | N(t) = n\} \Pr\{N(t) = n\}$$
$$= \Pr\{Y_1 + Y_2 + \cdots + Y_n \leq x\} \Pr\{X_1 + \cdots + X_n \leq t - x < X_1 + \cdots + X_{n+1}\}$$
$$= G^{(n)}(x)[F^{(n)}(t-x) - F^{(n+1)}(t-x)]$$

we have [82]

$$\Pr\{D(t) \leq x\} = \begin{cases} \sum_{n=0}^{\infty} G^{(n)}(x)[F^{(n)}(t-x) - F^{(n+1)}(t-x)] & \text{for } t > x \\ 1 & \text{for } t \leq x. \end{cases} \quad (1.35)$$

Thus, the distribution of $T_\delta \equiv \min_t \{D(t) > \delta\}$ for a specified $\delta > 0$, which is the first time that the total amount of off time has exceeded δ, is given by $\Pr\{D(t) > \delta\}$.

Next, consider the first time that an amount of off time exceeds a fixed time $c > 0$, where c is called a critically allowed time for maintenance [83]. In general, it is assumed that c is a random variable U with distribution K. Let $\widetilde{Y}_i \equiv \{Y_i; Y_i \leq U\}$ and $\widetilde{U}_i \equiv \{U; U < Y_i\}$. If the process ends with the first event of $\{U < Y_N\}$ then the terminating process of interest is $\{X_1, \widetilde{Y}_1, X_2, \widetilde{Y}_2, \dots, X_{N-1}, \widetilde{Y}_{N-1}, X_N, \widetilde{U}_N\}$, the sum of random variables $W \equiv \sum_{i=1}^{N-1}(X_i + \widetilde{Y}_i) + X_N + \widetilde{U}_N$, and its distribution $L(t) \equiv \Pr\{W \leq t\}$.

The probability that Y_i is not greater than U and $Y_i \leq t$ is

$$B(t) \equiv \Pr\{Y_i \leq U, Y_i \leq t\} = \int_0^t \overline{K}(u) \, dG(u)$$

and Y_i is greater than U and $U \leq t$ is

$$\widetilde{B}(t) \equiv \Pr\{U < Y_i, U \leq t\} = \int_0^t \overline{G}(u) \, dK(u).$$

Thus, from the formula of the sum of independent random variables,

$$L(t) = \sum_{N=1}^{\infty} \Pr\left\{\sum_{i=1}^{N-1}(X_i + \widetilde{Y}_i) + X_N + \widetilde{U}_N \leq t\right\}$$
$$= \sum_{n=0}^{\infty} F^{(n)}(t) * B^{(n)}(t) * F(t) * \widetilde{B}(t). \quad (1.36)$$

Therefore, the LS transform of $L(t)$ is

$$L^*(s) = \frac{F^*(s)\widetilde{B}^*(s)}{1 - F^*(s)B^*(s)} \tag{1.37}$$

and its mean time is

$$l \equiv \lim_{s \to 0} \frac{1 - L^*(s)}{s} = \frac{\mu + \int_0^\infty \overline{G}(t)\overline{K}(t)\,\mathrm{d}t}{\int_0^\infty K(t)\,\mathrm{d}G(t)}. \tag{1.38}$$

In particular, when c is constant, the corresponding results of (1.37) and (1.38) are, respectively,

$$L^*(s) = \frac{F^*(s)e^{-sc}\overline{G}(c)}{1 - F^*(s)\int_0^c e^{-st}\,\mathrm{d}G(t)} \tag{1.39}$$

$$l = \frac{\mu + \int_0^c \overline{G}(t)\,\mathrm{d}t}{\overline{G}(c)}. \tag{1.40}$$

1.3.3 Markov Processes

When we analyze complex systems, it is essential to learn Markov processes. This section briefly explains the theory of Markov chains, semi-Markov processes, and Markov renewal processes.

(1) Markov Chain

Consider a discrete time stochastic process $\{X_n, n = 0, 1, 2, \ldots\}$ with a finite state set $\{0, 1, 2, \ldots, m\}$. If we suppose that

$$\Pr\{X_{n+1} = i_{n+1} | X_0 = i_0, X_1 = i_1, \ldots, X_n = i_n\} = \Pr\{X_{n+1} = i_{n+1} | X_n = i_n\}$$

for all states $i_0, i_1, \ldots, i_n$, and all $n \geq 0$, then the process $\{X_n, n = 0, 1, \ldots\}$ is said to be a *Markov chain*. This property shows that, given the value of X_n, the future value of X_{n+1} does not depend on the value of X_k for $0 \leq k \leq n-1$. If the probability of X_{n+1} being in state j, given that X_n is in state i, is independent of n, i.e.,

$$\Pr\{X_{n+1} = j | X_n = i\} = P_{ij} \tag{1.41}$$

then the process has a *stationary (one-step) transition probability*. We restrict ourselves to discrete time Markov chains with stationary transition probabilities. Manifestly, the transition probabilities P_{ij} satisfy $P_{ij} \geq 0$, and $\sum_{j=0}^m P_{ij} = 1$.

A Markov chain is completely specified by the transition probabilities P_{ij} and an initial probability distribution of X_0 at time 0. Let P_{ij}^n denote the

1.3 Stochastic Processes

probability that the process goes from state i to state j in n transactions; or formally,

$$P_{ij}^n \equiv \Pr\{X_{n+k} = j | X_k = i\}.$$

Then,

$$P_{ij}^n = \sum_{k=0}^m P_{ik}^r P_{kj}^{n-r} \qquad (r = 0, 1, \ldots, n), \qquad (1.42)$$

where $P_{ii}^0 = 1$ and otherwise $P_{ij}^0 = 0$ for convenience. This equation is known as the *Chapman–Kolmogorov equation*.

We define the first-passage time distribution as

$$F_{ij}^n \equiv \Pr\{X_n = j, X_k \neq j, k = 1, 2, \ldots, n-1 | X_0 = i\} \qquad (1.43)$$

which is the probability that starting in state i, the first transition into state j occurs at the nth transition, where we define $F_{ij}^0 \equiv 0$ for all i, j. Then,

$$P_{ij}^n \equiv \sum_{k=0}^n F_{ij}^k P_{jj}^{n-k} \qquad (n = 1, 2, \ldots) \qquad (1.44)$$

and hence, the probability F_{ij}^k of the first passage from state i to state j at the kth transition is determined uniquely by the above equation.

Furthermore, let M_{ij}^n denote the expected number of visits to state j in the nth transition if the process starts in state i, not including the first at time 0. Then,

$$M_{ij}^n = \sum_{k=1}^n P_{ij}^k \qquad (n = 1, 2, \ldots), \qquad (1.45)$$

where we define $M_{ij}^0 \equiv 0$ for all i, j.

We next introduce the following generating functions.

$$P_{ij}^*(z) \equiv \sum_{n=0}^{\infty} z^n P_{ij}^n, \qquad F_{ij}^*(z) \equiv \sum_{n=0}^{\infty} z^n F_{ij}^n, \qquad M_{ij}^*(z) \equiv \sum_{n=0}^{\infty} z^n M_{ij}^n$$

for $|z| < 1$. Then, forming the generating operations of (1.44) and (1.45), we have [59]

$$P_{ii}^*(z) = \frac{1}{1 - F_{ii}^*(z)}, \qquad P_{ij}^*(z) = F_{ij}^*(z) P_{jj}^*(z) \quad (i \neq j) \qquad (1.46)$$

$$M_{jj}^*(z) = \frac{P_{jj}^*(z) - 1}{1 - z}, \qquad M_{ij}^*(z) = \frac{P_{ij}^*(z)}{1 - z} \quad (i \neq j). \qquad (1.47)$$

Two states i and j are said to communicate if and only if there exist integers $n_1 \geq 0$ and $n_2 \geq 0$ such that $P_{ij}^{n_1} > 0$ and $P_{ji}^{n_2} > 0$. The period $d(i)$ of states i is defined as the greatest common divisor of all integers $n \geq 1$ for which $P_{ii}^n > 0$. If $d(i) = 1$ then state i is said to be *nonperiodic*.

Consider a Markov chain in which all states communicate. Such a chain is called *irreducible*. We only consider the nonperiodic case. Then, the following limiting results of such a Markov chain are known.

$$\sum_{n=1}^{\infty} F_{ij}^n = 1, \qquad \mu_{jj} = \sum_{n=1}^{\infty} n F_{jj}^n < \infty \tag{1.48}$$

$$\lim_{n \to \infty} M_{ij}^n = \sum_{n=1}^{\infty} P_{ij}^n = \infty, \quad \lim_{n \to \infty} \frac{M_{ij}^n}{n} = \lim_{n \to \infty} P_{ij}^n = \frac{1}{\mu_{jj}} < \infty \tag{1.49}$$

for all i, j. Furthermore, $\lim_{n \to \infty} P_{ij}^n = \pi_j$ ($j = 0, 1, 2, \ldots, m$) are uniquely determined by a set of equations:

$$\pi_j = \sum_{i=0}^{m} \pi_i P_{ij} \quad (j = 0, 1, 2, \ldots, m), \qquad \sum_{j=0}^{m} \pi_j = 1. \tag{1.50}$$

(2) Semi-Markov and Markov Renewal Processes

Consider a stochastic process with a finite state set $\{0, 1, 2, \ldots, m\}$ that makes transitions from state to state in accordance with a Markov chain with stationary transition probabilities. However, in the process the amount of time spent in each state until the next transition is not always constant but random.

Let $Q_{ij}(t)$ denote the probability that after making a transition into state i, the next process makes a transition into state j, in an amount of time less than or equal to t. Clearly, we have $Q_{ij}(t) \geq 0$ and $\sum_{j=0}^{m} Q_{ij}(\infty) = 1$, where $Q_{ij}(\infty)$ represents the probability that the next process makes a transition into state j, given that the process goes into state i. We call the probability $Q_{ij}(t)$ a *mass function*. Letting

$$G_{ij}(t) = \frac{Q_{ij}(t)}{Q_{ij}(\infty)} \quad \text{for } Q_{ij}(\infty) > 0$$

then $G_{ij}(t)$ represents the conditional probability that the process makes a transition in an amount of time less than or equal to t, given that the process goes from state i to state j at the next transition.

Let J_n denote the state of the process immediately after the nth transition has occurred for $n \geq 1$ and let J_0 denote the initial state of the process. Then, the stochastic process $\{J_n, n = 0, 1, 2, \ldots\}$ is called an *embedded Markov chain*. If the process makes a transition from one state to another with one unit of time, i.e., $G_{ij}(t) = 0$ for $t < 1$, and 1 for $t \geq 1$, then an embedded Markov chain becomes a Markov chain. Furthermore, if an amount of time spent in state i depends only on state i and is exponential independent of the next state; $G_{ij}(t) = 1 - e^{-\lambda_i t}$ for constant $\lambda_i > 0$, the process is said to be a *continuous time parameter Markov process*. In addition, the process becomes a *renewal process* if it is only one state. If we let $Z(t)$ denote the state of the process at time t, then the stochastic process $\{Z(t), t \geq 0\}$ is called a *semi-Markov process*. Let $N_i(t)$ denote the number of times that the process visits state i in $(0, t]$. It follows from renewal theory that with probability 1, $N_i(t) < \infty$ for $t \geq 0$. The stochastic process $\{N_0(t), N_1(t), N_2(t), \ldots, N_m(t)\}$ is called a *Markov renewal process*.

An embedded Markov chain records the state of the process at each transition point, a semi-Markov process records the state of the process at each time point, and a Markov renewal process records the total number of times that each state has been visited.

Let $H_i(t)$ denote the distribution of an amount of time spent in state i until the process makes a transition to the next state;

$$H_i(t) \equiv \sum_{j=0}^{m} Q_{ij}(t)$$

which is called the *unconditional distribution* for state i. We suppose that $H_i(0) < 1$ for all i. Denoting

$$\eta_i \equiv \int_0^\infty t \, dH_i(t), \qquad \mu_{ij} \equiv \int_0^\infty t \, dG_{ij}(t)$$

it is easily seen that

$$\eta_i = \sum_{j=0}^{m} Q_{ij}(\infty) \mu_{ij}$$

which represents the mean time spent in state i.

We define transition probabilities, first-passage time distributions, and renewal functions as, respectively,

$$P_{ij}(t) \equiv \Pr\{Z(t) = j | Z(0) = i\}$$
$$F_{ij}(t) \equiv \Pr\{N_j(t) > 0 | Z(0) = i\}$$
$$M_{ij}(t) \equiv E\{N_j(t) | Z(0) = i\}.$$

We have the following relationships for $P_{ij}(t)$, $F_{ij}(t)$, and $M_{ij}(t)$ in terms of the mass functions $Q_{ij}(t)$.

$$P_{ii}(t) = 1 - H_i(t) + \sum_{k=0}^{m} \int_0^t P_{ki}(t-u) \, dQ_{ik}(u) \qquad (1.51)$$

$$P_{ij}(t) = \sum_{k=0}^{m} \int_0^t P_{kj}(t-u) \, dQ_{ik}(u) \qquad \text{for } i \neq j \qquad (1.52)$$

$$F_{ij}(t) = Q_{ij}(t) + \sum_{\substack{k=0 \\ k \neq j}}^{m} \int_0^t F_{kj}(t-u) \, dQ_{ik}(u) \qquad (1.53)$$

$$M_{ij}(t) = Q_{ij}(t) + \sum_{k=0}^{m} \int_0^t M_{kj}(t-u) \, dQ_{ik}(u). \qquad (1.54)$$

Therefore, the mass functions $Q_{ij}(t)$ determine $P_{ij}(t)$, $F_{ij}(t)$, and $M_{ij}(t)$ uniquely. Furthermore, we have

$$P_{ii}(t) = 1 - H_i(t) + \int_0^t P_{ii}(t-u)\,dF_{ii}(u) \tag{1.55}$$

$$P_{ij}(t) = \int_0^t P_{jj}(t-u)\,dF_{ij}(u) \quad \text{for } i \neq j \tag{1.56}$$

$$M_{ij}(t) = F_{ij}(t) + \int_0^t M_{jj}(t-u)\,dF_{ij}(u). \tag{1.57}$$

Thus, forming the LS transforms of the above equations,

$$P_{ii}^*(s) = \frac{1 - H_i^*(s)}{1 - F_{ii}^*(s)} \tag{1.58}$$

$$P_{ij}^*(s) = F_{ij}^*(s) P_{jj}^*(s) \quad \text{for } i \neq j \tag{1.59}$$

$$M_{ij}^*(s) = F_{ij}^*(s)[1 + M_{jj}^*(s)], \tag{1.60}$$

where the asterisk denotes the LS transform of the function with itself.

Consider the process in which all states communicate, $G_{ii}(\infty) = 1$, and $\mu_{ii} < \infty$ for all i. It is said that the process consists of one positive recurrent class. Further suppose that each $G_{jj}(t)$ is a nonlattice distribution. Then, we have

$$G_{ij}(\infty) = 1, \quad \mu_{ij} < \infty$$
$$\mu_{ij} = \eta_i + \sum_{k \neq j} Q_{ik}(\infty)\mu_{kj}. \tag{1.61}$$

Furthermore,

$$\lim_{t\to\infty} M_{ij}(t) = \int_0^\infty P_{ij}(u)\,du = \infty$$
$$\lim_{t\to\infty} \frac{M_{ij}(t)}{t} = \frac{1}{\mu_{jj}} < \infty, \quad \lim_{t\to\infty} P_{ij}(t) = \frac{\eta_j}{\mu_{jj}} < \infty. \tag{1.62}$$

In this case, because there exist $\lim_{t\to\infty} M_{ij}(t)/t$ and $\lim_{t\to\infty} P_{ij}(t)$, we also have, from a Tauberian theorem that if for some nonnegative integer n, $\lim_{s\to 0} s^n \Phi^*(s) = C$ then $\lim_{t\to\infty} \Phi(t)/t^n = C/n!$,

$$\lim_{t\to\infty} \frac{M_{ij}(t)}{t} = \lim_{s\to 0} s M_{ij}^*(s)$$
$$\lim_{t\to\infty} P_{ij}(t) = \lim_{t\to\infty} \frac{1}{t}\int_0^t P_{ij}(u)\,du = \lim_{s\to 0} P_{ij}^*(s). \tag{1.63}$$

1.3.4 Markov Renewal Process with Nonregeneration Points

This section explains unique modifications of Markov renewal processes and applies them to redundant repairable systems including some nonregeneration points [84]. It has already been shown that such modifications give powerful plays for analyzing two-unit redundant systems [85] and communication systems [86]. In this book, this is used for the one-unit system with repair in Section 2.1, and the two-unit standby system with preventive maintenance in Section 6.2.

It is assumed that the Markov renewal process under consideration has only one positive recurrent class, because we restrict ourselves to applications to reliability models. Consider the case where epochs at which the process enters some states are not regeneration points. Then, we partition a state space S into $S = S^* \bigcup S^\dagger$ ($S^* \bigcap S^\dagger = \phi$), where S^* is the portion of the state space such that the epoch entering state i ($i \in S^*$) is not a regeneration point, and $S^\dagger$ is such that the epoch entering state i ($i \in S^\dagger$) is a regeneration point, where S^* and $S^\dagger$ are assumed not to be empty.

Define the mass function $Q_{ij}(t)$ from state i ($i \in S^\dagger$) to state j ($j \in S$) by the probability that after entering state i, the process makes a transition into state j, in an amount of time less than or equal to t. However, it is impossible to define mass functions $Q_{ij}(t)$ for $i \in S^*$, because the epoch entering state i is not a regeneration point. We define the new mass function $Q_{ij}^{(k_1,k_2,\ldots,k_m)}(t)$ which is the probability that after entering state i ($i \in S^\dagger$), the process next makes transitions into states $k_1, k_2, \ldots, k_m$ ($k_1, k_2, \ldots, k_m \in S^*$), and finally, enters state j ($j \in S$), in an amount of time less than or equal to t. Moreover, we define that $H_i(t) \equiv \sum_{j \in S} Q_{ij}(t)$ for $i \in S^\dagger$, which is the unconditional distribution of the time elapsed from state i to the next state entered, possibly i itself.

(1) Type 1 Markov Renewal Process

Consider a Markov renewal process with $m+1$ states, that consists of $S^\dagger = \{0\}$ and $S^* = \{1, 2, \ldots, m\}$ in Figure 1.5. The process starts in state 0, *i.e.*, $Z(0) = 0$, and makes transitions into state 1, 2, ..., m, and comes back to state 0. Then, from straightforward renewal arguments, the first-passage time distributions are

$$F_{01}(t) = Q_{01}(t)$$
$$F_{0j}(t) = Q_{0j}^{(1,2,\ldots,j-1)}(t) \quad (j = 2, 3, \ldots, m) \quad (1.64)$$
$$F_{00}(t) = Q_{00}^{(1,2,\ldots,m)}(t),$$

the renewal functions are

$$M_{01}(t) = Q_{01}(t) + Q_{00}^{(1,2,\ldots,m)}(t) * M_{01}(t) = Q_{01}(t) + F_{00}(t) * M_{01}(t)$$
$$M_{0j}(t) = Q_{0j}^{(1,2,\ldots,j-1)}(t) + F_{00}(t) * M_{0j}(t) \quad (j = 2, 3, \ldots, m) \quad (1.65)$$
$$M_{00}(t) = F_{00}(t) + F_{00}(t) * M_{00}(t),$$

and the transition probabilities are

$$P_{01}(t) = Q_{01}(t) - Q_{02}^{(1)}(t) + Q_{00}^{(1,2,\ldots,m)}(t) * P_{01}(t)$$
$$= Q_{01}(t) - Q_{02}^{(1)}(t) + F_{00}(t) * P_{01}(t)$$
$$P_{0j}(t) = Q_{0j}^{(1,2,\ldots,j-1)}(t) - Q_{0j+1}^{(1,2,\ldots,j)}(t) + F_{00}(t) * P_{0j}(t) \quad (j=2,3,\ldots,m)$$
$$P_{00}(t) = 1 - Q_{01}(t) + F_{00}(t) * P_{00}(t), \tag{1.66}$$

where $Q_{0m+1}^{(1,2,\ldots,m)}(t) = Q_{00}^{(1,2,\ldots,m)}(t)$.

Taking the LS transforms on both sides of (1.65) and (1.66),

$$M_{01}^*(s) = \frac{Q_{01}^*(s)}{1 - F_{00}^*(s)}$$
$$M_{0j}^*(s) = \frac{Q_{0j}^{(1,2,\ldots,j-1)}(s)}{1 - F_{00}^*(s)} \quad (j=2,3,\ldots,m) \tag{1.67}$$
$$M_{00}^*(s) = \frac{F_{00}^*(s)}{1 - F_{00}^*(s)}$$

$$P_{01}^*(s) = \frac{Q_{01}^*(s) - Q_{02}^{*(1)}(s)}{1 - F_{00}^*(s)}$$
$$P_{0j}^*(s) = \frac{Q_{0j}^{*(1,2,\ldots,j-1)}(s) - Q_{0j+1}^{*(1,2,\ldots,j)}(s)}{1 - F_{00}^*(s)} \quad (j=2,3,\ldots m) \tag{1.68}$$
$$P_{00}^*(s) = \frac{1 - Q_{01}^*(s)}{1 - F_{00}^*(s)},$$

where note that $\sum_{j=0}^{m} P_{0j}^*(s) = 1$. From (1.67) and (1.68), the renewal functions and the transition probabilities are computed explicitly upon inversion, however, this might not be easy except in simple cases.

Example 1.6. Consider an n-unit parallel redundant system: When at least one of n units is operating, the system is operating. When all units are down simultaneously, the system fails and will begin to operate again immediately by replacing all failed units with new ones. Each unit operates independently and has an identical failure distribution $F(t)$. The states are denoted by the total number of failed units. When all units begin to operate at time 0, the mass functions are

$$Q_{01}(t) = 1 - [\overline{F}(t)]^n$$
$$Q_{0j}^{(1,2,\ldots,j-1)}(t) = \sum_{i=j}^{n} \binom{n}{i} [F(t)]^i [\overline{F}(t)]^{n-i} \quad (j=2,3,\ldots,n). \tag{1.69}$$

Thus, substituting the above equations into (1.67) and (1.68), we can obtain the renewal functions and the transition probabilities. ∎

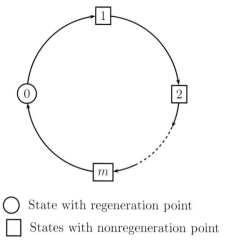

Fig. 1.5. State transition diagram for Type 1 Markov renewal process

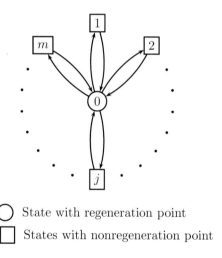

Fig. 1.6. State transition diagram for Type 2 Markov renewal process

(2) Type 2 Markov Renewal Process

Consider a Markov renewal process with $S^\dagger = \{0\}$ and $S^* = \{1, 2, \ldots, m\}$ in Figure 1.6. The process starts in state 0 and only is permitted to make a transition into one state $j \in S^*$, and then return to 0.

The LS transforms of the first-passage time distributions, the renewal functions, and the transition probabilities are

1 Introduction

$$F_{00}^*(s) = \sum_{i=1}^{m} Q_{00}^{*(i)}(s)$$

$$F_{0j}^*(s) = Q_{0j}^*(s) \bigg/ \left[1 - \sum_{\substack{i=1 \\ i \neq j}}^{m} Q_{00}^{*(i)}(s)\right] \qquad (j = 1, 2, \ldots m) \qquad (1.70)$$

$$M_{00}^*(s) = \frac{F_{00}^*(s)}{1 - F_{00}^*(s)}$$

$$M_{0j}^*(s) = \frac{Q_{0j}^*(s)}{1 - F_{00}^*(s)} \qquad (j = 1, 2, \ldots, m) \qquad (1.71)$$

$$P_{00}^*(s) = \frac{1 - \sum_{j=1}^{m} Q_{0j}^*(s)}{1 - F_{00}^*(s)}$$

$$P_{0j}^*(s) = \frac{Q_{0j}^*(s) - Q_{00}^{*(j)}(s)}{1 - F_{00}^*(s)} \qquad (j = 1, 2, \ldots, m). \qquad (1.72)$$

The process corresponds to a special one of Type 1 when $m = 1$. That is, it is the simplest state space with one nonregeneration point. The process takes two alternate states 0 and 1. When the epoch entering state 1 is also a regeneration point, the process becomes an alternating renewal process (see Section 1.3.2).

Example 1.7. Consider a two-unit standby redundant system with repair maintenance [85]. The failure distribution of an operating unit is $F(t)$ and the repair distribution of a failed unit is $G(t)$. When an operating unit fails and the other unit is on standby, the failed unit undergoes repair immediately and the unit on standby takes over the operation. However, when an operating unit fails while the other unit is under repair, the failed unit has to wait for repair until a repairperson is free.

Define the following states.

State 0: One unit is operating and other unit is under repair.
State 1: One unit is operating and the other unit is on standby.
State 2: One unit is under repair and the other unit waits for repair.

The system generates a Markov renewal process with $S^\dagger = \{0\}$ and $S^* = \{1, 2\}$. Then, the mass functions are

$$Q_{01}(t) = \int_0^t \overline{F}(u) \, dG(u), \qquad Q_{02}(t) = \int_0^t \overline{G}(u) \, dF(u)$$

$$Q_{00}^{(1)}(t) = \int_0^t G(u) \, dF(u), \qquad Q_{00}^{(2)}(t) = \int_0^t F(u) \, dG(u).$$

Thus, we can obtain the reliability quantities of the system by using the results of the Type 2 process.

Many redundant systems and stochastic models can be described by the Type 1 or Type 2 process, or mixtures and linkages of Type 1 and Type 2, and the usual Markov renewal processes. ∎

References

1. Hudson WR, Haas R, Uddin W (1997) Infrastructure Management. McGraw-Hill, New York.
2. Blanche KM, Shrisvastava AB (1994) Defining failure of manufacturing machinery and equipment. In: Proceedings Annual Reliability and Maintainability Symposium:69–75.
3. Rausand M, Høyland A (2004) System Reliability Theory. J Wiley & Sons, Hoboken, NJ.
4. Gertsbakh I (2000) Reliability Theory with Applications to Preventive Maintenance. Springer, New York.
5. Duchesne T, Lawless JF (2000) Alternative time scales and failure time models. Life Time Data Analysis 6:157–179.
6. Finkelstein MS (2004) Alternative time scales for systems with random usage. IEEE Trans Reliab 53:261–264.
7. Barlow RE, Proschan F (1965) Mathematical Theory of Reliability. J Wiley & Sons, New York.
8. Barlow RE, Proschan F (1975) Statistical Theory of Reliability and Life Testing Probability Models. Holt, Rinehart & Winston, New York.
9. Jorgenson DW, McCall JJ, Radner R (1967) Optimal Replacement Policy. North-Holland, Amsterdam.
10. Gnedenko BV, Belyayev YK, Solovyev AD (1969) Mathematical Methods of Reliability Theory. Academic, New York.
11. Gertsbakh I (1977) Models of Preventive Maintenance. North-Holland, Amsterdam.
12. Osaki S, Nakagawa T (1976) Bibliography for reliability and availability of stochastic systems. IEEE Trans Reliab R-25:284–287.
13. Pierskalla WP, Voelker JA (1976) A survey of maintenance models: The control and surveillance of deteriorating systems. Nav Res Logist Q 23:353–388.
14. Sherif YS, Smith ML (1981) Optimal maintenance models for systems subject to failure – A review. Nav Res Logist Q 28:47–74.
15. Thomas LC (1986) A survey of maintenance and replacement models for maintainability and reliability of multi-item systems. Reliab Eng 16:297–309.
16. Valdez-Flores C, Feldman RM (1989) A survey of preventive maintenance models for stochastically deteriorating single-unit system. Nav Res Logist Q 36:419–446.
17. Cho DI, Palar M (1991) A survey of maintenance models for multi-unit systems. Eur J Oper Res 51:1–23.
18. Dekker R (1996) Applications of maintenance optimization models: A review and analysis. Reliab Eng Syst Saf 51:229–240.
19. Wang H (2002) A survey of maintenance policies of deteriorating systems. Eur J Oper Res 139:469–489.

20. Özekici S (ed) (1996) Reliability and Maintenance of Complex Systems. Springer, Berlin.
21. Christer AH, Osaki S, Thomas LC (eds) (1997) Stochastic Modelling in Innovative Manufacturing. Lecture Notes in Economics and Mathematical Systems 445, Springer, Berlin.
22. Ben-Daya M, Duffuaa SO, Raouf A (eds) (2000) Maintenance, Modeling and Optimization. Kluwer Academic, Boston.
23. Rahin MA, Ben-Daya M (2001) Integrated Models in Production Planning, Inventory, Quality, and Maintenance. Kluwer Academic, Boston.
24. Osaki S (ed) (2002) Stochastic Models in Reliability and Maintenance. Springer, New York.
25. Pham H (2003) Handbook of Reliability Engineering. Springer, London.
26. Naresky JJ (1970) Reliability definitions. IEEE Trans Reliab R-19:198–200.
27. Hosford JE (1960) Measures of dependability. Oper Res 8:53–64.
28. Lai CD, Xie M (2003) Concepts and applications of stochastic aging in reliability. In: Pham H (ed) Handbook of Reliability Engineering. Springer, London:165–180.
29. Gumbel EJ (1958) Statistics of Extremes. Columbia University Press, New York.
30. Bartholomew DJ (1973) Stochastic Models for Social Processes. J Wiley & Sons, New York.
31. Gross AJ, Clark VA (1975) Survival Distributions: Reliability Applications in the Biomedical Sciences. J Wiley & Sons, New York.
32. Badenius D (1970) Failure rate/MTBF. IEEE Trans Reliab R-19:66–67.
33. Kalbfleisch JD, Prentice RL (1980) The Statistical Analysis of Failure Time Data. J Wiley & Sons, New York.
34. Lane WR, Looney ST, Wansley JW (1986) An application of the Cox proportional hazard model to bank failure. J Banking Insurance 10:511–532.
35. Kumar D, Klefsjö B (1994) Proportional hazard model: A review. Reliab Eng Syst Saf 44:177–188.
36. Block HW, Savits TH, Singh H (1998) The reversed hazard rate function. Prob Eng Inform Sci 12:69–90.
37. Finkelstein MS (2002) On the reversed hazard rate. Reliab Eng Syst Saf 78:71–75.
38. Berg M (1996) Towards rational age-based failure modelling. In: Özekici S (ed) Reliability and Maintenance of Complex Systems. Springer, New York:107–113.
39. Ichida T (1968) Introduction to Maintainability Engineering. Nikkagiren, Tokyo.
40. Nakagawa T (1978) Some inequalities for failure distributions. IEEE Trans Reliab R-27:58–59.
41. Shaked M, Shanthikumar JG (1994) Stochastic Orders and Their Applications. Academic, Boston.
42. Sheikh AK, Younas M, Raouf A (2000) Reliability based spare parts forecasting and procurement strategies. In: Ben-Daya M, Duffuaa SO, Raouf A (eds) Maintenance, Modeling and Optimization. Kluwer Academic, Boston:81–110.
43. Kumar UD, Crocker J, Knezevic J, El-Haram M (2000) Reliability, Maintenance and Logistics Support. Kluwer Academic, Boston.
44. Jiang R, Ji P (2002) Age replacement policy: A multi-attribute value model. Reliab Eng Syst Saf 76:311–318.
45. Xie M, Gaudoin O, Bracquemond C (2002) Redefining failure rate function for discrete distributions. Inter J Reliab Qual Saf Eng 9:275–285.

46. Sandler GH (1963) System Reliability Engineering. Prentice-Hall, Englewood Cliffs, NJ.
47. Kabak IW (1969) System availability and some design implications. Oper Res 17:827–837.
48. Martz Jr HF (1971) On single-cycle availability. IEEE Trans Reliab R-20:21–23.
49. Nakagawa T, Goel AL (1973) A note on availability for a finite interval. IEEE Trans Reliab R-22:271–272.
50. Lie CH, Hwang CL, Tillman FA (1977) Availability of maintained systems: A state-of-the-art survey. AIIE Trans 9:247–259.
51. Aven T (1996) Availability analysis of monotone systems. In: Özekichi S (ed) Reliability and Maintenance of Complex Systems. Springer, New York:206–223.
52. Finkelstein MS, Zarudnij VI (2002) Laplace-transforms and fast-repair approximations for multiple availability and its generations. IEEE Trans Reliab 51:168–176.
53. Finekelstein MS (2003) Modeling the observed failure rate. In: Pham H (ed) Handbook of Reliability Engineering. Springer, London:117–139.
54. Lee KW (2003) Random-request availability. In: Pham (ed) Handbook of Reliability Engineering. Springer, London:643–652.
55. Barlow RE, Hunter LC (1961) Reliability analysis of a one-unit system. Oper Res 9:200–208.
56. Pinedo M (2002) Scheduling Theory, Algorithms, and Systems. Prentice-Hall, Upper Saddle River NJ.
57. Johnson NL, Kotz S (1972) Distributions in Statistics. Vols I, II, III. J Wiley & Sons, New York.
58. Tsokos CP (1972) Probability Distributions: An Introduction to Probability Theory with Applications. Duxbury, Belmont CA.
59. Osaki S (1992) Applied Stochastic System Modeling. Springer, New York.
60. Lawless JF (1983) Statistical methods in reliability. Technometrics 25:305–316.
61. Murthy DNP, Xie M, Jiang R (2004) Weibull Models. J Wiley & Sons, Hoboken, NJ.
62. Nakagawa T, Yoda H (1977) Relationships among distributions. IEEE Trans Reliab R-26:352–353.
63. Nakagawa T, Osaki S (1975) The discrete Weibull distribution. IEEE Trans Reliab R-24:300–301.
64. Ali Khan MS, Khalique A, Abouammoh AM (1989) On estimating parameters in a discrete Weibull distribution. IEEE Trans Reliab 38:348–350.
65. Stein WE, Dattero R (1984) A new discrete Weibull distribution. IEEE Trans Reliab 33:196–197.
66. Padgett WJ, Spurrier JD (1985) Discrete failure models. IEEE Trans Reliab 34:253–256.
67. Råde L (1976) Reliability systems in random environment. J Appl Prob 13:407–410.
68. Feller W (1957) An Introduction to Probability Theory and Its Applications Vol 1. J Wiley & Sons, New York.
69. Cox DR (1962) Renewal Theory. Methuen, London.
70. Smith WL (1958) Renewal theory and its ramifications. J Roy Statist Soc Ser B 20:243–302.
71. Pyke R (1961) Markov renewal processes: Definitions and preliminary properties. Ann Math Statist 32:1231–1242.

72. Pyke R (1961) Markov renewal processes with finitely many states. Ann Math Statist 32:1243–1259.
73. Ross SM (1970) Applied Probability Models with Optimization Applications. Holden-Day, San Francisco.
74. Karlin S, Taylor HM (1975) A First Course in Stochastic Processes. Academic, New York.
75. Çinlar E (1975) Introduction to Stochastic Processes. Prentice-Hall, Englewood Cliffs, NJ
76. Davies B, Martin BL (1979) Numerical inversion of the Laplace transform: A survey and comparison of methods. J Comput Physics 33:1–32.
77. Abate J, Whitt W (1995) Numerical inversion of Laplace transforms of probability distributions. ORSA J Comput 7:36–43.
78. Lorden G (1970) On excess over the boundary. Ann Math Statist 41:520–527.
79. Belzunce F, Ortega EM, Ruiz JM (2001) A note on stochastic comparisons of excess lifetimes of renewal processes. J Appl Prob 38:747–753.
80. Esary JD, Marshall AW, Proschan F (1973) Shock models and wear processes. Ann Prob 1:627–649.
81. Nakagawa T (1974) The expected number of visits to state k before a total system failure of a complex system with repair maintenance. Oper Res 22:108–116.
82. Takács L (1957) On certain sojourn time problems in the theory of stochastic processes. Acta Math Acad Sci Hungary 8:161–191.
83. Calabro SR (1962) Reliability Principles and Practices. McGraw-Hill, New York.
84. Nakagawa T, Osaki S (1976) Markov renewal processes with some non-regeneration points and their applications to reliability theory. Microelectron Reliab 15:633–636.
85. Nakagawa T (2002) Two-unit redundant models. In: Osaki S (ed) Stochastic Models in Reliability and Maintenance. Springer, New York:165–185.
86. Yasui K, Nakagawa T, Sandoh H (2002) Reliability models in data communication systems. In: Osaki S (ed) Stochastic Models in Reliability and Maintenance. Springer, New York:281–306.

2
Repair Maintenance

The most basic maintenance policy for units is to do some maintenance of failed units which is called *corrective maintenance*; *i.e.,* when units fail, they may undergo repair or may be scrapped and replaced. After the repair completion, units can operate again. A system with several units forms semi-Markov processes and Markov renewal processes in stochastic processes. Such reliability models are called *repairman problems* [1], and some useful expressions of reliability measures of many redundant systems were summarized in [2,3]. Early results of two-unit systems and their maintenance (see Section 6.2) were surveyed in [4]. Furthermore, imperfect repair models that do not always become like new after repair were proposed in [5,6] (see Chapter 7).

In this chapter, we are concerned only with reliability characteristics of repairable systems such as mean time to system failure, availability, and expected number of system failures. Such reliability measures are obtained by using the techniques of stochastic processes as described in Section 1.3.

In Section 2.1, we consider the most fundamental one-unit system and survey its reliability quantities such as transition probabilities, downtime distribution, and availabilities. Another point of interest is the *repair limit policy* where the repair of a failed unit is stopped if it is not completed within a planned time T [7]. It is shown that there exists an optimum repair limit time T^* that minimizes the expected cost rate when the repair cost is proportional to time. In Section 2.2, we consider a system with a main unit supported by n spare units, and obtain the mean time to system failure and the expected number of failed spare units [8]. Using these results, we propose several optimization problems. Finally, in Section 2.3, we consider $(n+1)$-unit standby and parallel systems, and derive transition probabilities and first-passage time distributions.

2.1 One-Unit System

An operating unit is repaired or replaced when it fails. When the failed unit undergoes repair, it takes a certain time which may not be negligible. When the repair is completed, the unit begins to operate again. If the failed unit cannot be repaired and spare units are not on hand, it takes a replacement time which may not be negligible.

We consider one operating unit that is repaired immediately when it fails. The failed unit is returned to the operating state when its repair is completed and becomes as good as new. It is assumed that the switchover time from the operating state to the repair state and from the repair state to the operating state are instantaneous. The successive operating times between failures are independently and identically distributed. The successive repair times are also independently, identically distributed and independent of the operating times. Of course, we can consider the repair time as the time required to make a replacement. In this case, the failed unit is replaced with a new one, and its unit operates as same as the failed one.

This system is the most fundamental system that repeats up and down states alternately. The process of such a system can be described by a Markov renewal process with two states, *i.e.*, an alternating renewal process given in Section 1.3 [9]. Many of the known results were summarized in [1, 10].

This section surveys the reliability quantities of a one-unit system and considers a repair limit policy in which the unit under repair is replaced with a new one when the repair is not completed by a fixed time.

2.1.1 Reliability Quantities

(1) Renewal Functions and Transition Probabilities

In the analysis of stochastic models, we are interested in the expected number of system failures during $(0, t]$ and the probability that the system is operating at time t. We obtain the stochastic behavior of a one-unit system by using the techniques in Markov renewal processes.

Assume that the failure time of an operating unit has a general distribution $F(t)$ with finite mean μ and the repair time of failed units has a general distribution $G(t)$ with finite mean β, where $\overline{\Phi} \equiv 1 - \Phi$ for any function Φ, where, in general, μ and β are referred to as *mean time to failure* (MTTF) and *mean time to repair* (MTTR), respectively. To analyze the system, we define the following states.

State 0: Unit is operating.
State 1: Unit is under repair.

Suppose that the unit begins to operate at time 0. The system forms a Markov renewal or semi-Markov process with two states of up and down as shown in Figure 1.4 of Section 1.3.2.

2.1 One-Unit System

Define the mass function $Q_{ij}(t)$ from state i to state j by the probability that after making a transition into state i, the system next makes a transition into state j ($i, j = 0, 1$), in an amount of time less than or equal to time t. Then, from a Markov renewal process, we can easily have

$$Q_{01}(t) = F(t), \qquad Q_{10}(t) = G(t).$$

Let $M_{ij}(t)$ denote the expected number of occurrences of state j during $(0, t]$ when the system goes into state i at time 0, where the first visit to state j is not counted when $i = j$. Then, from Section 1.3, we have the following renewal equations:

$$M_{01}(t) = Q_{01}(t) * [1 + M_{11}(t)], \quad M_{10}(t) = Q_{10}(t) * [1 + M_{00}(t)],$$

and $M_{11}(t) = Q_{10}(t) * M_{01}(t)$, $M_{00}(t) = Q_{01}(t) * M_{10}(t)$, where the asterisk denotes the pairwise Stieltjes convolution; i.e., $a(t) * b(t) \equiv \int_0^t a(t-u) db(u)$. Thus, forming the Laplace–Stieltjes (LS) transforms of both sides of these equations and solving them, we have

$$M_{01}^*(s) = \frac{Q_{01}^*(s)}{1 - Q_{01}^*(s) Q_{10}^*(s)} = \frac{F^*(s)}{1 - F^*(s) G^*(s)} \qquad (2.1)$$

$$M_{10}^*(s) = \frac{Q_{10}^*(s)}{1 - Q_{01}^*(s) Q_{10}^*(s)} = \frac{G^*(s)}{1 - F^*(s) G^*(s)} \qquad (2.2)$$

and $M_{11}^*(s) = G^*(s) M_{01}^*(s) = M_{00}^*(s) = F^*(s) M_{10}^*(s)$, where the asterisk of the function denotes the LS transform with itself; i.e., $\Phi^*(s) \equiv \int_0^\infty e^{-st} d\Phi(t)$ for any function $\Phi(t)$.

Furthermore, let $P_{ij}(t)$ denote the probability that the system is in state j at time t if it starts in state i at time 0. Then, from Section 1.3,

$$P_{00}(t) = 1 - Q_{01}(t) + Q_{01}(t) * P_{10}(t)$$
$$P_{11}(t) = 1 - Q_{10}(t) + Q_{10}(t) * P_{01}(t)$$

and $P_{10}(t) = Q_{10}(t) * P_{00}(t)$, $P_{01}(t) = Q_{01}(t) * P_{11}(t)$. Thus, again forming the LS transforms,

$$P_{00}^*(s) = \frac{1 - Q_{01}^*(s)}{1 - Q_{01}^*(s) Q_{10}^*(s)} = \frac{1 - F^*(s)}{1 - F^*(s) G^*(s)} \qquad (2.3)$$

$$P_{11}^*(s) = \frac{1 - Q_{10}^*(s)}{1 - Q_{01}^*(s) Q_{10}^*(s)} = \frac{1 - G^*(s)}{1 - F^*(s) G^*(s)} \qquad (2.4)$$

and $P_{10}^*(s) = G^*(s) P_{00}^*(s)$, $P_{01}^*(s) = F^*(s) P_{11}^*(s)$. Thus, from (2.1) to (2.4), we have the following relations.

$$P_{01}(t) = M_{01}(t) - M_{00}(t), \qquad P_{10}(t) = M_{10}(t) - M_{11}(t).$$

Moreover, we have

$$P_{01}^*(s) = \frac{F^*(s)[1-G^*(s)]}{1-F^*(s)G^*(s)} = \int_0^\infty e^{-st}\overline{G}(t-u)\,\mathrm{d}M_{01}(u)$$

$$P_{10}^*(s) = \frac{G^*(s)[1-F^*(s)]}{1-F^*(s)G^*(s)} = \int_0^\infty e^{-st}\overline{F}(t-u)\,\mathrm{d}M_{10}(u);$$

i.e.,

$$P_{01}(t) = \int_0^t \overline{G}(t-u)\,\mathrm{d}M_{01}(u), \qquad P_{10}(t) = \int_0^t \overline{F}(t-u)\,\mathrm{d}M_{10}(u).$$

These relations with renewal functions and transition probabilities would be useful for the analysis of more complex systems.

Next, let $h(t)$ and $r(t)$ be the failure rate and the repair rate of the unit, respectively; i.e., $h(t) \equiv f(t)/\overline{F}(t)$ and $r(t) \equiv g(t)/\overline{G}(t)$, where f and g are the respective density functions of F and G. Then, from (2.1) to (2.4), we also have

$$\min_{x \le t} h(x) \int_0^t P_{00}(u)\,\mathrm{d}u \le M_{01}(t) \le \max_{x \le t} h(x) \int_0^t P_{00}(u)\,\mathrm{d}u$$

$$\min_{x \le t} r(x) \int_0^t P_{11}(u)\,\mathrm{d}u \le M_{10}(t) \le \max_{x \le t} r(x) \int_0^t P_{11}(u)\,\mathrm{d}u.$$

All inequalities equal when both F and G are exponential, which is shown in Example 2.1.

There exist $P_j \equiv \lim_{t\to\infty} P_{ij}(t)$ and $M_j \equiv \lim_{t\to\infty} M_{ij}(t)/t$, independent of an initial state i, because the system forms a Markov renewal process with one positive recurrent. Thus, from (1.63) we have

$$M_0 = \lim_{s\to 0} sM_{00}^*(s) = \frac{1}{\mu+\beta} = M_1 \tag{2.5}$$

$$P_0 = \lim_{s\to 0} P_{00}^*(s) = \frac{\mu}{\mu+\beta} = 1 - P_1. \tag{2.6}$$

In general, it is often impossible to invert explicitly the LS transforms of $M_{ij}^*(s)$ and $P_{ij}^*(s)$ in (2.1) to (2.4), and it is very difficult even to invert them numerically [11,12]. However, we can state the following asymptotic describing behaviors for small t and large t.

First, we consider the approximation calculation for small t. Reliability calculations for small t are needed in considering the near-term future security of an operating bulk power system [13]. We can rewrite (2.3) as

$$P_{00}^*(s) = 1 - F^*(s) + F^*(s)G^*(s) - [F^*(s)]^2 G^*(s) + \cdots.$$

Because the probability that the process makes more than two transitions in a short time is very small, by dropping the terms with higher degrees than $F^*(s)G^*(s)$, we have

$$P_{00}^*(s) \approx 1 - F^*(s) + F^*(s)G^*(s);$$

i.e.,
$$P_{00}(t) \approx \overline{F}(t) + \int_0^t G(t-u)\,dF(u). \tag{2.7}$$

Similarly,
$$P_{01}(t) \approx \int_0^t \overline{G}(t-u)\,dF(u) \tag{2.8}$$

$$M_{00}(t) \approx \int_0^t G(t-u)\,dF(u), \qquad M_{01}(t) \approx F(t). \tag{2.9}$$

Next, we obtain the asymptotic forms for large t [9]. By expanding e^{-st} in a Taylor series on the LS transforms of $F^*(s)$ and $G^*(s)$ as $s \to 0$, it follows that

$$F^*(s) = 1 - \mu s + \frac{1}{2}(\mu^2 + \sigma_\mu^2)s^2 + o(s^2)$$

$$G^*(s) = 1 - \beta s + \frac{1}{2}(\beta^2 + \sigma_\beta^2)s^2 + o(s^2),$$

where σ_μ^2 and σ_β^2 are the variances of F and G, respectively, and $o(s)$ is an infinite decimal higher than s. Thus, substituting these equations into (2.1), we have

$$M_{01}^*(s) = \frac{1}{\mu+\beta}\frac{1}{s} - \frac{\mu}{\mu+\beta} + \frac{1}{2} + \frac{\sigma_\mu^2 + \sigma_\beta^2}{2(\mu+\beta)^2} + o(1).$$

Formal inversion of $M_{01}^*(s)$ gives that for large t,

$$M_{01}(t) = \frac{t}{\mu+\beta} - \frac{\mu}{\mu+\beta} + \frac{1}{2} + \frac{\sigma_\mu^2 + \sigma_\beta^2}{2(\mu+\beta)^2} + o(1). \tag{2.10}$$

Similarly,
$$M_{00}(t) = \frac{t}{\mu+\beta} - \frac{1}{2} + \frac{\sigma_\mu^2 + \sigma_\beta^2}{2(\mu+\sigma)^2} + o(1) \tag{2.11}$$

$$P_{00}(t) = \frac{\mu}{\mu+\beta} + o(1), \qquad P_{01}(t) = \frac{\beta}{\mu+\beta} + o(1). \tag{2.12}$$

Example 2.1. Suppose that $F(t) = 1 - e^{-\lambda t}$ and $G(t) = 1 - e^{-\theta t}$ ($\theta \neq \lambda$). Then, it is easy to invert the LS transforms of $P_{01}^*(s)$ and $M_{01}^*(s)$,

$$P_{01}(t) = \frac{\lambda}{\lambda+\theta}[1 - e^{-(\lambda+\theta)t}]$$

$$M_{01}(t) = \frac{\lambda\theta t}{\lambda+\theta} + \left(\frac{\lambda}{\lambda+\theta}\right)^2 [1 - e^{-(\lambda+\theta)t}].$$

Furthermore, for small t,

$$P_{01}(t) \approx \frac{\lambda}{\theta-\lambda}(e^{-\lambda t} - e^{-\theta t}), \qquad M_{01}(t) \approx 1 - e^{-\lambda t}$$

44   2 Repair Maintenance

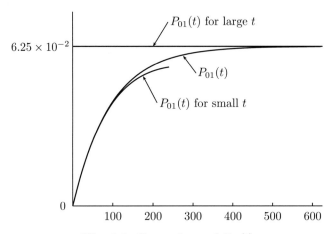

Fig. 2.1. Comparisons of $P_{01}(t)$

and for large t,

$$P_{01}(t) \approx \frac{\lambda}{\lambda + \theta}, \qquad M_{01}(t) \approx \frac{\lambda \theta t}{\lambda + \theta} + \left(\frac{\lambda}{\lambda + \theta}\right)^2.$$

Figure 2.1 shows the value of $P_{01}(t)$ and the approximate values of $P_{01}(t)$ for small t and large t when $1/\lambda = 1500$ hours and $1/\theta = 100$ hours. In this case, we can use these approximate values for about fewer than 100 hours and more than 500 hours. This indicates that these approximations are comparatively fitted for a long interval of time t. ∎

Example 2.2. When $F(t) = 1 - (1 + \lambda t)e^{-\lambda t}$ and the time for repair is constant β,

$$P_{01}^*(s) = \frac{\lambda^2(1 - e^{-\beta s})}{(s + \lambda)^2 - \lambda^2 e^{-\beta s}}.$$

Furthermore, for small t,

$$P_{01}(t) \approx 1 - (1 + \lambda t)e^{-\lambda t} - \begin{cases} 0 & \text{for } t < \beta \\ 1 - [1 + \lambda(t - \beta)]e^{-\lambda(t - \beta)} & \text{for } t \geq \beta \end{cases}$$

and for large t,

$$P_{01}(t) \approx \frac{\beta}{2/\lambda + \beta}. \quad \blacksquare$$

(2) Downtime Distribution

A quantity of most interest is the behavior of system down or system failure. It is of great importance to know how long and how many times the system is down during $(0, t]$, because the system down is sometimes costly and/or dangerous. It was shown in [10] that the downtime distribution of a one-unit system is given from the result of a stochastic process [14]. The excess time which is the time spent in t due to failures was proposed and its stochastic properties were reviewed in [15, 16]. Furthermore, the downtime distribution was derived in the case where failure and repair times are dependent [17].

We have already derived in (1): the probability $P_{01}(t)$ that the system is down at time t, the mean downtime $\int_0^t P_{01}(u)du$ during $(0, t]$, and the expected number $M_{01}(t)$ of system down during $(0, t]$. Of other interest is to show (i) the downtime distribution, (ii) the mean time that the total downtime during $(0, t]$ exceeds a specified level $\delta > 0$ for the first time, and (iii) the first time that an amount of downtime exceeds a specified level c.

Suppose that the unit begins to operate at time 0. Let $D(t)$ denote the total amount of downtime during $(0, t]$. Then, the distribution of downtime $D(t)$ is, from (1.35) in Section 1.3,

$$\Omega(t, x) \equiv \Pr\{D(t) \leq x\}$$
$$= \begin{cases} \sum_{n=0}^{\infty} G^{(n)}(x)[F^{(n)}(t-x) - F^{(n+1)}(t-x)] & \text{for } t > x \\ 1 & \text{for } t \leq x, \end{cases} \quad (2.13)$$

where $F^{(n)}(t)$ ($G^{(n)}(t)$) denotes the n-fold Stieltjes convolution of F (G) with itself, and $F^{(0)}(t) = G^{(0)}(t) \equiv 1$ for $t \geq 0$ and 0 for $t < 0$. Equation (2.13) can also be written as

$$\Omega(t+x, x) = \Pr\{D(t+x) \leq x\}$$
$$= \sum_{n=0}^{\infty} G^{(n)}(x)[F^{(n)}(t) - F^{(n+1)}(t)] \quad (2.14)$$

which is called *excess time* [15]. Furthermore, the survival distribution of downtime is

$$1 - \Omega(t, x) = \Pr\{D(t) > x\}$$
$$= \begin{cases} \sum_{n=0}^{\infty} [G^{(n)}(x) - G^{(n+1)}(x)]F^{(n+1)}(t-x) & \text{for } t > x \\ 0 & \text{for } t \leq x. \end{cases} \quad (2.15)$$

Takács also proved the following important theorem.

Theorem 2.1. Suppose that μ, β and σ_μ^2, σ_β^2 are the means and variances of distributions $F(t)$ and $G(t)$, respectively. If $\sigma_\mu^2 < \infty$ and $\sigma_\beta^2 < \infty$ then

$$\lim_{t\to\infty}\Pr\left\{\frac{D(t)-\beta t/(\mu+\beta)}{\sqrt{[(\beta\sigma_\mu)^2+(\mu\sigma_\beta)^2]t/(\mu+\beta)^3}}\le x\right\}=\frac{1}{\sqrt{2\pi}}\int_{-\infty}^{x}e^{-u^2/2}du. \quad (2.16)$$

That is, if the means and variances of F and G are statistically estimated then the probability of the amount of $D(t)$ is approximately obtained for large t, by using a standard normal distribution.

Next, let $T_\delta \equiv \min_t\{D(t) > \delta\}$ be the first time that the total downtime exceeds a specified level $\delta > 0$. Then, from (2.15),

$$J_\delta(t) \equiv \Pr\{T_\delta \le t\} = \Pr\{D(t) > \delta\}$$
$$= \sum_{n=0}^{\infty}[G^{(n)}(\delta) - G^{(n+1)}(\delta)]F^{(n+1)}(t-\delta) \qquad \text{for } t > \delta. \quad (2.17)$$

The mean time that the total time first exceeds δ is

$$l_\delta \equiv \int_0^\infty \overline{J}_\delta(t)\,dt = \delta + \mu\sum_{n=0}^{\infty}G^{(n)}(\delta). \quad (2.18)$$

Example 2.3. Suppose that $F(t) = 1 - e^{-\lambda t}$ and the time for repair is constant β [1, pp. 78–79]. Then, the downtime distribution is

$$\Omega(t,x) = \sum_{n=0}^{[x/\beta]}\frac{[\lambda(t-x)]^n}{n!}e^{-\lambda(t-x)} \qquad \text{for } t > x$$

and

$$l_\delta = \delta + \frac{1}{\lambda}\left\{\left[\frac{\delta}{\beta}\right]+1\right\},$$

where $[x]$ denotes the greatest integer contained in x. In addition, the expected number of systems down during $(0, t]$ is

$$M_{01}(t) = \left[\frac{t}{\beta}\right]+1-\sum_{j=0}^{[t/\beta]}\sum_{k=0}^{j}\frac{\lambda^k(t-\beta j)^k e^{-\lambda(t-\beta j)}}{k!}$$

and the probability that the system is down at time t is

$$P_{01}(t) = 1 - \sum_{j=0}^{[t/\beta]}\frac{\lambda^j(t-\beta j)^j e^{-\lambda(t-\beta j)}}{j!}. \quad \blacksquare$$

Finally, we consider the first time that an amount of a single downtime exceeds a fixed time $c > 0$, where c is considered to be a critically allowed time for repair [18]. For example, we can give a fuel charge and discharge system for a nuclear reactor that shuts down spontaneously when the system has failed more than time c [19]. The distribution $L(t)$ of the first time that an amount of downtime first exceeds time c is given by applying a terminating renewal process. Then, from (1.39) and (1.40), the LS transform of $L(t)$ and its mean time l are, respectively,

$$L^*(s) = \frac{F^*(s)e^{-sc}\overline{G}(c)}{1 - F^*(s)\int_0^c e^{-st}\,dG(t)}, \qquad l = \frac{\mu + \int_0^c \overline{G}(t)\,dt}{\overline{G}(c)}. \qquad (2.19)$$

(3) Availability

We derive the exact expressions of availabilities for a one-unit system with repair introduced in Section 1.1. Suppose that the unit begins to operate at time 0.

(i) *Pointwise availability*: From (2.3),

$$A(t) = P_{00}(t) = \overline{F}(t) * [1 + F(t)*G(t) + F(t)*G(t)*F(t)*G(t) + \cdots]; \quad (2.20)$$

i.e.,

$$A(t) = \overline{F}(t) + \int_0^t \overline{F}(t-u)\,dM_{00}(u)$$

and its LS transform is

$$A^*(s) = \frac{1 - F^*(s)}{1 - F^*(s)G^*(s)}. \qquad (2.21)$$

Furthermore, when $m_{01}(t) \equiv dM_{01}(t)/dt$ exists, from the results **(1)** of Section 2.1.1, we have

$$\min_{x \le t} h(x) A(t) \le m_{01}(t) \le \max_{x \le t} h(x) A(t)$$

$$\overline{A}(t) \equiv 1 - A(t) = \int_0^t \overline{G}(t-u) m_{01}(u)\,du.$$

Thus, we have the inequality [20, p. 107]

$$\overline{A}(t) \le \max_{x \le t} h(x) \int_0^t \overline{G}(t-u) A(u)\,du$$

$$\le \max_{x \le t} h(x) \int_0^t \overline{G}(u)\,du \le \max_{x \le t} h(x)\beta \qquad (2.22)$$

which give the upper bounds of the unavailability at time t.

(ii) *Interval availability*:

$$\frac{1}{t}\int_0^t A(u)\,du = \frac{1}{t}\int_0^t P_{00}(u)\,du. \qquad (2.23)$$

48   2 Repair Maintenance

(iii) *Limiting interval availability*:

$$A = \lim_{t \to \infty} P_{00}(t) = \frac{\mu}{\mu + \beta} = \frac{\text{MTTF}}{\text{MTTF} + \text{MTTR}} \quad (2.24)$$

which is sometimes called simply *availability*.

(iv) *Multiple cycle availability*:

$$A(n) = \int_0^\infty \int_0^\infty \frac{x}{x+y} \, dF^{(n)}(x) \, dG^{(n)}(y) \quad (n = 1, 2, \dots). \quad (2.25)$$

(v) *Multiple cycle availability with probability*: Because

$$\Pr\left\{a \sum_{i=1}^n X_i \geq \sum_{i=1}^n Y_i\right\} = \int_0^\infty G^{(n)}(ax) \, dF^{(n)}(x) \quad \text{for } a > 0$$

$$\Pr\left\{\frac{\sum_{i=1}^n X_i}{\sum_{i=1}^n (X_i + Y_i)} \geq y\right\} = \int_0^\infty G^{(n)}\left(\frac{x}{y} - x\right) dF^{(n)}(x) \quad \text{for } 0 < y \leq 1.$$

Thus, putting $y = A_\nu(n)$ in the above equation,

$$\int_0^\infty G^{(n)}\left(\frac{x}{A_\nu(n)} - x\right) dF^{(n)}(x) = \nu \quad (n = 1, 2, \dots). \quad (2.26)$$

(vi) *Interval availability with probability*: Let $U(t)$ denote the total amount of uptime during $(0, t]$; i.e., $U(t) \equiv t - D(t)$. Then, from the downtime distribution in (2.13),

$$\Pr\left\{\frac{U(t)}{t} \geq y\right\} = \Pr\{D(t) \leq t - ty\}$$

$$= \sum_{n=0}^\infty G^{(n)}(t - ty)[F^{(n)}(ty) - F^{(n+1)}(ty)] \quad \text{for } 0 < y \leq 1.$$

Thus, it is given by solving

$$\sum_{n=0}^\infty G^{(n)}(t - tA_\nu(t))[F^{(n)}(tA_\nu(t)) - F^{(n+1)}(tA_\nu(t))] = \nu. \quad (2.27)$$

Furthermore, the interval reliability is, from (1.14),

$$R(x; t) = \overline{F}(t + x) + \int_0^t \overline{F}(t + x - u) \, dM_{00}(u) \quad (2.28)$$

and its Laplace transform is

$$R^*(x; s) = \int_0^\infty e^{-st} R(x; t) \, dt = \frac{e^{sx} \int_x^\infty e^{-st} \overline{F}(t) \, dt}{1 - F^*(s) G^*(s)}. \quad (2.29)$$

Thus, the limiting interval reliability is [21, 22]

$$R(x) \equiv \lim_{t \to \infty} R(x; t) = \lim_{s \to 0} s R^*(x; s) = \frac{\int_x^\infty \overline{F}(t) \, dt}{\mu + \beta}. \quad (2.30)$$

2.1 One-Unit System

We give the exact expressions of the above availabilities for two particular cases [10, 23–26].

Example 2.4. When $F(t) = 1 - e^{-\lambda t}$ and $G(t) = 1 - e^{-\theta t}$, the availabilities are given as follows.

(i) $A(t) = \dfrac{\theta}{\lambda + \theta} + \dfrac{\lambda}{\lambda + \theta} e^{-(\lambda+\theta)t}$

$\overline{A}(t) = \dfrac{\lambda}{\lambda + \theta}(1 - e^{-(\lambda+\theta)t}) \leq \dfrac{\lambda}{\lambda + \theta} < \dfrac{\lambda}{\theta}.$

(ii) $\dfrac{1}{t}\displaystyle\int_0^t A(u)\,du = \dfrac{\theta}{\lambda + \theta} + \dfrac{\lambda}{(\lambda + \theta)^2 t}(1 - e^{-(\lambda+\theta)t})$

$\dfrac{1}{t}\displaystyle\int_0^t \overline{A}(u)\,du = \dfrac{\lambda}{\lambda + \theta} - \dfrac{\lambda}{(\lambda + \theta)^2 t}(1 - e^{-(\lambda+\theta)t}) \leq \dfrac{\lambda t}{2}.$

(iii) $A = \dfrac{\theta}{\lambda + \theta}.$

(iv) $A(n) = \displaystyle\int_0^\infty \dfrac{n(\lambda\theta)^n}{(n-1)!} y^{2n-1} \Gamma(-n, \lambda y) e^{(\lambda-\theta)y}\,dy,$

where $\Gamma(\alpha, x) \equiv \int_x^\infty u^{\alpha-1} e^{-u}\,du$. In particular,

$$A(1) = \begin{cases} \dfrac{\theta}{\theta - \lambda} + \dfrac{\lambda\theta}{(\theta-\lambda)^2}\log\dfrac{\lambda}{\theta} & \text{for } \lambda \neq \theta \\ \dfrac{1}{2} & \text{for } \lambda = \theta. \end{cases}$$

(v) $A_\nu(n)$ is given by solving

$$\sum_{j=0}^{n-1}\binom{n+j-1}{j}\left\{\dfrac{\theta[(1/A_\nu(n))-1]}{\lambda+\theta[(1/A_\nu(n))-1]}\right\}^j \left\{\dfrac{\lambda}{\lambda+\theta[1/(A_\nu(n))-1]}\right\}^n = 1 - \nu.$$

In particular,

$$A_\nu(1) = \dfrac{(1-\nu)\theta}{\lambda\nu + (1-\nu)\theta}.$$

(vi) $A_\nu(t)$ is given by solving

$$e^{-\lambda t A_\nu(t)}\left[1 + \sqrt{\lambda\theta t A_\nu(t)}\int_0^{t(1-A_\nu(t))} e^{-\theta y} y^{-1/2} I_1(2\sqrt{\lambda\theta y A_\nu(t)})\,dy\right] = \nu,$$

where $I_1(x) \equiv \sum_{j=0}^\infty (x/2)^{2j+1}/[j!(j+1)!]$.

The interval reliability is

$$R(x; t) = \left[\dfrac{\theta}{\lambda+\theta} + \dfrac{\lambda}{\lambda+\theta} e^{-(\lambda+\theta)t}\right] e^{-\lambda x} = A(t)\overline{F}(x)$$

and its limiting interval reliability is

$$R(x) = \frac{\theta}{\lambda+\theta}e^{-\lambda x} = A\overline{F}(x). \blacksquare$$

Example 2.5. Suppose that $F(t) = 1 - e^{-\lambda t}$ and the time for repair is constant β.

(i) $A(t) = \sum_{j=0}^{[t/\beta]} \frac{\lambda^j(t-\beta j)^j}{j!} e^{-\lambda(t-\beta j)}.$

(ii) $\frac{1}{t}\int_0^t A(u)\, du = \frac{1}{\lambda t}\left\{\left[\frac{t}{\beta}\right]+1-\sum_{j=0}^{[t/\beta]}\sum_{k=0}^{j}\frac{\lambda^k(t-\beta j)^k}{k!}e^{-\lambda(t-\beta j)}\right\}.$

(iii) $A = \frac{1/\lambda}{1/\lambda + \beta}.$

(iv) $A(n) = n(n\lambda\beta)^n e^{n\lambda\beta}\Gamma(-n, n\lambda\beta).$
In particular,
$$A(1) = 1 - \lambda\beta e^{\lambda\beta}\int_{\lambda\beta}^\infty u^{-1}e^{-u}\, du.$$

(v) $A_\nu(n)$ is given by solving
$$\sum_{j=0}^{n-1}\frac{[n\lambda\beta A_\nu(n)/(1-A_\nu(n))]^j}{j!}\exp[-n\lambda\beta A_\nu(n)/(1-A_\nu(n))] = \nu.$$

In particular,
$$A_\nu(1) = \frac{\log(1/\nu)}{\lambda\beta + \log(1/\nu)}.$$

(vi) $A_\nu(t)$ is given by solving
$$\sum_{j=0}^{[t(1-A_\nu(t))/\beta]}\frac{[\lambda t A_\nu(t)]^j}{j!}\exp[-\lambda t A_\nu(t)] = \nu. \blacksquare$$

Finally, we give the example of asymptotic behavior shown in [1, 26].

Example 2.6. We wish to compute the time T when the probability that the system is down more than T in $t = 10{,}000$ hours of operation is given by 0.90, and the availability $A_\nu(t)$ when $\nu = 0.90$. The failure and repair distributions are unknown, but from the sample data, the estimates of means and variances are:
$$\mu = 1{,}000, \qquad \sigma_\mu^2 = 100{,}000, \qquad \beta = 100, \qquad \sigma_\beta^2 = 400.$$
Then, from Theorem 2.1, when $t = 10{,}000$,
$$\frac{D(t) - \beta t/(\mu+\beta)}{\sqrt{[(\beta\sigma_\mu)^2 + (\mu\sigma_\beta)^2]t/(\mu+\beta)^3}} = \frac{D(10{,}000) - 909.09}{102.56}$$

is approximately normally distributed with mean 0 and variance 1. Thus,

$$\Pr\{D(t) > T\} = \Pr\left\{\frac{D(10,000) - 909.09}{102.56} > \frac{T - 909.09}{102.56}\right\}$$

$$\approx \frac{1}{\sqrt{2\pi}} \int_{(T-909.09)/102.56}^{\infty} e^{-u^2/2}\, du = 0.90.$$

Because $u_0 = -1.28$ such that $(1/\sqrt{2\pi}) \int_{u_0}^{\infty} e^{-u^2/2} du = 0.90$, we have

$$T = 909.09 - 102.56 \times 1.28 = 777.81.$$

Moreover, from the relation $U(t) = t - D(t)$, we have

$$\Pr\left\{\frac{U(t)/t - \mu/(\mu+\beta)}{\sqrt{[(\beta\sigma_\mu)^2 + (\mu\sigma_\beta)^2]/[t(\mu+\beta)^3]}} > -y\right\} \approx \frac{1}{\sqrt{2\pi}} \int_{-\infty}^{y} e^{-u^2/2}\, du$$

$$= 0.90.$$

Thus, we have approximately

$$A_\nu(t) = \frac{\mu}{\mu+\beta} + u_0 \sqrt{\frac{(\beta\sigma_\mu)^2 + (\mu\sigma_\beta)^2}{t(\mu+\beta)^3}} = 0.896.$$

In this case, it can be said that with probability 0.90 the system will operate for at least 89.6 percent of the time interval 10,000 hours. ∎

2.1.2 Repair Limit Policy

Until now, we have analyzed a one-unit system which is repaired upon failure and then returns to operation without having any preventive maintenance (PM). The first PM policy for an operating unit, in which it is repaired at failure or at time T, whichever occurs first, was defined in [27]. The optimum PM policy that maximizes the availability was derived in [10]. We discuss some PM policies in Chapters 6 and 7.

An alternative considered here is to repair a failed unit if the repair time is short or to replace it if the repair time is long. This is achieved by stopping the repair if it is not completed within a repair limit time, and the unit is replaced. This policy is optimum over both deterministic and random repair limit time policies [28]. We discuss optimum repair limit policies that minimize the expected cost rates for an infinite time span. An optimum repair limit time is analytically obtained in the case where the repair cost is proportional to time.

Similar repair limit problems can be applied to army vehicles [29–33]. When a unit requires repair, it is first inspected and its repair cost is estimated. If the estimated cost exceeds a certain amount, the unit is not repaired but is replaced. The authors further derived the repair limiting value, in which the expected future cost per vehicle-year when the failed vehicle is repaired is equal to the cost when the failed vehicle is scrapped and a new one is substituted. They used three methods of optimizing the repair limit policies such as simulation, hill-climbing, and dynamic programming. More general forms of repair costs were given in [34]. Using the nonparametric and graphical methods, several problems were solved in [35, 36].

Consider a one-unit system that is repaired or replaced if it fails. Let μ denote the finite mean failure time of the unit and $G(t)$ denote the repair distribution of the failed unit with finite mean β. It is assumed that a failure of the unit is immediately detected, and it is repaired or replaced and becomes as good as new upon repair or replacement.

When the unit fails, its repair is started immediately, and when the repair is not completed within time T ($0 \leq T \leq \infty$), which is called the *repair limit time*, it is replaced with a new one. Let c_1 be the replacement cost of a failed unit that includes all costs caused by failure and replacement. Let $c_r(t)$ be the expected repair cost during $(0, t]$, which also includes all costs incurred due to repair and downtime during $(0, t]$, and be bounded on a finite interval.

Consider one cycle from the beginning of an operative unit to the repair or replacement completion. Each cycle is independently and identically distributed, and hence, a sequence of cycles forms a renewal process. Then, the expected cost of one cycle is

$$[c_1 + c_r(T)]\overline{G}(T) + \int_0^T c_r(t)\,\mathrm{d}G(t) = c_1\overline{G}(T) + \int_0^T \overline{G}(t)\,\mathrm{d}c_r(t)$$

and the mean time of one cycle is

$$\mu + T\overline{G}(T) + \int_0^T t\,\mathrm{d}G(t) = \mu + \int_0^T \overline{G}(t)\,\mathrm{d}t.$$

Thus, from Theorem 1.6, the expected cost rate for an infinite span (see (3.3) in Chapter 3) is

$$C(T) = \frac{c_1\overline{G}(T) + \int_0^T \overline{G}(t)\,\mathrm{d}c_r(t)}{\mu + \int_0^T \overline{G}(t)\,\mathrm{d}t}. \tag{2.31}$$

It is evident that

$$C(0) \equiv \lim_{T \to 0} C(T) = \frac{c_1}{\mu} \tag{2.32}$$

$$C(\infty) \equiv \lim_{T \to \infty} C(T) = \frac{\int_0^\infty \overline{G}(t)\,\mathrm{d}c_r(t)}{\mu + \beta} \tag{2.33}$$

which represent the expected cost rates with only replacement and only repair maintenance, respectively.

Consider the special case where the repair cost is proportional to time; i.e., $c_r(t) = at^b$ for $a > 0$ and $b \geq 0$. The repair cost would be dependent on downtime and repairpersons, both of which are approximately proportional to time. In this case, the expected cost rate is

$$C(T) = \frac{c_1 \overline{G}(T) + ab \int_0^T t^{b-1} \overline{G}(t)\,dt}{\mu + \int_0^T \overline{G}(t)\,dt}. \tag{2.34}$$

If $\int_0^\infty t^b dG(t) \equiv \beta_b < \infty$ then

$$C(\infty) = \frac{a\beta_b}{\mu + \beta}. \tag{2.35}$$

We find an optimum repair limit time T^* that minimizes $C(T)$. It is assumed that there exists a density function $g(t)$ of $G(t)$ and let $r(t) \equiv g(t)/\overline{G}(t)$ be the repair rate. Then, differentiating $C(T)$ with respect to T and setting it equal to zero yield

$$r(T)\left[\mu + \int_0^T \overline{G}(t)\,dt\right] + \overline{G}(T)$$
$$= \frac{ab}{c_1}\left\{T^{b-1}\left[\mu + \int_0^T \overline{G}(t)\,dt\right] - \int_0^T t^{b-1}\overline{G}(t)\,dt\right\}. \tag{2.36}$$

If there exists a finite and positive T^* that minimizes $C(T)$, it has to satisfy (2.36). Otherwise, an optimum repair limit time is $T^* = 0$ or $T^* = \infty$.

Consider the particular case of $b = 1$; i.e., $c_r(t) = at$. Let

$$k \equiv \frac{a\mu - c_1}{c_1 \mu}, \qquad K \equiv \frac{a\mu}{c_1(\mu + \beta)},$$

where k might be negative. Substituting $b = 1$ into (2.36),

$$r(T)\left[\mu + \int_0^T \overline{G}(t)\,dt\right] + \overline{G}(T) = \frac{a\mu}{c_1}. \tag{2.37}$$

Letting $Q(T)$ be the left-hand side of (2.37), we have

$$Q(0) \equiv \mu r(0) + 1, \qquad Q(\infty) = (\mu + \beta)r(\infty)$$

and furthermore, $Q(T)$ and $r(T)$ are monotonic together. Hence, if $r(t)$ is strictly decreasing and $Q(0) > a\mu/c_1 > Q(\infty)$; i.e., $r(0) > k$ and $r(\infty) < K$, there exists uniquely a finite and positive T^* that minimizes $C(T)$, and

$$C(T^*) = a - c_1 r(T^*). \tag{2.38}$$

If $r(0) \leq k$ then $Q(T) < a\mu/c_1$ and $dC(T)/dT > 0$ for any $T > 0$. Thus, the optimum time is $T^* = 0$; i.e., no repair should be made. If $r(\infty) \geq K$ then $Q(T) > a\mu/c_1$ and $dC(T)/dT < 0$ for any $T < \infty$. Thus, the optimum time is $T^* = \infty$; i.e., no replacement should be made.

From the above discussions, we have the following optimum policy when $r(t)$ is continuous and strictly decreasing.

2 Repair Maintenance

Table 2.1. Optimum repair limit time T^* and expected cost rate $C(T^*)$ when $a = 3$, $\mu = 10$, and $c_1 = 10$

θ	T^*	$C(T^*)$
0.1	0.062	0.989
0.2	0.239	0.953
0.3	0.510	0.900
0.4	0.854	0.836
0.5	1.252	0.766
0.6	1.693	0.694
0.7	2.170	0.624
0.8	2.682	0.557
0.9	3.229	0.496
1.0	3.813	0.439

(i) If $r(0) > k$ and $r(\infty) < K$ then there exists a finite and unique T^* ($0 < T^* < \infty$) that satisfies (2.37), and the resulting cost rate is given in (2.38).
(ii) If $r(0) \le k$ then $T^* = 0$ and the expected cost rate is given in (2.32).
(iii) If $r(\infty) \ge K$ then $T^* = \infty$ and the expected cost rate is given in (2.35).

It is evident in the above result that if $r(t)$ is not decreasing then $T^* = 0$ or $T^* = \infty$. In this case, if $a/c_1 > 1/\mu + 1/\beta$ then $T^* = 0$, and conversely, if $a/c_1 < 1/\mu + 1/\beta$ then $T^* = \infty$. In other cases of $b \ne 1$, it is, in general, difficult to discuss an optimum repair limit policy. However, it could compute an optimum time T^* that satisfies (2.36) if the parameters a, b, and $G(t)$ are specified.

Example 2.7. Suppose that $c_r(t) = at$ and $G(t) = 1 - e^{-\theta\sqrt{t}}$. Then, $r(t) = \theta/(2\sqrt{t})$ which is strictly decreasing from infinity to zero. Then, from (2.37), there exists a unique solution T^* that satisfies

$$\frac{a\mu}{c_1}\sqrt{T} - \frac{1}{\theta}(1 - e^{-\theta\sqrt{T}}) = \frac{\theta\mu}{2}$$

and from (2.38), the expected cost rate is $C(T^*) = a - c_1\theta/(2\sqrt{T^*})$. Table 2.1 shows a numerical example of the optimum repair limit time T^* and the resulting cost rate $C(T^*)$ for $\theta = 0.1 \sim 1.0$ when $a = 3$, $\mu = 10$, and $c_1 = 10$. ∎

Example 2.8. Suppose that $c_r(t) = at^2$ and $G(t) = 1 - e^{-\theta t}$. Then, from (2.36), there exists a unique solution T^* that satisfies

$$T - \frac{1 - e^{-\theta T}}{\theta(\mu\theta + 1)} = \frac{c_1\theta}{2a}$$

because the left-hand side is strictly increasing from 0 to ∞, and from (2.34), the expected cost rate is $C(T^*) = 2aT^* - c_1\theta$. Table 2.2 shows a numerical example of T^* and $C(T^*)$ for θ when $a = 3$, $\mu = 10$, and $c_1 = 10$. ∎

Table 2.2. Optimum repair limit time T^* and expected cost rate $C(T^*)$ when $a = 3$, $\mu = 10$, and $c_1 = 10$

θ	T^*	$C(T^*)$
0.1	0.330	0.981
0.2	0.489	0.931
0.3	0.647	0.883
0.4	0.804	0.826
0.5	0.961	0.763
0.6	1.116	0.697
0.7	1.272	0.632
0.8	1.428	0.568
0.9	1.584	0.507
1.0	1.742	0.450

Until now, we have discussed the case where the repair cost is not estimated when an operating unit fails. However, if the repair cost can be previously estimated when an operating unit fails and the decision can be made as to whether the failed unit should be repaired or replaced, the expected cost rate is easily given by

$$C(T) = \frac{c_1 \overline{G}(T) + \int_0^T c_r(t) \, dG(t)}{\mu + \int_0^T t \, dG(t)}. \tag{2.39}$$

Finally, we introduce the following earnings in specifying the repair limit policy. Let e_0 be a net earning per unit of time made by the production of an operating unit, e_1 be an earning gained for replacing a failed unit, and e_2 be an earning rate per unit of time while the unit is under repair, where both e_1 and e_2 would usually be negative. Then, by the similar method to that of obtaining (2.31), the expected earning rate is

$$C(T) = \frac{e_0 \mu + e_1 \overline{G}(T) + e_2 \int_0^T \overline{G}(t) \, dt}{\mu + \int_0^T \overline{G}(t) \, dt}. \tag{2.40}$$

Checking up on these models with actual systems, modifying, and extending them, we could get an optimum repair limit policy.

2.2 Standby System with Spare Units

Most standby systems with spare units have been discussed only for the case where any failed units are repaired and become as good as new upon the repair completion. In the real world, it may be worthwhile to scrap some failed units without repairing, depending on the nature of the failed units. For instance, we have scrapped failed units according to the repair limit policy proposed in Section 2.1.2.

Consider the system with a main unit and n spare subunits that are statistically not identical to each other, but any spare ones have the same function as the main unit if they take over operation. The system functions as follows. When the main unit fails, it undergoes repair immediately and one of the spare units replaces it. As soon as the repair of the main unit is completed, it begins to operate and the operating spare unit is available for further use. Any failed spare units are scrapped. The system functions until the nth spare unit fails; *i.e.*, system failure occurs when the last spare unit fails while the main unit is under repair. This model often occurs when something is broken or lost, and we temporarily use a substitute until it is repaired or replaced. We believe that this could be applicable to other practical fields.

We are interested in the following operating characteristics of the system.

(i) The distribution and the mean time to first system failure, given that n spare units are provided at time 0.
(ii) The probability that the number of failed spare units is exactly equal to n and its expected number during $(0, t]$.

These quantities are derived by forming renewal equations, and using them, two optimization problems to determine an initial number of spares to stock are considered.

We adopt the expected cost per unit of time for an infinite time span; *i.e.*, the expected cost rate (see Section 3.1) as an appropriate objective function. First, we compare two systems with (1) both main and spare units and (2) only unrepairable spare units. Secondly, we do the preventive maintenance (PM) of the main unit. When the main unit works for a specified time T $(0 \leq T \leq \infty)$ without failure, its operation is stopped and one of the spare units takes over operation. The main unit is serviced on failure or its age T, whichever occurs first. The costs incurred for each failed unit and each PM are introduced. Then, we derive an optimum PM policy that minimizes the expected cost rate under suitable conditions.

2.2.1 Reliability Quantities

Suppose that the failure time of the main unit has a general distribution $F(t)$ with finite mean μ and its repair time has a general distribution $G(t)$ with finite mean β, where $\overline{\Phi} \equiv 1 - \Phi$ for any function. The failure time of each spare unit also has a general distribution $F_s(t)$ with finite mean μ_s, even if it has been used before; *i.e.*, the life of spare units is not affected by past operation. It is assumed that all random variables considered here are independent, and all units are good at time 0. Furthermore, any failures are instantly detected and repaired or scrapped, and each switchover is perfect and its time is instantaneous.

Let $L_j(t)$ $(j = 1, 2, \ldots, n)$ denote the first-passage time distribution to system failure when j spares are provided at time 0. Then, we have the following renewal equation.

$$L_n(t) = F(t) * \left\{ \int_0^t \overline{G}(u) \, dF_s^{(n)}(u) \right.$$
$$\left. + \sum_{j=0}^{n-1} L_{n-j}(t) * \int_0^t [F_s^{(j)}(u) - F_s^{(j+1)}(u)] \, dG(u) \right\} \quad (n = 1, 2, \ldots), \quad (2.41)$$

where the asterisk represents the Stieltjes convolution, and $F_s^{(j)}(t)$ ($j = 1, 2, \ldots$) represents the j-fold Stieltjes convolution of $F_s(t)$ with itself and $F_s^{(0)}(t) \equiv 1$ for $t \geq 0$. The first term of the bracket on the right-hand side is the time distribution that all of n spares have failed before the first repair completion of the failed main unit, and the second term is the time distribution that j ($j = 0, 1, \ldots, n-1$) spares fail exactly before the first repair completion, and then, the main unit with $n-j$ spares operates again.

The first-passage time distribution $L_n(t)$ to system failure can be calculated recursively and determined from (2.41). To obtain $L_n(t)$ explicitly, we introduce the notation of the generating function of LS transforms;

$$L^*(z, s) \equiv \sum_{j=1}^{\infty} z^j \int_0^{\infty} e^{-st} \, dL_j(t) \quad \text{for } |z| < 1.$$

Then, taking the LS transform on both sides of (2.41) and using the generating function $L^*(z, s)$, we have

$$L^*(z, s) = \frac{F^*(s) \sum_{j=1}^{\infty} z^j \int_0^{\infty} e^{-st} \overline{G}(t) \, dF_s^{(j)}(t)}{1 - F^*(s) \sum_{j=0}^{\infty} z^j \int_0^{\infty} e^{-st} [F_s^{(j)}(t) - F_s^{(j+1)}(t)] \, dG(t)}, \quad (2.42)$$

where $F^*(s) \equiv \int_0^{\infty} e^{-st} dF(t)$.

Moreover, let l_n denote the mean first-passage time to system failure. Then, by a similar method to that of (2.41), we easily have

$$l_n = \mu + \int_0^{\infty} [1 - F_s^{(n)}(t)] \overline{G}(t) \, dt + \sum_{j=0}^{n-1} l_{n-j} \int_0^{\infty} [F_s^{(j)}(t) - F_s^{(j+1)}(t)] \, dG(t)$$
$$(n = 1, 2, \ldots) \quad (2.43)$$

and hence, the generating function is

$$l^*(z) \equiv \sum_{j=1}^{\infty} z^j l_j = \frac{\mu[z/(1-z)] + \sum_{j=1}^{\infty} z^j \int_0^{\infty} [1 - F_s^{(j)}(t)] \overline{G}(t) \, dt}{1 - \sum_{j=0}^{\infty} z^j \int_0^{\infty} [F_s^{(j)}(t) - F_s^{(j+1)}(t)] \, dG(t)}. \quad (2.44)$$

In a similar way, we obtain the expected number of failed spares during $(0, t]$. Let $p_n(t)$ be the probability that the total number of failed spares during $(0, t]$ is exactly n. Then, we have

2 Repair Maintenance

$$p_0(t) = \overline{F}(t) + F(t) * \left[\overline{F}_s(t)\overline{G}(t) + p_0(t) * \int_0^t \overline{F}_s(u) \, dG(u) \right] \tag{2.45}$$

$$p_n(t) = F(t) * \left\{ \left[F_s^{(n)}(t) - F_s^{(n+1)}(t) \right] \overline{G}(t) \right.$$

$$\left. + \sum_{j=0}^n p_{n-j}(t) * \int_0^t \left[F_s^{(j)}(u) - F_s^{(j+1)}(u) \right] dG(u) \right\} \quad (n = 1, 2, \ldots).$$
$$\tag{2.46}$$

Introducing the notation

$$p^*(z, s) \equiv \sum_{n=0}^{\infty} z^n \int_0^{\infty} e^{-st} \, dp_n(t) \quad \text{for } |z| < 1$$

we have, from (2.45) and (2.46),

$$p^*(z, s) = \frac{1 - F^*(s)\left[1 - \sum_{j=0}^{\infty} z^j \int_0^{\infty} e^{-st} \, d\{[F_s^{(j)}(t) - F_s^{(j+1)}(t)]\overline{G}(t)\}\right]}{1 - F^*(s) \sum_{j=0}^{\infty} z^j \int_0^{\infty} e^{-st} [F_s^{(j)}(t) - F_s^{(j+1)}(t)] \, dG(t)},$$
$$\tag{2.47}$$

where note that $p^*(1, s) \equiv \lim_{z \to 1} p^*(z, s) = 1$. Thus, the LS transform of the expected number $M(t) \equiv \sum_{n=1}^{\infty} np_n(t)$ of failed spares during $(0, t]$ is

$$M^*(s) \equiv \sum_{n=1}^{\infty} \int_0^{\infty} e^{-st} \, dM(t) = \lim_{z \to 1} \frac{\partial p^*(z, s)}{\partial z}$$

$$= \frac{F^*(s) \int_0^{\infty} e^{-st} \overline{G}(t) \, dM_s(t)}{1 - F^*(s) G^*(s)}, \tag{2.48}$$

where $M_s(t) \equiv \sum_{j=1}^{\infty} F_s^{(j)}(t)$ is the renewal function of $F_s(t)$. Furthermore, the limit of the expected number of failed spares per unit of time is

$$M \equiv \lim_{t \to \infty} \frac{M(t)}{t} = \lim_{s \to 0} sM^*(s) = \frac{\int_0^{\infty} M_s(t) \, dG(t)}{\mu + \beta}. \tag{2.49}$$

The result of M can be intuitively derived because the numerator represents the total expected number of failed spares during the repair time of the main unit and the denominator represents the mean time from the operation to the repair completion of the main unit.

Example 2.9. Suppose that $G(t) = 1 - e^{-\theta t}$. In this case, from (2.44), when n spares are provided at time 0, the mean time to system failure is

$$l_n = \mu + n\left(\mu + \frac{1}{\theta}\right) \frac{1 - F_s^*(\theta)}{F_s^*(\theta)}.$$

2.2 Standby System with Spare Units

Note that by adding one spare unit to the system, the mean time increases a constant $\alpha \equiv (\mu + 1/\theta)[1 - F_s^*(\theta)]/F_s^*(\theta)$. Furthermore, the LS transform of the expected number of failed spares during $(0, t]$ is

$$M^*(s) = \frac{F^*(s) F_s^*(s + \theta)}{\{1 - [\theta/(s+\theta)] F^*(s)\}[1 - F_s^*(s+\theta)]}$$

and its limit per unit of time is

$$M = \frac{F_s^*(\theta)}{(\mu + 1/\theta)[1 - F_s^*(\theta)]}$$

which is equal to $1/\alpha$; i.e., $l_n = \mu + n/M$. ∎

2.2.2 Optimization Problems

First, we obtain the expected cost rate, by introducing costs incurred for each failed main unit and spare unit. This expected cost rate is easily deduced from the expected number of failed units. We compare two expected costs of the system with both main unit and spares and the system with only spares, and determine which of the systems is more economical.

Cost c_1 is incurred for each failed main unit, which includes all costs resulting from its failure and repair, and cost c_s is incurred for each failed spare, which includes all costs resulting from its failure, replacement, and cost of itself. Let $C(t)$ be the total expected cost during $(0, t]$. Then, the expected cost rate is, from Theorems 1.2 and 1.6 in Section 1.3,

$$C \equiv \lim_{t \to \infty} \frac{C(t)}{t} = c_1 M_1 + c_s M, \tag{2.50}$$

where M_1 is the expected number of the failed main unit per unit of time, and from (2.5), $M_1 = 1/(\mu + \beta)$.

Thus, from (2.49) the expected cost rate is

$$C = \frac{c_1 + c_s \int_0^\infty M_s(t) \, dG(t)}{\mu + \beta} \tag{2.51}$$

which is also equal to the expected cost per one cycle from the beginning of the operation to the repair completion of the main unit. If only spare units are allowed then the expected cost rate is

$$C_s \equiv \frac{c_s}{\mu_s}. \tag{2.52}$$

Therefore, comparing (2.51) and (2.52), we have $C \leq C_s$ if and only if

$$c_1 \leq c_s \left[\frac{\mu + \beta}{\mu_s} - \int_0^\infty M_s(t) \, dG(t) \right]$$

and *vice versa*.

In general, it is hard to compute the above costs directly. However, simple results that would be useful in practical fields can be obtained in the following particular cases. Because $M_s(t) \geq t/\mu_s - 1$ [1, p. 53], if $c_1 > c_s(\mu/\mu_s + 1)$ then $C > C_s$. In the case of Example 2.9, we have the relation $C \leq C_s$ if and only if

$$c_1 \leq c_s \left[\frac{\mu + 1/\theta}{\mu_s} - \frac{F_s^*(\theta)}{1 - F_s^*(\theta)} \right]$$

and *vice versa*.

Next, consider the PM policy where the operating main unit is preventively maintained at time T ($0 \leq T \leq \infty$) after its installation or is repaired at failure, whichever occurs first. The several PM policies are discussed fully in Chapter 6. In this model, spare units work temporarily during the interval of repair or PM time of the main unit. It is assumed that the PM time has the same distribution $G(t)$ with finite mean β as the repair time. The main unit becomes as good as new upon repair or PM, and begins to operate immediately. The costs incurred for each failed main unit and each failed spare are the same as c_1 and c_s, respectively, as those in the previous model. The PM cost c_2 with $c_2 < c_1$ incurs for each nonfailed main unit that is preventively maintained.

The total expected cost of one cycle from the operation to the repair or PM completion of the main unit is

$$F(T) \left[c_1 + c_s \int_0^\infty M_s(t) \, dG(t) \right] + \overline{F}(T) \left[c_2 + c_s \int_0^\infty M_s(t) \, dG(t) \right]$$

and the mean time of one cycle is

$$\int_0^T (t + \beta) \, dF(t) + \overline{F}(T)(T + \beta) = \int_0^T \overline{F}(t) \, dt + \beta.$$

Thus, in a similar way to that of obtaining (2.51), the expected cost rate is

$$C(T) = \frac{\tilde{c}_1 F(T) + \tilde{c}_2 \overline{F}(T)}{\int_0^T \overline{F}(t) \, dt + \beta}, \qquad (2.53)$$

where $\tilde{c}_1 \equiv c_1 + c_s \int_0^\infty M_s(t) \, dG(t)$ and $\tilde{c}_2 \equiv c_2 + c_s \int_0^\infty M_s(t) \, dG(t)$, and $\tilde{c}_1 > \tilde{c}_2$ from $c_1 > c_2$.

We find an optimum PM time T^* that minimizes $C(T)$. Clearly, $C(0) = \tilde{c}_2/\beta$ is the expected cost in the case where the main unit is always under PM, and $C(\infty)$ is the expected cost of the main unit with no PM and is given in (2.51). Let $h(t) \equiv f(t)/\overline{F}(t)$ be the failure rate of $F(t)$ with $h(0) \equiv \lim_{t \to 0} h(t)$ and $h(\infty) \equiv \lim_{t \to \infty} h(t)$, and $k \equiv \tilde{c}_2/[\beta(\tilde{c}_1 - \tilde{c}_2)]$ and $K \equiv \tilde{c}_1/[(\mu + \beta)(\tilde{c}_1 - \tilde{c}_2)]$. Then, we have the following optimum policy.

Theorem 2.2. Suppose that the failure rate $h(t)$ is continuous and strictly increasing.

2.2 Standby System with Spare Units

(i) If $h(0) < k$ and $h(\infty) > K$ then there exists a finite and unique T^* $(0 < T^* < \infty)$ that satisfies

$$h(T)\left[\int_0^T \overline{F}(t)\,dt + \beta\right] - F(T) = \frac{\tilde{c}_2}{\tilde{c}_1 - \tilde{c}_2} \qquad (2.54)$$

and the resulting expected cost rate is

$$C(T^*) = (c_1 - c_2)h(T^*). \qquad (2.55)$$

(ii) If $h(0) \geq k$ then $T^* = 0$.
(iii) If $h(\infty) \leq K$ then $T^* = \infty$.

Proof. Differentiating $C(T)$ in (2.53) with respect to T and putting it equal to zero, we have (2.54). Letting $Q(T)$ be the left-hand side of (2.54), it is easily proved that $Q(0) = \beta h(0)$, $Q(\infty) = (\mu + \beta)h(\infty) - 1$, and $Q(T)$ is strictly increasing because $h(t)$ is strictly increasing. Thus, if $h(0) < k$ and $h(\infty) > K$ then $Q(0) < \tilde{c}_2/(\tilde{c}_1 - \tilde{c}_2) < Q(\infty)$, and hence, there exists a finite and unique T^* that satisfies (2.54) and minimizes $C(T)$. Furthermore, from (2.54), we have (2.55).

If $h(0) \geq k$ then $Q(0) \geq \tilde{c}_2/(\tilde{c}_1 - \tilde{c}_2)$. Thus, $C(T)$ is strictly increasing, and hence, $T^* = 0$. Finally, if $h(\infty) \leq K$ then $Q(\infty) \leq \tilde{c}_2/(\tilde{c}_1 - \tilde{c}_2)$. Thus, $C(T)$ is strictly decreasing, and $T^* = \infty$. ∎

It is easily noted in Theorem 2.2 that if the failure rate $h(t)$ is nonincreasing then $T^* = 0$ or $T^* = \infty$. Similar theorems are derived in Section 3.1.

Until now, it has been assumed that the time to the PM completion has the same repair distribution $G(t)$. In reality, the PM time might be smaller than the repair time. So that, suppose that the repair time is $G_1(t)$ with mean β_1 and the PM time is $G_2(t)$ with mean β_2. Then, the expected cost rate is similarly given by

$$C(T) = \frac{\left[c_1 + c_s \int_0^\infty M_s(t)\,dG_1(t)\right]F(T) + \left[c_2 + c_s \int_0^\infty M_s(t)\,dG_2(t)\right]\overline{F}(T)}{\int_0^T \overline{F}(t)\,dt + \beta_1 F(T) + \beta_2 \overline{F}(T)}. \qquad (2.56)$$

Example 2.10. Consider the optimization problem of ensuring that sufficient numbers of spares are initially provided to protect against shortage. If the probability α of occurrences of no shortage during $(0, t]$ is given *a priori*, we can find a minimum number of spares to maintain this level of confidence. One solution of this problem can be shown by computing a minimum n such that $\sum_{i=0}^n p_i(t) \geq \alpha$. If we need a minimum number of initial stocks during $(0, t]$ on *the average* without probabilistic guarantee, we might compute a minimum n such that $l_n \geq t$, or $M(t) \leq n$.

Table 2.3. Optimum PM time T^*, its cost rates $C(T^*)$, and C when $1/\lambda_s = 1$, $1/\theta = 5$, $c_1 = 10$, $c_2 = 1$, and $c_s = 2$

$2/\lambda$	T^*	$C(T^*)$	C
1	0.06	2.18	3.33
2	0.31	2.13	2.86
3	0.78	2.06	2.50
4	1.54	1.94	2.22
5	2.63	1.84	2.00
6	4.08	1.72	1.82
7	5.91	1.61	1.67
8	8.14	1.50	1.54
9	10.78	1.41	1.43
10	13.88	1.32	1.33

Next, compare two systems with main and spare units, and with only spares, when $F(t) = 1 - (1 + \lambda t)e^{-\lambda t}$, $F_s(t) = 1 - \exp(-\lambda_s t)$ and $G(t) = 1 - e^{-\theta t}$. Then, from (2.51) and (2.52), the expected cost rates are

$$C = \frac{c_1 + c_s(\lambda_s/\theta)}{2/\lambda + 1/\theta}, \qquad C_s = \lambda_s c_s.$$

Thus, $C \leq C_s$ if and only if $c_1 \leq c_s(2\lambda_s/\lambda)$ and *vice versa*.

Furthermore, when $F(t) = 1 - (1 + \lambda t)e^{-\lambda t}$, the failure rate is $h(t) = \lambda^2 t/(1 + \lambda t)$ which is strictly increasing from 0 to λ. Thus, from (i) of Theorem 2.2, if $\lambda(\tilde{c}_1 - \tilde{c}_2) > \theta(2\tilde{c}_2 - \tilde{c}_1)$ then there exists a finite and unique T^* $(0 < T^* < \infty)$ that satisfies

$$\frac{1}{1 + \lambda T}\left[\frac{\lambda^2 T}{\theta} + \lambda T - (1 - e^{-\lambda T})\right] = \frac{\tilde{c}_2}{\tilde{c}_1 - \tilde{c}_2}$$

and the expected cost rate is

$$C(T^*) = \frac{\lambda^2 T^*}{1 + \lambda T^*}(c_1 - c_2).$$

Table 2.3 gives the optimum PM time T^*, its cost rates $C(T^*)$, and C with no PM for $2/\lambda$ when $1/\lambda_s = 1$, $1/\theta = 5$, $c_1 = 10$, $c_2 = 1$, and $c_s = 2$. This indicates that when the mean failure time $2/\lambda$ is small, the PM time T^* is small and it is very effective. In this case, because $C_s = 2$, we have that $C \geq C_s$ for $2/\lambda \leq 5$ and $C(T^*) > C_s$ for $2/\lambda \leq 3$. ∎

2.3 Other Redundant Systems

In this section, we briefly mention redundant systems with repair maintenance without detailed derivations [37–40]. For the analysis of redundant systems,

2.3 Other Redundant Systems

it is of great importance to know the behavior of system failure; *i.e.*, the probability that the system will be in system failure, the mean time to system failure, and the expected number of system failures. For instance, if the system failure is catastrophic, we have to make the time to system failure as long as possible, by doing the PM and providing standby units.

2.3.1 Standby Redundant System

Consider an $(n+1)$-unit standby redundant system with $n+1$ repairpersons and one operating unit supported by n identical spares (refer to [40] for s ($1 \le s \le n+1$) repairpersons). Each unit fails according to a general distribution $F(t)$ with finite mean μ and undergoes repair immediately. When the repair is completed, the unit rejoins the spares. It is also assumed that the repair time of each failed unit is an independent random variable with an exponential distribution $(1 - e^{-\theta t})$ for $0 < \theta < \infty$. Let $\xi(t)$ denote the number of units under repair at time t. The system is said to be in state k at time t if $\xi(t) = k$. In particular, it is also said that system failure occurs when the system is in state $n+1$. Furthermore, let $0 \equiv t_0 < t_1 < \cdots < t_m \cdots$ be the failure times of an operating unit. If we define $\xi_m \equiv \xi(t_m - 0)$ $(m = 0, 1, \dots)$ then ξ_m represents the number of units under repair immediately before the mth failure occurs. Then, we present only the results of transition probabilities and first-passage time distributions.

The Laplace transform of the binomial moment of transition probabilities $p_{ik}(t) \equiv \Pr\{\xi(t) = k | \xi_0 = i\}$ ($i = 0, 1, \dots, n; k = 0, 1, \dots, n+1$) is

$$\Psi_{ir}(s) \equiv \sum_{k=r}^{n+1} \binom{k}{r} \int_0^\infty e^{-st} p_{ik}(t)\, dt$$

$$= \frac{B_{r-1}(s)}{s+r\theta} \left\{ \sum_{j=0}^{r} \binom{i+1}{j} \frac{1}{B_{j-1}(s)} - \sum_{j=0}^{i+1} \binom{i+1}{j} \frac{1}{B_{j-1}(s)} \right.$$

$$\left. \times \frac{\sum_{j=0}^{r-1} \binom{n+1}{j}(s+j\theta)/B_{j-1}(s)}{\sum_{j=0}^{n+1} \binom{n+1}{j}(s+j\theta)/B_{j-1}(s)} \right\} \quad (r = 0, 1, \dots, n+1)$$

and the limiting probability $p_k \equiv \lim_{t \to \infty} p_{ik}$ ($k = 0, 1, \dots, n+1$) is

$$\Psi_r \equiv \sum_{k=r}^{n+1} \binom{k}{r} p_k$$

$$= \frac{(n+1)B_{r-1}(0)}{r} \frac{\sum_{j=r-1}^{n} \binom{n}{j}/B_j(0)}{1+(n+1)(\mu\theta)\sum_{j=0}^{n} \binom{n}{j}/B_j(0)} \quad (r = 1, 2, \dots, n+1)$$

and $\Psi_0 \equiv 1$, where $\sum_{j=0}^{-1} \equiv 0$, $B_{-1}(s) = B_0(0) \equiv 1$ and

$$B_r(s) \equiv \prod_{j=0}^{r} \frac{F^*(s+j\theta)}{1-F^*(s+j\theta)} \qquad (r=0,1,2,\dots)$$

$$B_r(0) \equiv \prod_{j=1}^{r} \frac{F^*(j\theta)}{1-F^*(j\theta)} \qquad (r=1,2,\dots).$$

Thus, by the inversion formula of binomial moments,

$$p_{ik}^*(s) \equiv \int_0^\infty e^{-st} p_{ik}(t)\,dt = \sum_{r=k}^{n+1} (-1)^{r-k} \binom{r}{k} \Psi_{ir}(s)$$
$$(i=0,1,\dots,n;\, k=0,1,\dots,n+1) \qquad (2.57)$$

$$p_k = \sum_{r=k}^{n+1} (-1)^{r-k} \binom{r}{k} \Psi_r \quad (k=0,1,\dots,n+1). \qquad (2.58)$$

It was shown in [41] that there exists the limiting probability p_k for $\mu < \infty$.

Next, the LS transform of the first-passage time distribution $F_{ik}(t) \equiv \sum_{m=1}^{\infty} \Pr\{\xi_m = k, \xi_j \neq k \text{ for } j=1,2,\dots,m-1, t_m \leq t \mid \xi_0 = i\}$ is, for $i < k$,

$$F_{ik}^*(s) \equiv \int_0^\infty e^{-st}\,dF_{ik}(t) = \frac{\sum_{j=0}^{i+1} \binom{i+1}{j}/B_{j-1}(s)}{\sum_{j=0}^{k+1} \binom{k+1}{j}/B_{j-1}(s)} \quad (k=0,1,\dots,n) \quad (2.59)$$

and its mean time is

$$l_{ik} \equiv \int_0^\infty t\,dF_{ik}(t) = \mu \sum_{j=1}^{k+1} \left[\binom{k+1}{j} - \binom{i+1}{j}\right] \frac{1}{B_{j-1}(0)}$$
$$(k=0,1,\dots,n), \qquad (2.60)$$

where $\binom{i}{j} \equiv 0$ for $j > i$. The mean time l_{ik} when $i = -1$ and $k = n$ agrees with the result of [37], where state -1 means the initial condition that one unit begins to operate and n units are on standby at time 0.

The expected number M_k ($k=0,1,\dots,n-1$) of visits to state k before system failure is

$$M_k = \sum_{r=k}^{n} (-1)^{r-k} \binom{r}{k} B_r(0) \sum_{j=r+1}^{n+1} \binom{n+1}{j} \frac{1}{B_{j-1}(0)} \quad (k=0,1,\dots,n-1).$$
$$(2.61)$$

Thus, the total expected number M of unit failures before system failure from state 0 is

$$M \equiv 1 + \sum_{k=0}^{n-1} M_k = \sum_{j=1}^{n+1} \binom{n+1}{j} \frac{1}{B_{j-1}(0)} \qquad (2.62)$$

and the expected number of repairs before system failure is $M - (n+1)$. It is noted that μM is also the mean time to system failure $l_{-1\,n}$ in (2.60).

In the case of one repairperson, the first-passage time from state i to state k for $i < k$ coincides with that of queue $G/M/1$. Thus, for $i < k$ [42],

$$F_{ik}^*(s) = \frac{1 + [1 - F^*(s)][A_{i+1}(s) - \delta_{i+1\,0}]}{1 + [1 - F^*(s)]A_{k+1}(s)} \quad (k = 0, 1, \ldots, n), \tag{2.63}$$

where $\delta_{ik} = 1$ for $i = k$ and 0 for $i \neq k$,

$$\sum_{j=0}^{\infty} A_j(s) z^j \equiv z^2 \Big[(1-z)\{F^*[s + \theta(1-z)] - z\}\Big] \quad \text{for } |z| < 1.$$

From the relation of transition probability and first-passage time distribution, we easily have

$$p_{ik}(t) = \int_0^t p_{k-1\,k}(t-u)\,\mathrm{d}F_{i\,k-1}(u)$$

$$p_{n\,n+1}(t) = e^{-\theta t} + \int_0^t p_{n\,n+1}(t-u)\,\mathrm{d}F_{nn}(u)$$

$$F_{nn}(t) = \int_0^t F_{n-1\,n}(t-u)\theta e^{-\theta u}\,\mathrm{d}u.$$

Thus, forming the Laplace transforms of the above equations and using the result of $F_{ik}^*(s)$,

$$p_{i\,n+1}^*(s) = \frac{1 + [1 - F^*(s)][A_{i+1}(s) - \delta_{i+1\,0}]}{s + [1 - F^*(s)]\{sA_{n+1}(s) + \theta[A_{n+1}(s) + \delta_{n\,0} - A_n(s)]\}} \tag{2.64}$$

$$p_{n+1} = \frac{1}{1 + (\mu\theta)[A_{n+1}(0) + \delta_{n\,0} - A_n(0)]}. \tag{2.65}$$

2.3.2 Parallel Redundant System

Consider an $(n+1)$-unit parallel redundant system with one repairperson. Then, it can be easily seen that this system is equivalent to a standby system with $n+1$ repairpersons as described in Section 2.3.1 wherein the notations of failure and repair change one another. For instance, the transition probability p_{ik} in (2.57) becomes the transition probability for the number of units under operation. The LS transform of the busy period of a repairperson is

$$F_{n-1\,n}^*(s) = \frac{\sum_{j=0}^{n} \binom{n}{j}/B_{j-1}(s)}{\sum_{j=0}^{n+1} \binom{n+1}{j}/B_{j-1}(s)} \tag{2.66}$$

and its mean time is

$$l_{n-1\,n} = \mu \sum_{j=0}^{n} \binom{n}{j} \frac{1}{B_j(0)}. \tag{2.67}$$

In addition, when a system has $n+1$ repairpersons (*i.e.*, there are as many repairpersons as the number of units), we may consider only n one-unit systems [1, p. 145]. In this model, we have

$$p_{ik}(t) = \sum_{j_1}\sum_{j_2} \binom{i}{j_1}\binom{n-i}{j_2}[P_{11}(t)]^{j_1}[P_{10}(t)]^{i-j_1}[P_{01}(t)]^{j_2}[P_{00}(t)]^{n-i-j_2}, \tag{2.68}$$

where the summation takes over $j_1 + j_2 = k$, $j_1 \leq i$, and $j_2 \leq n-i$, and $P_{ij}(t)$ ($i,j=0,1$) are given in (2.3) and (2.4).

Finally, consider n parallel units in which system failure occurs where k ($1 \leq k \leq n$) out of n units are down simultaneously. The LS transform of the distribution of time to system failure and its mean time were obtained in [43], by applying a birth and death process, and 2-out-of-n systems were discussed in [4].

References

1. Barlow RE and Proschan F (1965) Mathematical Theory of Reliability. J Wiley & Sons, New York.
2. Ushakov IA (1994) Handbook of Reliability Engineering. J Wiley & Sons, New York.
3. Birolini A (1999) Reliability Engineering Theory and Practice. Springer, New York.
4. Nakagawa T (2002) Two-unit redundant models. In: Osaki S (ed) Stochastic Models in Reliability and Maintenance. Springer, New York:165–185.
5. Brown M, Proschan F (1983) Imperfect repair. J Appl Prob 20:851–859.
6. Fontenot RA, Proschan F (1984) Some imperfect maintenance models. In: Abdel-Hameed MS, Çinlar E, Quinn J (eds) Reliability Theory and Models. Academic, Orlando, FL:83–101.
7. Nakagawa T, Osaki S (1974) The optimum repair limit replacement policies. Oper Res Q 25:311–317.
8. Nakagawa T, Osaki S (1976) Reliability analysis of a one-unit system with unrepairable spare units and its optimization applications. Oper Res Q 27:101–110.
9. Cox DR (1962) Renewal Theory. Methuen, London.
10. Barlow RE, Hunter LC (1961) Reliability analysis of a one-unit system. Oper Res 9:200–208.
11. Bellman R, Kalaba RE, Lockett J (1966) Numerical Inversion of the Laplace Transform. American Elsevier, New York.
12. Abate J, Choudury G, Whitt W (1999) An introduction to numerical transform inversion and its application to probability models. In: Grassmann WK (ed) Commutational Probability. Kluwer Academic, The Netherlands:257–323.
13. Patton AD (1972) A probability method for bulk power system security assessment, I-basic concepts. IEEE Trans Power Apparatus Syst PAS-91:54–61.
14. Takács L (1957) On certain sojourn time problems in the theory of stochastic processes. Acta Math Acad Sci Hungary 8:169–191.
15. Muth EJ (1968) A method for predicting system downtime. IEEE Trans Reliab R-17:97–102.

16. Muth EJ (1970) Excess time, a measure of system repairability. IEEE Trans Reliab R-19:16–19.
17. Suyona, Van der Weide JAM (2003) A method for computing total downtime distributions in repairable systems. J Appl Prob 40:643–653.
18. Calabro SR (1962) Reliability Principles and Practices. McGraw-Hill, New York.
19. Buzacott JA (1973) Reliability analysis of a nuclear reactor fuel charging system. IEEE Trans Reliab R-22:88–91.
20. Aven T, Jensen U (1999) Stochastic Models in Reliability. Springer, New York.
21. Baxter LA (1981) Availability measures for a two-state system. J Appl Prob 18:227–235.
22. Mi J (1998) Some comparison results of system availability. Nav Res Logist 45:205–218.
23. Welker EL (1966) System effectiveness. In: Ireson WG (ed) Reliability Handbook Section 1. McGraw-Hill, New York.
24. Kabak IW (1969) System availability and some design implications. Oper Res 17:827–837.
25. Martz Jr HF (1971) One single-cycle availability. IEEE Trans Reliab R-20:21–23.
26. Nakagawa T and Goel AL (1973) A note on availability for a finite interval. IEEE Trans Reliab R-22:271–272.
27. Morse PM (1958) Queues, Inventories, and Maintenance. J Wiley & Sons, New York.
28. Nguyen DG, Murthy DNP (1980) A note on the repair limit replacement policy. J Oper Res Soc 31:103–104.
29. Drinkwater RW, Hastings NAJ (1967) An economic replacement model. Oper Res Q 18:121–138.
30. Hastings NAJ (1968) Some notes on dynamic programming and replacement. Oper Res Q 19:453–464.
31. Hastings NAJ (1969) The repair limit replacement method. Oper Res Q 20:337–349.
32. Love CE, Rodger R, Blazenko G (1982) Repair limit policies for vehicle replacement. INFOR 20:226–237.
33. Love CE, Guo R (1996) Utilizing Weibull failure rates in repair limit analysis for equipment replacement/preventive maintenance decision. J Oper Res Soc 47:1366–1376.
34. Choi CH, Yun WY (1998) A note on pseudodynamic cost limit replacement model. Int J Reliab Qual Saf Eng 5:287–292.
35. Dohi T, Matsushima N, Kaio N, Osaki S (1996) Nonparametric repair limit replacement policies with imperfect repair. Eur J Oper Res 96:260–273.
36. Dohi T, Takeshita K, Osaki S (2000) Graphical methods for determining/estimating optimal repair-limit replacement policies. Int J Reliab Qual Saf Eng 7:43–60.
37. Srinivasan VS (1968) First emptiness in the spare parts problem for repairable components. Oper Res 16:407–415.
38. Natarajan R (1968) A reliability problem with spares and multiple repair facilities. Oper Res 16:1041–1057.
39. Bhat UN (1973) Reliability of an independent component, s-spare system with exponential life times and general repair times. Technometrics 15:529–539.

40. Nakagawa T (1974) The expected number of visits to state k before a total system failure of a complex system with repair maintenance. Oper Res 22:108–116.
41. Takács, L (1962) Introduction to the Theory of Queues. Oxford University Press, New York.
42. Cohen JW (1969) The Single Server Queue. North-Holland, Amsterdam.
43. Downton F (1966) The reliability of multiplex systems with repair. J Roy Stat Soc B 28:459–476.

3
Age Replacement

Failures of units are roughly classified into two failure modes: catastrophic failure in which a unit fails suddenly and completely, and degraded failure in which a unit fails gradually with time by its performance deterioration. In the former, failures during actual operation might sometimes be costly or dangerous. It is an important problem to determine when to replace or preventively maintain a unit before failure. In the latter, maintenance costs of a unit increase with its age, and inversely, its performance suffers some deterioration. In this case, it is also required to measure some performance parameters and to determine when to replace or preventively maintain a unit before it has been degraded into failure state.

In this chapter, we consider the replacement of a single unit with catastrophic failure mode, where its failure is very serious, and sometimes may incur a heavy loss. Some electronic and electric parts or equipment are typical examples. We introduce a high cost incurred for failure during operation and a low cost incurred for replacement before failure. The replacements after failure and before failure are called *corrective replacement* and *preventive replacement*, respectively. It is assumed that the distribution of failure time of a unit is known *a priori* by investigating its life data, and the planning horizon is infinite. It is also assumed that an operating unit is supplied with unlimited spare units. In Section 9.4 we discuss the optimization problem of maximizing the mean time to failure in the case of limited spare units. We may consider the age of a unit as the real operating time or the number of uses.

The most reasonable replacement policy for such a unit is based on its age, which is called *age replacement* [1]. A unit is always replaced at failure or time T if it has not failed up to time T, where T $(0 < T \leq \infty)$ is constant. In this case, it is appropriate to adopt the expected cost per unit of time as an objective function because the planning horizon is infinite. Of course, it is reasonable to adopt the total expected cost for a finite time span (see Sections 8.6 and 9.2) and in consideration of discounted cost. It is theoretically shown in Chapter 6 that the policy maximizing the availability is the same one as formally minimizing the expected cost.

Age replacement policies have been studied theoretically by many author. The known results were summarized and the optimum policies were studied in detail in [1]. First, a sufficient condition for a finite optimum time to exist was shown in [2]. The replacement times for the cases of truncated normal, gamma, and Weibull failure distributions were computed in [3]. Furthermore, more general models and cost structures were provided successively in [4–12]. An age replacement with continuous discounting was proposed in [13,14], and the comparison between age and block replacements was made in [15]. For the case of unknown failure distribution, the statistical confidence interval of the optimum replacement time was shown in [16–18]. Fuzzy set theory was applied to age replacement policies in [19]. The time scale that combines the age and usage times was given in [20,21]. Some chapters [22–24] of the recently published books summarized the basic results of age and the other replacement policies. Opportunistic replacement policies [1], in which a maintenance action is taken to depend on states of systems, are needed for the maintenance of complex systems. This area is omitted in this book (for example, see [25]).

In Section 3.1, we consider an age replacement policy in which a unit is replaced at failure or at age T, whichever occurs first. When the failure rate is strictly increasing, it is shown that there exists an optimum replacement time that minimizes the expected cost [26]. Furthermore, we give the upper and lower limits of the optimum replacement time [27]. Also, the optimum time is compared with other replacement times in a numerical example. In Section 3.2, we show three modified models of age replacement with discounting [26], age replacement in discrete time [28], and age replacement of a parallel system [29]. In Section 3.3, we suggest extended age replacement policies in which a unit is replaced at time T and at number N of uses, and discuss their optimum policies [30]. Furthermore, some replacement models where a unit is replaced at discrete times, and is replaced at random times are proposed in Sections. 9.1 and 9.3, respectively.

3.1 Replacement Policy

Consider an age replacement policy in which a unit is replaced at constant time T after its installation or at failure, whichever occurs first. We call a specified time T the *planned replacement time* which ranges over $(0, \infty]$. Such an age replacement policy is optimum among all reasonable policies [31, 32]. The event $\{T = \infty\}$ represents that no replacement is made at all. It is assumed that failures are instantly detected and each failed unit is replaced with a new one, where its replacement time is negligible, and so, a new installed unit begins to operate instantly. Furthermore, suppose that the failure time X_k ($k = 1, 2, \ldots$) of each unit is independent and has an identical distribution

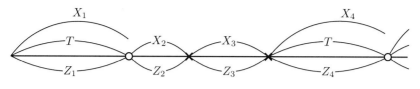

○ Planned replacement at time T ✗ Replacement at failure

Fig. 3.1. Process of age replacement with planned time T

$F(t) \equiv \Pr\{X_k \leq t\}$ with finite mean μ, where $\overline{F} \equiv 1 - F$ throughout this chapter; i.e., $\mu \equiv \int_0^\infty \overline{F}(t)dt < \infty$.

A new unit is installed at time $t = 0$. Then, an age replacement procedure generates a renewal process as follows. Let $\{X_k\}_{k=1}^\infty$ be the failure times of successive operating units. Define a new random variable $Z_k \equiv \min\{X_k, T\}$ $(k = 1, 2, \dots)$. Then, $\{Z_k\}_{k=1}^\infty$ represents the intervals between replacements caused by either failures or planned replacements such as shown in Figure 3.1. A sequence of random variables $\{Z_k\}_{k=1}^\infty$ is independently and identically distributed, and forms a renewal process as described in Section 1.3, and has an identical distribution

$$\Pr\{Z_k \leq t\} = \begin{cases} F(t) & \text{for } t < T \\ 1 & \text{for } t \geq T. \end{cases} \tag{3.1}$$

We consider the problem of minimizing the expected cost per unit of time for an infinite time span. Introduce the following costs. Cost c_1 is incurred for each failed unit that is replaced; this includes all costs resulting from a failure and its replacement. Cost c_2 ($< c_1$) is incurred for each nonfailed unit that is exchanged. Also, let $N_1(t)$ denote the number of failures during $(0, t]$ and $N_2(t)$ denote the number of exchanges of nonfailed units during $(0, t]$. Then, the expected cost during $(0, t]$ is given by

$$\widehat{C}(t) \equiv c_1 E\{N_1(t)\} + c_2 E\{N_2(t)\}. \tag{3.2}$$

When the planning is infinite, it is appropriate to adopt the expected cost per unit of time $\lim_{t \to \infty} \widehat{C}(t)/t$ as an objective function [1].

We call the time interval from one replacement to the next replacement as one cycle. Then, the pairs of time and cost on each cycle are independently and identically distributed, and both have finite means. Thus, from Theorem 1.6, the expected cost per unit of time for an infinite time span is

$$C(T) \equiv \lim_{t \to \infty} \frac{\widehat{C}(t)}{t} = \frac{\text{Expected cost of one cycle}}{\text{Mean time of one cycle}}. \tag{3.3}$$

We call $C(T)$ the *expected cost rate* and generally adopt it as the objective function of an optimization problem.

When we set a planned replacement at time T ($0 < T \leq \infty$) for a unit with failure time X, the expected cost of one cycle is

$$c_1 \Pr\{X \leq T\} + c_2 \Pr\{X > T\} = c_1 F(T) + c_2 \overline{F}(T)$$

and the mean time of one cycle is

$$\int_0^T t\, \mathrm{d}\Pr\{X \leq t\} + T \Pr\{X > T\} = \int_0^T t\, \mathrm{d}F(t) + T\overline{F}(T)$$

$$= \int_0^T \overline{F}(t)\, \mathrm{d}t.$$

Thus, the expected cost rate is, from (3.3),

$$C(T) = \frac{c_1 F(T) + c_2 \overline{F}(T)}{\int_0^T \overline{F}(t)\, \mathrm{d}t}, \tag{3.4}$$

where $c_1 = $ cost of replacement at failure and $c_2 = $ cost of replacement at planned time T with $c_2 < c_1$.

If $T = \infty$ then the policy corresponds to the replacement only at failure, and the expected cost rate is

$$C(\infty) \equiv \lim_{T \to \infty} C(T) = \frac{c_1}{\mu}. \tag{3.5}$$

The expected cost rate is generalized on the following form,

$$C(T) = \frac{\int_0^T c(t)\, \mathrm{d}F(t) + c_2}{\int_0^T \overline{F}(t)\, \mathrm{d}t}, \tag{3.6}$$

where $c(t) = $ marginal cost of replacement at time t [33, 34].

Furthermore, the expected cost per the operating time in one cycle is [35, 36]

$$C(T) = \int_0^T \frac{c_1}{t}\, \mathrm{d}F(t) + \int_T^\infty \frac{c_2}{T}\, \mathrm{d}F(t). \tag{3.7}$$

In this case, an optimum time that minimizes $C(T)$ is given by a solution of $Th(T) = c_2/(c_1 - c_2)$, where $h(t)$ is the failure rate of $F(t)$.

Putting that $F(T_p) = p$ ($0 < p \leq 1$), i.e., denoting T_p by a pth percentile point, the expected cost rate in (3.4) is rewritten as

$$C(p) = \frac{c_1 p + c_2 (1 - p)}{\int_0^{F^{-1}(p)} \overline{F}(t)\, \mathrm{d}t}, \tag{3.8}$$

where $F^{-1}(p)$ is the inverse function of $F(T_p) = p$. Then, the problem of minimizing $C(T)$ with respect to T becomes the problem of minimizing $C(p)$ with respect to a pth percentile point [37]. Using a graphical method based

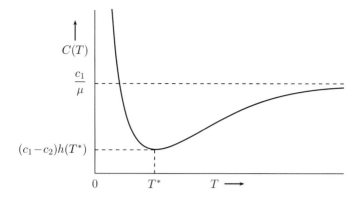

Fig. 3.2. Expected cost rate $C(T)$ of age replacement with planned time T

on the total time on test (TTT) plot, an optimum time that minimizes $C(p)$ was derived in [38–41].

Our aim is to derive an optimum planned replacement time T^* that minimizes the expected cost rate $C(T)$ in (3.4) as shown in Figure 3.2. It is assumed that there exists a density function $f(t)$ of the failure distribution $F(t)$ with finite mean μ. Let $h(t) \equiv f(t)/\overline{F}(t)$ be the failure rate and $K \equiv c_1/[\mu(c_1-c_2)]$.

Theorem 3.1. *Suppose that there exists the limit of the failure rate $h(\infty) \equiv \lim_{t \to \infty} h(t)$, possibly infinite, as $t \to \infty$. A sufficient condition that $C(\infty) > C(T)$ for some T is that $h(\infty) > K$.*

Proof. Differentiating $\log C(T)$ with respect to T yields

$$\frac{\mathrm{d} \log C(T)}{\mathrm{d}T} = \overline{F}(T)\left[\frac{(c_1-c_2)h(T)}{c_1 F(T) + c_2 \overline{F}(T)} - \frac{1}{\int_0^T \overline{F}(t)\,\mathrm{d}t}\right]$$

$$\approx \overline{F}(T)\left[\frac{(c_1-c_2)h(\infty)}{c_1} - \frac{1}{\mu}\right] \quad \text{for large } T.$$

Thus, if the quantity within in the bracket of the right-hand side is positive, i.e., $h(\infty) > K$, then there exists at least some finite T such that $C(\infty) > C(T)$ [2, p. 119]. ∎

In the above theorem, it has been assumed that there exists only the limit of the failure rate. Next, consider the case that the failure rate $h(t)$ is strictly increasing.

Theorem 3.2. *Suppose that the failure rate $h(t)$ is continuous and strictly increasing.*

(i) *If $h(\infty) > K$ then there exists a finite and unique T^* $(0 < T^* < \infty)$ that satisfies*

3 Age Replacement

$$h(T)\int_0^T \overline{F}(t)\,dt - F(T) = \frac{c_2}{c_1 - c_2} \tag{3.9}$$

and the resulting expected cost rate is

$$C(T^*) = (c_1 - c_2)h(T^*). \tag{3.10}$$

(ii) If $h(\infty) \le K$ then $T^* = \infty$; i.e., a unit is replaced only at failure, and the expected cost rate is given in (3.5).

Proof. Differentiating $C(T)$ in (3.4) with respect to T and putting it equal to zero imply (3.9). Letting

$$Q_1(T) \equiv h(T)\int_0^T \overline{F}(t)\,dt - F(T)$$

it is proved that $\lim_{T \to 0} Q_1(T) = 0$, $Q_1(\infty) \equiv \lim_{T \to \infty} Q_1(T) = \mu h(\infty) - 1$, and $Q_1(T)$ is strictly increasing because for any $\Delta T > 0$,

$$h(T+\Delta T)\int_0^{T+\Delta T} \overline{F}(t)\,dt - F(T+\Delta T) - h(T)\int_0^T \overline{F}(t)\,dt + F(T)$$

$$\ge h(T+\Delta T)\int_0^{T+\Delta T} \overline{F}(t)\,dt - h(T+\Delta T)\int_T^{T+\Delta T} \overline{F}(t)\,dt - h(T)\int_0^T \overline{F}(t)\,dt$$

$$= [h(T+\Delta T) - h(T)]\int_0^T \overline{F}(t)\,dt > 0$$

because $h(T+\Delta T) \ge [F(T+\Delta T) - F(T)]/\int_T^{T+\Delta T} \overline{F}(t)\,dt$.

If $h(\infty) > K$ then $Q_1(\infty) > c_2/(c_1-c_2)$. Thus, from the monotonicity and the continuity of $Q_1(T)$, there exists a finite and unique T^* $(0 < T^* < \infty)$ that satisfies (3.9) and it minimizes $C(T)$. Furthermore, from (3.9), we easily have (3.10).

If $h(\infty) \le K$ then $Q_1(\infty) \le c_2/(c_1-c_2)$, i.e., $Q_1(T) < c_2/(c_1-c_2)$, which implies $dC(T)/dT < 0$ for any finite T. Thus, the optimum time is $T^* = \infty$; i.e., a unit is replaced only at failure. ∎

It is easily noted from Theorem 3.2 that if the failure rate is nonincreasing then the optimum replacement time is $T^* = \infty$. It is intuitively apparent because a used unit tends to have a longer remaining life than its replacement unit. Such an intuition is made in the case of $c_1 \le c_2$.

In the case (i) of Theorem 3.2, we can get the following upper and lower limits of the optimum replacement time T^*.

Theorem 3.3. Suppose that the failure rate $h(t)$ is continuous, strictly increasing, and $h(\infty) > K$. Then, there exists a finite and unique $\overline{T}$ that satisfies $h(T) = K$, and a finite and unique $\underline{T}$ that satisfies

$$Th(T) - \int_0^T h(t)\,dt = \frac{c_2}{c_1 - c_2}$$

and consequently, $\underline{T} < T^* < \overline{T}$.

Proof. It is evident that $h(T) < \overline{F}(T)/\int_T^\infty \overline{F}(t)\mathrm{d}t$ for $0 \leq T < \infty$ from (1.7) because $h(t)$ is strictly increasing. Thus, we have

$$Q_1(T) > \mu h(T) - 1. \tag{3.11}$$

If $h(t)$ is continuous, strictly increasing, $h(0) < K$, and $h(\infty) > K$, then there exists a finite and unique $\overline{T}$ that satisfies $\mu h(T) - 1 = c_2/(c_1 - c_2)$; i.e., $h(T) = K$. Therefore, we have $T^* < \overline{T}$ from (3.11). If $h(0) \geq K$ then we may put that $\overline{T} = \infty$.

Also, letting

$$Q_2(T) \equiv Th(T) - \int_0^T h(t)\,\mathrm{d}t$$

we have that $Q_2(0) = 0$ and

$$Q_2(T) - Q_1(T) = Th(T) - \int_0^T h(t)\,\mathrm{d}t - h(T)\int_0^T \overline{F}(t)\,\mathrm{d}t + F(T)$$

$$= \int_0^T [h(T)F(t) - h(t) + f(t)]\,\mathrm{d}t$$

$$> \int_0^T [f(t) - h(t)\overline{F}(t)]\,\mathrm{d}t = 0$$

and hence, $Q_2(T) > Q_1(T)$ for $0 < T < \infty$. Thus, there exists a finite and unique $\underline{T}$ that satisfies $Q_2(T) = c_2/(c_1 - c_2)$, and $T^* > \underline{T}$. ∎

Note that the function $Q_2(T)$ plays an important role in analyzing the periodic replacement with minimal repair (see Section 4.2).

We have two advantages of introducing two such limits of $\overline{T}$ and $\underline{T}$: One is to use a suboptimum replacement time instead of T^*, and $\overline{T}$ becomes sharp if T goes to large. Furthermore, if the failure rate were estimated from actual data, we might replace a unit approximately before its failure rate reaches a level K. The other is to use an initial guess for computing an optimum T^* in Newton's method or the successive approximations.

Next, let $H(t)$ be the cumulative hazard function of $F(t)$; i.e., $H(t) \equiv \int_0^t h(u)\mathrm{d}u$. Then, from Figure 4.2 in Chapter 4 and Theorem 3.3, we have approximately $H(T) = c_2/(c_1 - c_2)$. Thus, if $\widetilde{T}$ is a solution of $H(T) = c_2/(c_1 - c_2)$ then it might be one approximation of optimum time T^*.

Another simple method of age replacement is to balance the cost of replacement at failure against that at nonfailure; i.e., $c_1 F(T) = c_2 \overline{F}(T)$. In this case,

$$F(T) = \frac{c_2}{c_1 + c_2} \tag{3.12}$$

and a solution T_p to satisfy it represents a p $(= c_2/(c_1 + c_2))$th percentile point of distribution $F(t)$.

Example 3.1. In this numerical example, we show how two limits give better approximations and compare them with other replacement times. When the failure time has a Weibull distribution with a shape parameter m ($m > 1$), i.e., $\overline{F}(t) \equiv \exp[-(\lambda t)^m]$, we have

$$\mu = \frac{1}{\lambda}\Gamma\left(\frac{1}{m}+1\right), \qquad h(t) = m\lambda^m t^{m-1}$$

which is strictly increasing from 0 to ∞. Thus, an optimum replacement time T^* is given by unique solution of the equation:

$$m\lambda^m T^{m-1} \int_0^T \exp[-(\lambda t)^m]\,dt + \exp[-(\lambda T)^m] = \frac{c_1}{c_1 - c_2}$$

and

$$\overline{T} = \frac{1}{\lambda}\left[\frac{1}{m\Gamma(1/m+1)}\frac{c_1}{c_1-c_2}\right]^{1/(m-1)}$$

$$\underline{T} = \frac{1}{\lambda}\left[\frac{1}{m-1}\frac{c_2}{c_1-c_2}\right]^{1/m}$$

$$\tilde{T} = \frac{1}{\lambda}\left[\frac{c_2}{c_1-c_2}\right]^{1/m}.$$

It can be easily seen that $\tilde{T} < \underline{T}$ for $m < 2$, $\tilde{T} = \underline{T}$ for $m = 2$, and $\tilde{T} > \underline{T}$ for $m > 2$.

Table 3.1 gives an optimum time T^* and its upper limit $\overline{T}$, lower limit $\underline{T}$, $\tilde{T}$ and T_p for $m = 1.2, 1.6, 2.0, 2.4, 3.0, 3.4$ when $1/\lambda = 100$. This indicates that $\underline{T}$ becomes much better when m and c_1/c_2 are large. On the other hand, $\overline{T}$ is good when m and c_1/c_2 are small. It is of great interest that the computation of T_p is very simple, however, it is a good approximation to T^* for $2 \leq m \leq 3$ and $c_1/c_2 \geq 6$. Near $m = 2.4$, it might be sufficient in actual fields to replace a unit at a $c_2/(c_1+c_2)$th percentile point for $c_1/c_2 \geq 4$. When the failure distribution is uncertain, if the failure rate $h(t)$, cumulative hazard $H(t)$, or pth percentile is statistically estimated, we should usually examine whether such approximations can be used in practice.

Furthermore, when $F(t)$ is a gamma distribution; i.e., $f(t) = [\lambda(\lambda t)^\alpha/\Gamma(\alpha)] \times e^{-\lambda t}$, optimum T^* and the expected cost rate $C(T^*)$ are given in Table 9.1 of Chapter 9. ∎

3.2 Other Age Replacement Models

We show the following three modified models of age replacement: (1) age replacement with discounting, (2) age replacement in discrete time, and (3) age replacement of a parallel system. The detailed derivations are omitted and optimum policies are given directly.

Table 3.1. Comparative table of optimum time T^* and $F(T^*)$, its approximate values $\overline{T}$, $\underline{T}$, $\widetilde{T}$, and percentile T_p when $1/\lambda = 100$

c_1/c_2	T^*	$F(T^*)\times 100$	$\overline{T}$	$\underline{T}$	$\widetilde{T}$	T_p
\multicolumn{7}{c}{$m = 1.2$}
2	1746	100	1746	382	100	47
4	227	93	230	153	40	28
6	124	73	136	100	26	21
10	68	47	92	61	16	14
20	35	24	71	33	9	8
40	19	12	62	18	5	5
60	13	8	59	13	3	3
100	8	5	57	8	2	2

c_1/c_2	T^*	$F(T^*)\times 100$	$\overline{T}$	$\underline{T}$	$\widetilde{T}$	T_p
\multicolumn{7}{c}{$m = 1.6$}
2	173	91	174	138	100	57
4	74	46	89	69	50	39
6	53	30	74	50	37	31
10	36	18	65	35	25	23
20	22	9	60	22	16	15
40	14	4	57	14	10	10
60	11	3	56	11	8	8
100	8	2	56	8	6	6

c_1/c_2	T^*	$F(T^*)\times 100$	$\overline{T}$	$\underline{T}$	$\widetilde{T}$	T_p
\multicolumn{7}{c}{$m = 2.0$}
2	110	70	113	100	100	64
4	59	30	75	58	58	47
6	46	19	68	45	45	39
10	34	11	63	33	33	31
20	23	5	59	23	23	22
40	16	3	58	16	16	16
60	13	2	57	13	13	13
100	10	1	57	10	10	10

(continued on next page)

(Table 3.1 continued)

c_1/c_2	T^*	$F(T^*)\times 100$	$\overline{T}$	$\underline{T}$	$\widetilde{T}$	T_p
\multicolumn{7}{c}{$m = 2.4$}
2	91	55	96	87	100	69
4	56	22	72	55	63	54
6	45	14	67	44	51	46
10	35	8	63	35	40	38
20	26	4	60	25	29	28
40	19	2	59	19	22	21
60	16	1	59	16	18	18
100	13	1	59	13	15	15

c_1/c_2	T^*	$F(T^*)\times 100$	$\overline{T}$	$\underline{T}$	$\widetilde{T}$	T_p
\multicolumn{7}{c}{$m = 3.0$}
2	81	41	87	79	100	74
4	55	16	71	55	69	61
6	47	10	67	46	58	54
10	38	5	64	38	48	46
20	30	3	63	30	37	37
40	23	1	62	23	29	29
60	20	1	62	20	26	25
100	17	1	61	17	22	21

c_1/c_2	T^*	$F(T^*)\times 100$	$\overline{T}$	$\underline{T}$	$\widetilde{T}$	T_p
\multicolumn{7}{c}{$m = 3.4$}
2	78	35	84	77	100	77
4	56	13	71	56	72	64
6	48	8	68	48	62	57
10	41	5	66	41	52	50
20	33	2	64	33	42	41
40	26	1	64	26	34	34
60	23	1	63	23	30	30
100	20	0	63	20	26	26

(1) Age Replacement with Discounting

When we adopt the total expected cost as an appropriate objective function for an infinite time span, we should evaluate the present values of all replacement costs by using an appropriate discount rate. Suppose that a continuous discounting with rate α $(0 < \alpha < \infty)$ is used for the cost incurred at replacement time. That is, the present value of cost c at time t is $ce^{-\alpha t}$ at time 0. Then, the cost on one cycle starting at time t is

$$c_1 e^{-\alpha(t+X)} I_{(X<T)} + c_2 e^{-\alpha(t+T)} I_{(X\geq T)}, \qquad (3.13)$$

where X is a random variable that denotes the failure time of an operating unit on that stage and I_A is an indicator. The expected cost at each cycle is the same, except for a discount rate, and hence, the total expected cost is equal to the sum of discounted costs incurred on the individual stages.

We use the same notation as in the preceding policy except for a discount rate α. Let $C(T;\alpha)$ be the total expected cost for an infinite time span when the planned replacement time is set by T $(0 < T \leq \infty)$ at each stage. Then, from (3.13), we have the following renewal equation.

$$C(T;\alpha) = E\{[c_1 + C(T;\alpha)]e^{-\alpha X}I_{(X<T)} + [c_2 + C(T;\alpha)]e^{-\alpha T}I_{(X\geq T)}\}$$
$$= [c_1 + C(T;\alpha)]\int_0^T e^{-\alpha t}\,dF(t) + [c_2 + C(T;\alpha)]e^{-\alpha T}\overline{F}(T); \quad (3.14)$$

i.e.,

$$C(T;\alpha) = \frac{c_1 \int_0^T e^{-\alpha t}\,dF(t) + c_2 e^{-\alpha T}\overline{F}(T)}{\alpha \int_0^T e^{-\alpha t}\overline{F}(t)\,dt} \quad (3.15)$$

$$C(\infty;\alpha) \equiv \lim_{T\to\infty} C(T;\alpha) = \frac{c_1 F^*(\alpha)}{1 - F^*(\alpha)}, \quad (3.16)$$

where $F^*(s)$ is the Laplace–Stieltjes transform of $F(t)$; i.e., $F^*(s) \equiv \int_0^\infty e^{-st}\,dF(t)$ for $s > 0$. It is easy to see that $\lim_{\alpha\to 0} \alpha C(T;\alpha) = C(T)$ which represents the expected cost rate in (3.4).

Letting

$$K(\alpha) \equiv \frac{c_1 F^*(\alpha) + c_2[1 - F^*(\alpha)]}{(c_1 - c_2)[1 - F^*(\alpha)]/\alpha}$$

similar theorems corresponding to Theorems 3.1, 3.2, and 3.3 are given as follows.

Theorem 3.4. There exists the limit of the failure rate $h(\infty) \equiv \lim_{t\to\infty} h(t)$, possibly infinite, as $t \to \infty$. A sufficient condition that $C(\infty;\alpha) > C(T;\alpha)$ for some finite T is that $h(\infty) > K(\alpha)$.

Theorem 3.5. Suppose that the failure rate $h(t)$ is continuous and strictly increasing.

(i) If $h(\infty) > K(\alpha)$ then there exists a finite and unique T^* $(0 < T^* < \infty)$ that satisfies

$$h(T)\int_0^T e^{-\alpha t}\overline{F}(t)\,dt - \int_0^T e^{-\alpha t}\,dF(t) = \frac{c_2}{c_1 - c_2} \quad (3.17)$$

and the total expected cost is

$$C(T^*;\alpha) = \frac{1}{\alpha}(c_1 - c_2)h(T^*) - c_2. \quad (3.18)$$

(ii) If $h(\infty) \leq K(\alpha)$ then $T^* = \infty$, and the total expected cost is given in (3.16).

It is noted in (3.17) that its left-hand side is strictly decreasing in α, and hence, T^* is greater than an optimum time in Theorem 3.2 for any $\alpha > 0$. This means that the replacement time becomes larger for consideration of discount rates on future costs.

Theorem 3.6. Suppose that the failure rate $h(t)$ is continuous, strictly increasing, and $h(0) < K(\alpha) < h(\infty)$. Then, there exists a finite and unique $\overline{T}$ that satisfies $h(T) = K(\alpha)$, and a finite and unique $\underline{T}$ that satisfies

$$\frac{1 - e^{-\alpha T}}{\alpha} h(T) - \int_0^T e^{-\alpha t} h(t)\, dt = \frac{c_2}{c_1 - c_2}$$

and $\underline{T} < T^* < \overline{T}$.

Example 3.2. Consider a gamma distribution $\overline{F}(t) = (1 + \lambda t)e^{-\lambda t}$. Then,

$$h(t) = \frac{\lambda^2 t}{1 + \lambda t}, \qquad K(\alpha) = \frac{c_1 \lambda^2 + c_2(\alpha^2 + 2\lambda\alpha)}{(c_1 - c_2)(\alpha + 2\lambda)}.$$

The failure rate $h(t)$ is strictly increasing from 0 to λ. From Theorem 3.5, if $\lambda > K(\alpha)$, i.e., $c_1 \lambda > c_2(\alpha + 2\lambda)$, we make the planned replacement at time T^* which uniquely satisfies

$$\frac{(\alpha + \lambda)T - 1 + e^{-(\alpha + \lambda)T}}{1 + \lambda T} = \frac{c_2}{c_1 - c_2}\left(\frac{\alpha + \lambda}{\lambda}\right)^2$$

and

$$C(T^*; \alpha) = \frac{c_1 - c_2}{\alpha} \frac{\lambda^2 T^*}{1 + \lambda T^*} - c_2.$$

Also from Theorem 3.6, we have the inequality

$$\lambda T^* < \frac{c_1 \lambda^2 + c_2(\alpha^2 + 2\lambda\alpha)}{(\alpha + \lambda)[c_1 \lambda - c_2(\alpha + 2\lambda)]}.$$

For example, when $\alpha = 0.1$, $\lambda = 1$, $c_1 = 10$, and $c_2 = 1$, $\overline{T} = 1.17$, $T^* = 0.69$, $C(T^*; 0.1) = 35.75$, and $C(\infty; 0.1) = 47.62$. Thus, we have 25% reduction in cost by adopting the age replacement. ∎

(2) Age Replacement in Discrete Time

In failure studies for parts of airplanes, the time to unit failure is often measured by the number of cycles to failure. In actual situations, tires of jet fighters are replaced preventively at $4 \sim 14$ number of times of flights, which may depend on the kind of uses. In other cases, the lifetimes are sometimes not recorded at the exact instant of failure and are collected statistically per day, per month, or per year. For example, failure data of electric switching devices in electric power companies are recorded as the number of failures [23]. In any case, it would be interesting and possibly useful to consider discrete time processes [42].

Consider the time over an indefinitely long cycle n ($n = 1, 2, \ldots$) that a single unit should be operating. A unit is replaced at cycle N ($N = 1, 2, \ldots$) after its installation or at failure, whichever occurs first. Let $\{p_n\}_{n=1}^{\infty}$ denote the discrete failure distribution that a unit fails at cycle n. Cost c_1 is incurred for each failed unit that is replaced and cost c_2 ($< c_1$) is incurred for each nonfailed unit that is exchanged. Then, in a similar method to that of obtaining (3.4), the expected cost rate is given by

$$C(N) = \frac{c_1 \sum_{j=1}^{N} p_j + c_2 \sum_{j=N+1}^{\infty} p_j}{\sum_{j=1}^{N} \sum_{i=j}^{\infty} p_i} \quad (N = 1, 2, \ldots). \qquad (3.19)$$

Let $h_n \equiv p_n / \sum_{j=n}^{\infty} p_j$ ($n = 1, 2, \ldots$) be the failure rate of the discrete distribution and μ be the mean failure cycle; i.e., $\mu \equiv \sum_{n=1}^{\infty} n p_n < \infty$. Then, Theorem 3.2 in continuous time process is rewritten as the discrete time one of age replacement.

Theorem 3.7. Suppose that h_n is strictly increasing.

(i) If $h_\infty > K$ then there exists a finite and unique minimum N^* ($1 \leq N^* < \infty$) that satisfies

$$h_{N+1} \sum_{j=1}^{N} \sum_{i=j}^{\infty} p_i - \sum_{j=1}^{N} p_j \geq \frac{c_2}{c_1 - c_2} \quad (N = 1, 2, \ldots) \qquad (3.20)$$

and the resulting cost rate is

$$(c_1 - c_2) h_{N^*} \leq C(N^*) < (c_1 - c_2) h_{N^*+1}. \qquad (3.21)$$

(ii) If $h_\infty \leq K$ then $N^* = \infty$.

Note that $0 < h_n \leq 1$ from the definition of the failure rate in discrete time. Thus, if $K \geq 1$, i.e., $\mu \leq c_1/(c_1 - c_2)$, then we do not need to consider any planned replacement.

Example 3.3. Suppose that the failure distribution is a negative binomial one with a shape parameter 2; i.e., $p_n = n p^2 q^{n-1}$ ($n = 1, 2, \ldots$), where $q \equiv 1 - p$ ($0 < p < 1$). Then, $\mu = (1 + q)/p$, $h_n = np^2/(np + q)$ which is strictly increasing from p^2 to p. From Theorem 3.7, if $c_1 q > c_2 (1 + q)$, we should make the replacement cycle N^* ($1 \leq N^* < \infty$) which is a unique minimum such that

$$\frac{(N+1)p(1+q)+q^{N+2}}{Np+1} \geq \frac{c_1}{c_1-c_2}.$$

For example, when $c_1 = 10$, $c_2 = 1$ and $\mu = 9$; i.e., $p = 1/5$, $N^* = 4$. In this case, the expected cost rate is $C(N^*) = 0.92$ and that of no planned replacement is $C(\infty) = 1.11$. ∎

(3) Age Replacement of a Parallel System

Consider a parallel redundant system that consists of N ($N \geq 2$) identical units and fails when all units fail. Each unit has a failure distribution $F(t)$ with finite mean μ.

Suppose that the system is replaced at system failure or at planned time T ($0 < T \leq \infty$), whichever occurs first. Then, we give the expected cost rate as

$$C(T;N) = \frac{c_1 F(T)^N + c_2[1-F(T)^N] + Nc_0}{\int_0^T [1-F(t)^N]\,dt}, \tag{3.22}$$

where c_1 = cost of replacement at system failure, c_2 = cost of replacement at planned time T with $c_2 < c_1$, and c_0 = acquisition cost of one unit. When $N = 1$, this corresponds to the expected cost rate $C(T)$ in (3.4), formally replacing c_1 with $c_1 + c_0$ and c_2 with $c_2 + c_0$.

Let $h(t)$ be the failure rate of each unit, which is increasing and $h(\infty) \equiv \lim_{t\to\infty} h(t)$. We seek an optimum time T^* that minimizes $C(T;N)$ for $N \geq 2$. Differentiating $C(T;N)$ with respect to T and setting it equal to zero, we have

$$\lambda(T;N)\int_0^T [1-F(t)^N]\,dt - F(T)^N = \frac{c_2+Nc_0}{c_1-c_2}, \tag{3.23}$$

where

$$\lambda(t;N) \equiv \frac{Nh(t)[F(t)^{N-1}-F(t)^N]}{1-F(t)^N}.$$

It is easy to see that $\lambda(t;N)$ is strictly increasing when $h(t)$ is increasing, and $\lim_{t\to\infty}\lambda(t;N) = h(\infty)$ because

$$\frac{N[F(t)^{N-1}-F(t)^N]}{1-F(t)^N} = \frac{N[F(t)]^{N-1}}{\sum_{j=1}^N [F(t)]^{j-1}}.$$

Furthermore, it is clear from this result that the left-hand side of (3.23) is strictly increasing from 0 to $\mu_N h(\infty) - 1$, where $\mu_N \equiv \int_0^\infty [1-F(t)^N]\,dt$ is the mean time to system failure.

Therefore, we have the following optimum policy.

(i) If $\mu_N h(\infty) > (c_1 + Nc_0)/(c_1 - c_2)$ then there exists a finite and unique T^* ($0 < T^* < \infty$) that satisfies (3.23) and the resulting cost rate is

$$C(T^*;N) = (c_1-c_2)\lambda(T^*;N). \tag{3.24}$$

(ii) If $\mu_N h(\infty) \le (c_1+Nc_0)/(c_1-c_2)$ then $T^* = \infty$; i.e., the system is replaced only at system failure, and

$$C(\infty; N) = \frac{c_1 + Nc_0}{\mu_N}. \tag{3.25}$$

Moreover, the maintenance of k-out-of-n systems was analyzed in [43–45].

3.3 Continuous and Discrete Replacement

Almost all units deteriorate with age and use, and eventually, fail from either cause. If their failure rates increase with age and use, it may be wise to replace units when they reach a certain age or are used a certain number of times. This policy would be effective where units suffer great deterioration with both age and use, and are applied to the maintenance of some parts of large complex systems such as switching devices, car batteries, railroad pantographs, and printers.

This section suggests an extended age replacement model that combines the continuous replacement as described in Section 3.1 and the discrete replacement in **(2)** of Section 3.2 as follows. A unit should operate for an infinite time span and is replaced at failure. Furthermore, a unit begins to operate at time 0, and is used according to a renewal process with an arbitrary distribution $G(t)$ with finite mean $1/\theta \equiv \int_0^\infty [1 - G(t)]\mathrm{d}t < \infty$. The probability that a unit is used exactly j times during $(0, t]$ is $G^{(j)}(t) - G^{(j+1)}(t)$, where $G^{(j)}(t)$ ($j = 1, 2, \dots$) denotes the j-fold Stieltjes convolution of $G(t)$ with itself and $G^{(0)}(t) \equiv 1$ for $t \ge 0$. The continuous distribution of failures due to deterioration with age is $F(t)$, and the discrete distribution of failures due to use is $\{p_j\}_{j=1}^\infty$, where $F(t)$ and p_j are independent of each other, and the failure rates of both distributions are $h(t) \equiv f(t)/\overline{F}(t)$ and $h_j \equiv p_j/(1-P_{j-1})$ ($j = 1, 2, \dots$), respectively, where $P_j \equiv \sum_{i=1}^j p_i$ ($j = 1, 2, \dots$) and $P_0 \equiv 0$.

It is assumed that a unit is replaced before failure at time T ($0 < T \le \infty$) of age or at number N ($N = 1, 2, \dots$) of uses, whichever occurs first. Then, the probability that a unit is replaced at time T is

$$\overline{F}(T) \sum_{j=0}^{N-1} (1 - P_j)[G^{(j)}(T) - G^{(j+1)}(T)] \tag{3.26}$$

because the probability that the number of uses occurs exactly j times ($j = 0, 1, \dots, N-1$) until time T is $G^{(j)}(T) - G^{(j+1)}(T)$, and the probability that it is replaced at number N is

$$(1 - P_N) \int_0^T \overline{F}(t) \, \mathrm{d}G^{(N)}(t). \tag{3.27}$$

Thus, by adding (3.26) and (3.27), and rearranging them, the probability that a unit is replaced before failure is

$$(1 - P_N)\left\{1 - \int_0^T [1 - G^{(N)}(t)]\,dF(t)\right\} + \overline{F}(T)\sum_{j=1}^N p_j[1 - G^{(j)}(T)]. \quad (3.28)$$

The probability that a unit is replaced at time t $(0 < t \le T)$ by the failure due to continuous deterioration with age is

$$\sum_{j=0}^{N-1}(1 - P_j)\int_0^T [G^{(j)}(t) - G^{(j+1)}(t)]\,dF(t) \quad (3.29)$$

and the probability that it is replaced at number j of uses $(j = 1, 2, \ldots, N)$ is

$$\sum_{j=1}^N p_j\int_0^T \overline{F}(t)\,dG^{(j)}(t). \quad (3.30)$$

Thus, the probability that a unit is replaced at failure is

$$\sum_{j=0}^{N-1}\left\{(1 - P_j)\int_0^T [G^{(j)}(t) - G^{(j+1)}(t)]\,dF(t) + p_{j+1}\int_0^T \overline{F}(t)\,dG^{(j+1)}(t)\right\}. \quad (3.31)$$

It is evident that (3.28) + (3.31) = 1 because we have the relation

$$(1 - P_N)[1 - G^{(N)}(t)] + \sum_{j=1}^N p_j[1 - G^{(j)}(t)] = \sum_{j=0}^{N-1}(1 - P_j)[G^{(j)}(t) - G^{(j+1)}(t)].$$

The mean time to replacement is, referring to (3.26), (3.27), and (3.31),

$$T\overline{F}(T)\sum_{j=0}^{N-1}(1 - P_j)[G^{(j)}(T) - G^{(j+1)}(T)] + (1 - P_N)\int_0^T t\overline{F}(t)\,dG^{(N)}(t)$$

$$+ \sum_{j=0}^{N-1}\left\{(1 - P_j)\int_0^T t[G^{(j)}(t) - G^{(j+1)}(t)]\,dF(t) + p_{j+1}\int_0^T t\overline{F}(t)\,dG^{(j+1)}(t)\right\}$$

$$= \sum_{j=0}^{N-1}(1 - P_j)\int_0^T [G^{(j)}(t) - G^{(j+1)}(t)]\overline{F}(t)\,dt. \quad (3.32)$$

Therefore, the expected cost rate is, from (3.3),

$$C(T, N) = \frac{(c_1 - c_2)\sum_{j=0}^{N-1}\left\{(1 - P_j)\int_0^T [G^{(j)}(t) - G^{(j+1)}(t)]\,dF(t) + p_{j+1}\int_0^T \overline{F}(t)\,dG^{(j+1)}(t)\right\} + c_2}{\sum_{j=0}^{N-1}(1 - P_j)\int_0^T [G^{(j)}(t) - G^{(j+1)}(t)]\overline{F}(t)\,dt}, \quad (3.33)$$

3.3 Continuous and Discrete Replacement

where $c_1 =$ cost of replacement at failure and $c_2 =$ cost of planned replacement at time T or at number N with $c_2 < c_1$.

This includes some basic replacement models: when a unit is replaced before failure only at time T,

$$C(T) \equiv \lim_{N \to \infty} C(T, N) = \frac{c_1 - (c_1 - c_2)\overline{F}(T) \sum_{j=1}^{\infty} p_j \overline{G}^{(j)}(T)}{\sum_{j=1}^{\infty} p_j \int_0^T [1 - G^{(j)}(t)]\overline{F}(t)\, dt}. \qquad (3.34)$$

In particular, when $p_j \equiv 0$ ($j = 1, 2, \ldots$), i.e., a unit fails only by continuous deterioration with age, the expected cost rate $C(T)$ agrees with (3.4) of the standard age replacement.

On the other hand, when a unit is replaced before failure only at number N,

$$C(N) \equiv \lim_{T \to \infty} C(T, N) = \frac{c_1 - (c_1 - c_2)(1 - P_N)\int_0^{\infty} G^{(N)}(t)\, dF(t)}{\sum_{j=0}^{N-1}(1 - P_j)\int_0^{\infty}[G^{(j)}(t) - G^{(j+1)}(t)]\overline{F}(t)\, dt}. \qquad (3.35)$$

When $\overline{F}(t) \equiv 1$ for $t \geq 0$, i.e., a unit fails only from use, $C(N)/\theta$ agrees with (3.19) of the discrete age replacement. Finally, when $T = \infty$ and $N = \infty$, i.e., a unit is replaced only at failure,

$$C \equiv \lim_{N \to \infty} C(N) = \frac{c_1}{\sum_{j=1}^{\infty} p_j \int_0^{\infty}[1 - G^{(j)}(t)]\overline{F}(t)\, dt}. \qquad (3.36)$$

(1) Optimum T^*

Suppose that $G(t) = 1 - e^{-\theta t}$ and $G^{(j)}(t) = 1 - \sum_{i=0}^{j-1}[(\theta t)^i/i!]\, e^{-\theta t}$ ($j = 1, 2, \ldots$). Then, the expected cost rate $C(T)$ in (3.34) is rewritten as

$$C(T) = \frac{c_1 - (c_1 - c_2)\overline{F}(T)\sum_{j=0}^{\infty}(1 - P_j)[(\theta T)^j/j!]\, e^{-\theta T}}{\sum_{j=0}^{\infty}(1 - P_j)\int_0^T[(\theta t)^j/j!]\, e^{-\theta t}\overline{F}(t)\, dt}. \qquad (3.37)$$

We seek an optimum T^* that minimizes $C(T)$ when the failure rate $h(t)$ of $F(t)$ is continuous and strictly increasing with $h(\infty) \equiv \lim_{t \to \infty} h(t)$, and the failure rate h_j of $\{p_j\}_{j=1}^{\infty}$ is increasing with $h_\infty \equiv \lim_{j \to \infty} h_j$, where $h(\infty)$ may possibly be infinity.

Lemma 3.1. If the failure rate h_j is strictly increasing then

$$\frac{\sum_{j=0}^{N} p_{j+1}[(\theta T)^j/j!]}{\sum_{j=0}^{N}(1 - P_j)[(\theta T)^j/j!]} \qquad (3.38)$$

is strictly increasing in T and converges to h_{N+1} as $T \to \infty$ for any integer N.

86   3 Age Replacement

Proof.   Differentiating (3.38) with respect to T, we have

$$\frac{\theta}{\{\sum_{j=0}^{N}(1-P_j)[(\theta T)^j/j!]\}^2} \left\{ \sum_{j=1}^{N} P_{j+1} \frac{(\theta T)^{j-1}}{(j-1)!} \sum_{j=0}^{N}(1-P_j)\frac{(\theta T)^j}{j!} \right.$$

$$\left. - \sum_{j=0}^{N} P_{j+1} \frac{(\theta T)^j}{j!} \sum_{j=1}^{N}(1-P_j)\frac{(\theta T)^{j-1}}{(j-1)!} \right\}.$$

The expression within the bracket of the numerator is

$$\sum_{j=1}^{N} \frac{(\theta T)^{j-1}}{(j-1)!} \sum_{i=0}^{N} \frac{(\theta T)^i}{i!}(1-P_i)(1-P_j)(h_{j+1}-h_{i+1})$$

$$= \sum_{j=1}^{N} \frac{(\theta T)^{j-1}}{(j-1)!} \sum_{i=0}^{j-1} \frac{(\theta T)^i}{i!}(1-P_i)(1-P_j)(h_{j+1}-h_{i+1})$$

$$+ \sum_{j=1}^{N} \frac{(\theta T)^{j-1}}{(j-1)!} \sum_{i=j}^{N} \frac{(\theta T)^i}{i!}(1-P_i)(1-P_j)(h_{j+1}-h_{i+1})$$

$$= \sum_{j=1}^{N} \frac{(\theta T)^{j-1}}{j!} \sum_{i=0}^{j-1} \frac{(\theta T)^i}{i!}(1-P_i)(1-P_j)(h_{j+1}-h_{i+1})(j-i) > 0$$

which implies that (3.38) is strictly increasing in T. Furthermore, it is evident that this tends to h_{N+1} as $T \to \infty$. ∎

Lemma 3.2.   If the failure rate $h(t)$ is continuous and strictly increasing then

$$\frac{\int_0^T (\theta t)^N e^{-\theta t}\, dF(t)}{\int_0^T (\theta t)^N e^{-\theta t}\overline{F}(t)\, dt} \tag{3.39}$$

is strictly increasing in N and converges to $h(T)$ as $N \to \infty$ for all $T > 0$.

Proof.   Letting

$$q(T) \equiv \int_0^T (\theta t)^{N+1} e^{-\theta t}\, dF(t) \int_0^T (\theta t)^N e^{-\theta t}\overline{F}(t)\, dt$$

$$- \int_0^T (\theta t)^N e^{-\theta t}\, dF(t) \int_0^T (\theta t)^{N+1} e^{-\theta t}\overline{F}(t)\, dt,$$

it is easy to show that $\lim_{T \to 0} q(T) = 0$, and

$$\frac{dq(T)}{dT} = (\theta T)^N e^{-\theta T}\overline{F}(T) \int_0^T (\theta t)^N e^{-\theta t}\overline{F}(t)(\theta T - \theta t)[h(T) - h(t)]\, dt > 0$$

3.3 Continuous and Discrete Replacement

because $h(t)$ is strictly increasing. Thus, $q(T)$ is a strictly increasing function of T from 0, and hence, $q(T) > 0$ for all $T > 0$, which shows that the quantity in (3.39) is strictly increasing in N.

Next, from the assumption that $h(t)$ is increasing,

$$\frac{\int_0^T (\theta t)^N e^{-\theta t} \, dF(t)}{\int_0^T (\theta t)^N e^{-\theta t} \overline{F}(t) \, dt} \leq h(T).$$

On the other hand, we have, for any $\delta \in (0, T)$,

$$\frac{\int_0^T (\theta t)^N e^{-\theta t} \, dF(t)}{\int_0^T (\theta t)^N e^{-\theta t} \overline{F}(t) \, dt} = \frac{\int_0^{T-\delta} (\theta t)^N e^{-\theta t} \, dF(t) + \int_{T-\delta}^T (\theta t)^N e^{-\theta t} \, dF(t)}{\int_0^{T-\delta} (\theta t)^N e^{-\theta t} \overline{F}(t) \, dt + \int_{T-\delta}^T (\theta t)^N e^{-\theta t} \overline{F}(t) \, dt}$$

$$\geq \frac{h(T-\delta) \int_{T-\delta}^T (\theta t)^N e^{-\theta t} \overline{F}(t) \, dt}{\int_0^{T-\delta} (\theta t)^N e^{-\theta t} \overline{F}(t) \, dt + \int_{T-\delta}^T (\theta t)^N e^{-\theta t} \overline{F}(t) \, dt}$$

$$= \frac{h(T-\delta)}{1 + \left[\int_0^{T-\delta} (\theta t)^N e^{-\theta t} \overline{F}(t) \, dt \Big/ \int_{T-\delta}^T (\theta t)^N e^{-\theta t} \overline{F}(t) \, dt \right]}.$$

The quantity in the bracket of the denominator is

$$\frac{\int_0^{T-\delta} (\theta t)^N e^{-\theta t} \overline{F}(t) \, dt}{\int_{T-\delta}^T (\theta t)^N e^{-\theta t} \overline{F}(t) \, dt} \leq \frac{e^{\theta T}}{\delta \overline{F}(T)} \int_0^{T-\delta} \left(\frac{t}{T-\delta}\right)^N dt \to 0 \quad \text{as } N \to \infty.$$

Therefore, it follows that

$$h(T-\delta) \leq \lim_{N \to \infty} \frac{\int_0^T (\theta t)^N e^{-\theta t} \, dF(t)}{\int_0^T (\theta t)^N e^{-\theta t} \overline{F}(t) \, dt} \leq h(T)$$

which completes the proof because δ is arbitrary and $h(t)$ is continuous. ∎

Letting

$$Q(T) \equiv \left\{ h(T) + \frac{\theta \sum_{j=0}^\infty p_{j+1}[(\theta T)^j/j!]}{\sum_{j=0}^\infty (1-P_j)[(\theta T)^j/j!]} \right\} \sum_{j=0}^\infty (1-P_j) \int_0^T \frac{(\theta t)^j}{j!} e^{-\theta t} \overline{F}(t) \, dt$$

$$- \left[1 - \overline{F}(T) \sum_{j=0}^\infty (1-P_j) \frac{(\theta T)^j}{j!} e^{-\theta T} \right]$$

we have the following optimum policy that minimizes $C(T)$ in (3.37).

Theorem 3.8. Suppose that the failure rate $h(t)$ is strictly increasing and h_j is increasing.

88   3 Age Replacement

(i) If

$$Q(\infty) \equiv \lim_{T \to \infty} Q(T)$$
$$= [h(\infty) + \theta h_\infty] \sum_{j=0}^{\infty} (1 - P_j) \int_0^\infty \frac{(\theta t)^j}{j!} e^{-\theta t} \overline{F}(t) \, dt - 1$$
$$> \frac{c_2}{c_1 - c_2} \tag{3.40}$$

then there exists a finite and unique T^* $(0 < T^* < \infty)$ that satisfies

$$Q(T) = \frac{c_2}{c_1 - c_2} \tag{3.41}$$

and the resulting cost rate is

$$C(T^*) = (c_1 - c_2) \left\{ h(T^*) + \frac{\theta \sum_{j=0}^{\infty} p_{j+1}[(\theta T^*)^j / j!]}{\sum_{j=0}^{\infty} (1 - P_j)[(\theta T^*)^j / j!]} \right\}. \tag{3.42}$$

(ii) If $Q(\infty) \le c_2/(c_1 - c_2)$ then $T^* = \infty$; i.e., we should make no planned replacement and the expected cost rate is

$$C(\infty) \equiv \lim_{T \to \infty} C(T) = \frac{c_1}{\sum_{j=0}^{\infty} (1 - P_j) \int_0^\infty [(\theta t)^j / j!] e^{-\theta t} \overline{F}(t) \, dt}. \tag{3.43}$$

Proof. Differentiating $C(T)$ in (3.37) with respect to T and setting it equal to zero, we have (3.41). First, we note from Lemma 3.1 that when h_j is increasing, $\{\theta \sum_{j=0}^{\infty} p_{j+1}[(\theta T)^j / j!]\} / \{\sum_{j=0}^{\infty} (1 - P_j)[(\theta T)^j / j!]\}$ is increasing in T and converges to θh_∞ as $T \to \infty$. Thus, it is clearly seen that $dQ(T)/dT > 0$, and hence, $Q(T)$ is strictly increasing from 0 to $Q(\infty)$.

Therefore, if $Q(\infty) > c_2/(c_1 - c_2)$ then there exists a finite and unique T^* $(0 < T^* < \infty)$ that satisfies (3.41), and the expected cost rate is given in (3.42). Conversely, if $Q(\infty) \le c_2/(c_1 - c_2)$ then $C(T)$ is strictly decreasing to $C(\infty)$, and hence, we have (3.43) from (3.37). ∎

In particular, suppose that the discrete distribution of failure times is geometric; i.e., $p_j = pq^{j-1}$ $(j = 1, 2, \dots)$, and the Laplace–Stieltjes transform of $F(t)$ is $F^*(s) \equiv \int_0^\infty e^{-st} dF(t)$. In this case, if

$$h(\infty) > p\theta \left[\frac{c_1}{c_1 - c_2} \frac{1}{1 - F^*(p\theta)} - 1 \right]$$

then there exists a finite and unique T^* that satisfies

$$h(T) \int_0^T e^{-p\theta t} \overline{F}(t) \, dt - \int_0^T e^{-p\theta t} \, dF(t) = \frac{c_2}{c_1 - c_2}$$

and the resulting cost rate is

$$C(T^*) = (c_1 - c_2)[h(T^*) + p\theta]$$

which correspond to (3.9) and (3.10) in Section 3.1, respectively.

(2) Optimum N^*

The expected cost rate $C(N)$ in (3.35) when $G(t) = 1 - e^{-\theta t}$ is

$$C(N) = \frac{c_1 - (c_1 - c_2)(1 - P_N) \sum_{j=N}^{\infty} \int_0^{\infty} [(\theta t)^j / j!] e^{-\theta t} \, dF(t)}{\sum_{j=0}^{N-1} (1 - P_j) \int_0^{\infty} [(\theta t)^j / j!] e^{-\theta t} \overline{F}(t) \, dt}$$

$$(N = 1, 2, \ldots). \qquad (3.44)$$

Letting

$$L(N) \equiv \left\{ \frac{\int_0^{\infty} (\theta t)^N e^{-\theta t} \, dF(t)}{\int_0^{\infty} (\theta t)^N e^{-\theta t} \overline{F}(t) \, dt} + \theta h_{N+1} \right\} \sum_{j=0}^{N-1} (1 - P_j) \int_0^{\infty} \frac{(\theta t)^j}{j!} e^{-\theta t} \overline{F}(t) \, dt$$

$$- \left[1 - (1 - P_N) \sum_{j=N}^{\infty} \int_0^{\infty} \frac{(\theta t)^j}{j!} e^{-\theta t} \, dF(t) \right] \quad (N = 1, 2, \ldots)$$

we have the following optimum policy that minimizes $C(N)$.

Theorem 3.9. Suppose that $h(t)$ is increasing and h_j is strictly increasing.

(i) If

$$L(\infty) \equiv \lim_{N \to \infty} L(N)$$

$$= [h(\infty) + \theta h_{\infty}] \sum_{j=0}^{\infty} (1 - P_j) \int_0^{\infty} \frac{(\theta t)^j}{j!} e^{-\theta t} \overline{F}(t) \, dt - 1$$

$$> \frac{c_2}{c_1 - c_2}$$

then there exists a finite and unique minimum N^* that satisfies

$$L(N) \geq \frac{c_2}{c_1 - c_2} \qquad (N = 1, 2, \ldots). \qquad (3.45)$$

(ii) If $L(\infty) \leq c_2/(c_1 - c_2)$ then $N^* = \infty$, and the expected cost rate is given in (3.43).

Proof. From the inequality $C(N+1) \geq C(N)$, we have (3.45). Recalling that from Lemma 3.2 when $h(t)$ is increasing, $[\int_0^{\infty} (\theta t)^N e^{-\theta t} \, dF(t)] / [\int_0^{\infty} (\theta t)^N e^{-\theta t} \overline{F}(t) \, dt]$ is increasing in N and converges to $h(\infty)$ as $N \to \infty$, we can clearly see that $L(N)$ strictly increases to $L(\infty)$.

Therefore, if $L(\infty) > c_2/(c_1 - c_2)$ then there exists a finite and unique minimum N^* ($N^* = 1, 2, \ldots$) that satisfies (3.45). Conversely, if $L(\infty) \leq c_2/(c_1 - c_2)$ then $C(N)$ is decreasing in N, and hence, $N^* = \infty$. ∎

It is of great interest that the limit $L(\infty)$ is equal to $Q(\infty)$ in (3.40).

In particular, suppose that the failure distribution $F(t)$ is exponential, *i.e.*, $F(t) = 1 - e^{-\lambda t}$, and the probability generating function of $\{p_j\}$ is $p^*(z) \equiv \sum_{j=1}^{\infty} p_j z^j$ for $|z| < 1$. In this case, if

$$h_\infty > \frac{\lambda}{\theta} \left\{ \frac{c_1}{c_1 - c_2} \frac{1}{1 - p^*[\theta/(\theta + \lambda)]} - 1 \right\}$$

then there exists a finite and unique minimum that satisfies

$$h_{N+1} \sum_{j=0}^{N-1} (1 - P_j) \left(\frac{\theta}{\theta + \lambda} \right)^{j+1} - \sum_{j=1}^{N} p_j \left(\frac{\theta}{\theta + \lambda} \right)^j \geq \frac{c_2}{c_1 - c_2} \quad (N = 1, 2, \dots)$$

and the resulting cost rate is

$$\theta h_{N^*} \leq \frac{C(N^*)}{c_1 - c_2} - \lambda < \theta h_{N^*+1}$$

which corresponds to (3.20) and (3.21) in **(2)** of Section 3.2.

(3) Optimum T^* and N^*

When $G(t) = 1 - e^{-\theta t}$, the expected cost rate $C(T, N)$ in (3.33) is rewritten as

$$C(T, N) = \frac{(c_1 - c_2) \sum_{j=0}^{N-1} \left\{ (1 - P_j) \int_0^T [(\theta t)^j / j!] e^{-\theta t} \, dF(t) + p_{j+1} \int_0^T \theta [(\theta t)^j / j!] e^{-\theta t} \overline{F}(t) \, dt \right\} + c_2}{\sum_{j=0}^{N-1} (1 - P_j) \int_0^T [(\theta t)^j / j!] e^{-\theta t} \overline{F}(t) \, dt}. \quad (3.46)$$

We seek both optimum T^* and N^* that minimize $C(T, N)$ when $h(t)$ is continuous and strictly increasing to ∞ and h_j is strictly increasing. Differentiating $C(T, N)$ with respect to T and setting it equal to zero for a fixed N, we have

$$Q(T; N) = \frac{c_2}{c_1 - c_2}, \quad (3.47)$$

where

$$Q(T; N) \equiv \left\{ h(T) + \frac{\theta \sum_{j=0}^{N-1} p_{j+1} [(\theta T)^j / j!]}{\sum_{j=0}^{N-1} (1 - P_j)[(\theta T)^j / j!]} \right\} \sum_{j=0}^{N-1} (1 - P_j) \int_0^T \frac{(\theta t)^j}{j!} e^{-\theta t} \overline{F}(t) \, dt$$

$$- \sum_{j=0}^{N-1} \left\{ (1 - P_j) \int_0^T \frac{(\theta t)^j}{j!} e^{-\theta t} \, dF(t) + p_{j+1} \int_0^T \frac{\theta (\theta t)^j}{j!} e^{-\theta t} \overline{F}(t) \, dt \right\}.$$

It is evident that $\lim_{T \to 0} Q(T; N) = 0$ and $\lim_{T \to \infty} Q(T; N) = \infty$. Furthermore, we can easily prove from Lemma 3.1 that $Q(T; N)$ is strictly increasing

3.3 Continuous and Discrete Replacement

in T. Hence, there exists a finite and unique T^* that satisfies (3.47) for any $N \geq 1$, and the resulting cost rate is

$$C(T^*, N) = (c_1 - c_2)\left\{h(T^*) + \frac{\theta \sum_{j=0}^{N-1} p_{j+1}[(\theta T^*)^j/j!]}{\sum_{j=0}^{N-1}(1 - P_j)[(\theta T^*)^j/j!]}\right\}. \quad (3.48)$$

Next, from the inequality $C(T, N+1) - C(T, N) \geq 0$ for a fixed $T > 0$,

$$L(N; T) \geq \frac{c_2}{c_1 - c_2}, \quad (3.49)$$

where

$L(N; T) \equiv$

$$\left\{\theta h_{N+1} + \frac{\int_0^T (\theta t)^N e^{-\theta t}\, dF(t)}{\int_0^T (\theta t)^N e^{-\theta t} \overline{F}(t)\, dt}\right\} \sum_{j=0}^{N-1}(1 - P_j)\int_0^T \frac{(\theta t)^j}{j!} e^{-\theta t}\overline{F}(t)\, dt$$

$$- \sum_{j=0}^{N-1}\left\{(1 - P_j)\int_0^T \frac{(\theta t)^j}{j!} e^{-\theta t}\, dF(t) + p_{j+1}\int_0^T \frac{\theta(\theta t)^j}{j!} e^{-\theta t}\overline{F}(t)\, dt\right\}$$

$$(N = 1, 2, \dots).$$

From Lemma 3.2, $L(N; T)$ is strictly increasing in N because

$$L(N+1; T) - L(N; T) = \sum_{j=0}^{N}(1 - P_j)\int_0^T \frac{(\theta t)^j}{j!} e^{-\theta t}\overline{F}(t)\, dt$$

$$\times \left\{\theta(h_{N+2} - h_{N+1}) + \frac{\int_0^T (\theta t)^{N+1} e^{-\theta t}\, dF(t)}{\int_0^T (\theta t)^{N+1} e^{-\theta t}\overline{F}(t)\, dt} - \frac{\int_0^T (\theta t)^N e^{-\theta t}\, dF(t)}{\int_0^T (\theta t)^N e^{-\theta t}\overline{F}(t)\, dt}\right\} > 0.$$

In addition, because T and N have to satisfy (3.47), the inequality (3.49) can be rewritten as

$$\theta\left\{h_{N+1} - \frac{\sum_{j=0}^{N-1} p_{j+1}[(\theta T)^j/j!]}{\sum_{j=0}^{N-1}(1 - P_j)[(\theta T)^j/j!]}\right\} + \frac{\int_0^T (\theta t)^N e^{-\theta t}\, dF(t)}{\int_0^T (\theta t)^N e^{-\theta t}\overline{F}(t)\, dt} \geq h(T). \quad (3.50)$$

It is noted that the left-hand side of (3.50) is greater than $h(T)$ as $N \to \infty$ from Lemma 3.2. Thus, there exists a finite N^* that satisfies inequality (3.50) for all $T > 0$.

From the above discussions, we can specify the computing procedure for obtaining optimum T^* and N^*.

1. Compute a minimum N_1 to satisfy (3.45).
2. Compute T_k to satisfy (3.47) for N_k ($k = 1, 2, \dots$).
3. Compute a minimum N_{k+1} to satisfy (3.50) for T_k ($k = 1, 2, \dots$).
4. Continue the computation until $N_k = N_{k+1}$, and put $N_k = N^*$ and $T_k = T^*$.

Table 3.2. Optimum time T^* and its cost rate $C(T^*)$, optimum number N^* and its cost rate $C(N^*)$, optimum (T^*, N^*) and its cost rate $C(T^*, N^*)$ when $F(t) = 1 - \exp(-\lambda t^2)$, $p_j = jp^2 q^{j-1}$, and $\lambda = \pi p^2 / [4(1+q)^2]$

p	T^*	$C(T^*)$	N^*	$C(N^*)$	(T^*, N^*)	$C(T^*, N^*)$
0.1	5.0	0.5368	5	0.5370	(6.1, 5)	0.5206
0.05	9.8	0.2497	10	0.2495	(11.5, 10)	0.2457
0.02	24.3	0.0955	24	0.0954	(26.8, 25)	0.0947
0.01	48.3	0.0470	48	0.0470	(52.0, 50)	0.0469
0.005	96.5	0.0233	96	0.0233	(101.6, 99)	0.0233

Example 3.4. We give a numerical example when $G(t) = 1 - e^{-t}$, $F(t)$ is a Weibull distribution $[1 - \exp(-\lambda t^2)]$, and $\{p_j\}$ is a negative binomial distribution $p_j = jp^2 q^{j-1}$ $(j = 1, 2, \ldots)$, where $q \equiv 1 - p$. Furthermore, suppose that $\lambda = \pi p^2 / [4(1+q)^2]$; i.e., the mean time to failure caused by use is equal to that caused by deterioration with age.

Table 3.2 shows the optimum T^*, $C(T^*)$, N^*, $C(N^*)$ and (T^*, N^*), $C(T^*, N^*)$ for $c_1 = 10$, $c_2 = 1$, and $p = 0.1, 0.05, 0.02, 0.01, 0.005$. This indicates that expected cost rates $C(T^*)$ and $C(N^*)$ are almost the same, $C(T^*, N^*)$ is a little lower than these costs, and (T^*, N^*) are equal to or greater than each value of T^* and N^*, respectively. If failures due to continuous deterioration with age and discrete deterioration with use occur at the same mean time, we may make the planned replacement according to a time policy of either age or number of uses. ∎

References

1. Barlow RE and Proschan F (1965) Mathematical Theory of Reliability. J Wiley & Sons, New York.
2. Cox DR (1962) Renewal Theory. Methuen, London.
3. Glasser GJ (1967) The age replacement problem. Technometrics. 9:83–91.
4. Scheaffer RL (1971) Optimum age replacement policies with an increasing cost factor. Technometrics 13:139–144.
5. Cléroux R, Hanscom M (1974) Age replacement with adjustment and depreciation costs and interest charges. Technometrics 16:235–239.
6. Cléroux R, Dubuc S, Tilquin C (1979) The age replacement problem with minimal repair and random repair costs. Oper Res 27:1158–1167.
7. Subramanian R, Wolff MR (1976) Age replacement in simple systems with increasing loss functions. IEEE Trans on Reliab R-25:32–34.
8. Bergman B (1978) Optimal replacement under a general failure model. Adv Appl Prob 10:431–451.
9. Block HW, Borges WS, Savits TH (1988) A general age replacement model with minimal repair. Nav Res Logist 35:365–372.
10. Sheu SH, Kuo CM, Nakagawa T (1993) Extended optimal age replacement policy with minimal repair. RAIRO Oper Res 27:337–351.

11. Sheu SH, Griffith WS, Nakagawa T (1995) Extended optimal replacement model with random minimal repair costs. Eur J Oper Res 85:636–649.
12. Sheu SH, Griffith WS (2001) Optimal age-replacement policy with age-dependent minimal-repair and random-leadtime. IEEE Trans Reliab 50:302–309.
13. Fox B (1966) Age replacement with discounting. Oper Res 14:533–537.
14. Ran A, Rosenland SI (1976). Age replacement with discounting for a continuous maintenance cost model. Technometrics 18:459–465.
15. Berg M, Epstein B (1978) Comparison of age, block, and failure replacement policies. IEEE Trans Reliab R-27:25–28.
16. Ingram CR, Scheaffer RL (1976) On consistent estimation of age replacement intervals. Technometrics 18:213–219.
17. Frees EW, Ruppert D (1985) Sequential non-parametric age replacement policies. Ann Stat 13:650–662.
18. Léger C, Cléroux R (1992) Nonparametric age replacement: Bootstrap confidence intervals for the optimal costs. Oper Res 40:1062–1073.
19. Popova E, Wu HC (1999) Renewal reward processes with fuzzy rewards and their applications to T-age replacement policies. Eur J Oper Res 117:606–617.
20. Kordonsky KB, Gertsbakh I (1994) Best time scale for age replacement. Int J Reliab Qual Saf Eng 1:219–229.
21. Frickenstein SG, Whitaker LR (2003) Age replacement policies in two time scales. Nav Res Logist 50:592–613.
22. Dohi T, Kaio N, Osaki S (2000) Basic preventive maintenance policies and their variations. In: Ben-Daya M, Duffuaa SO, Raouf A (eds) Maintenance, Modeling and Optimization. Kluwer Academic, Boston:155–183.
23. Kaio N, Doshi T, Osaki S (2002) Classical maintenance models. In: Osaki S (ed) Stochastic Models in Reliability and Maintenance. Springer, New York:65–87.
24. Dohi T, Kaio N, Osaki S (2003) Preventive maintenance models: Replacement, repair, ordering, and inspection. In: Pham H (ed) Handbook of Reliability Engineering. Springer, London:349–366.
25. Zheng X (1995) All opportunity-triggered replacement policy for multiple-unit systems. IEEE Trans Reliab 44:648–652.
26. Osaki S, Nakagawa T (1975) A note on age replacement. IEEE Trans Reliab R-24:92–94.
27. Nakagawa T, Yasui K (1982) Bounds of age replacement time. Microelectron Reliab 22:603–609.
28. Nakagawa T, Osaki S (1977) Discrete time age replacement policies. Oper Res Q 28:881–885.
29. Yasui K, Nakagawa T, Osaki S (1988) A summary of optimum replacement policies for a parallel redundant system. Microelectron Reliab 28:635–641.
30. Nakagawa T (1985) Continuous and discrete age-replacement policies. J Oper Res Soc 36:147–154.
31. Berg M (1976) A proof of optimality for age replacement policies. J Appl Prob 13:751–759.
32. Bergman B (1980) On the optimality of stationary replacement strategies. J Appl Prob 17:178–186.
33. Berg M (1995) The marginal cost analysis and its application to repair and replacement policies. Eur J Oper Res 82:214–224.

34. Berg M (1996) Economics oriented maintenance analysis and the marginal cost approach. In: Özekici S (ed) Reliability and Maintenance of Complex Systems. Springer, New York:189–205.
35. Christer AH (1978) Refined asymptotic costs for renewal reward processes. J Oper Res Soc 29:577–583.
36. Ansell J, Bendell A, Humble S (1984) Age replacement under alternative cost criteria. Manage Sci 30:358–367.
37. Love CE, Guo R (1991) Using proportional hazard modelling in plant maintenance. Qual Reliab Eng Inter 7:7–17.
38. Bergman B, Klefsjö B (1982) A graphical method applicable to age replacement problems. IEEE Trans Reliab R-31:478–481.
39. Bergman B, Klefsjö B (1984) The total time on test concept and its use in reliability theory. Oper Res 32:596–606.
40. Bergman B (1985) On reliability theory and its applications. Scand J Statist 12:1–41.
41. Kumar D, Westberg U (1997) Maintenance scheduling under age replacement policy using proportional hazard model and TTT-policy. Eur J Oper Res 99:507–515.
42. Munter M (1971) Discrete renewal processes. IEEE Trans Reliab R-20:46–51.
43. Nakagawa T (1985) Optimization problems in k-out-of-n systems. IEEE Trans Reliab R-34:248–250.
44. Pham H (1992) On the optimal design of k-out-of-n: G subsystems. IEEE Trans Reliab R-41:572–574.
45. Kuo W, Zuo MJ (2003) Optimal Reliability Modeling. J Wiley & Sons, Hoboken NJ.

4
Periodic Replacement

When we consider large and complex systems that consist of many kinds of units, we should make only minimal repair at each failure, and make the planned replacement or do preventive maintenance at periodic times. We consider the following replacement policy which is called *periodic replacement with minimal repair at failures* [1]. A unit is replaced periodically at planned times kT ($k = 1, 2, \ldots$). Only minimal repair after each failure is made so that the failure rate remains undisturbed by any repair of failures between successive replacements.

This policy is commonly used with complex systems such as computers and airplanes. A practical procedure for applying the policy to large motors and small electrical parts was given in [2]. More general cost structures and several modified models were provided in [3–11]. On the other hand, the policy regarding the version that a unit is replaced at the Nth failure and ($N-1$)th previous failures are corrected with minimal repair proposed in [12]. The stochastic models to describe the failure pattern of repairable units subject to minimal maintenance are dealt with [13].

This chapter summarizes the periodic replacement with minimal repair based on our original work with reference to the book [1]. In Section 4.1, we make clear the theoretical definition of minimal repair, and give some useful theorems that can be applied to the analysis of optimum policies [14]. In Section 4.2, we consider the periodic replacement policy in which a unit is replaced at planned time T and any failed units undergo minimal repair between replacements. We obtain the expected cost rate as an objective function and analytically derive an optimum replacement time T^* that minimizes it [15]. In Section 4.3, we propose the extended replacement policy in which a unit is replaced at time T or at the Nth failure, whichever occurs first. Using the results in Section 4.1, we derive an optimum number N^* that minimizes the expected cost rate for a specified T [16–18]. Furthermore, in Section 4.4, we show five models of replacement with discounting and replacement in discrete time [15], replacement of a used unit [15], replacement with random and wearout failures, and replacement with threshold level [19]. Finally, in Sec-

4.1 Definition of Minimal Repair

Suppose that a unit begins to operate at time 0. If a unit fails then it undergoes minimal repair and begins to operate again. It is assumed that the time for repair is negligible. Let us denote by $0 \equiv Y_0 \leq Y_1 \leq \cdots \leq Y_n \leq \cdots$ the successive failure times of a unit. The times between failures $X_n \equiv Y_n - Y_{n-1}$ ($n = 1, 2, \dots$) are nonnegative random variables.

We define *to make minimal repair at failure* as follows.

Definition 4.1. Let $F(t) \equiv \Pr\{X_1 \leq t\}$ for $t \geq 0$. A unit undergoes minimal repair at failures if and only if

$$\Pr\{X_n \leq x | X_1 + X_2 + \cdots + X_{n-1} = t\} = \frac{F(t+x) - F(t)}{\overline{F}(t)} \quad (n = 2, 3, \dots) \quad (4.1)$$

for $x > 0$, $t \geq 0$ such that $F(t) < 1$, where $\overline{F} \equiv 1 - F$.

The function $[F(t+x) - F(t)]/\overline{F}(t)$ is called the failure rate and represents the probability that a unit with age t fails in a finite interval $(t, t+x]$. The definition means that the failure rate remains undisturbed by any minimal repair of failures; i.e., a unit after each minimal repair has the same failure rate as before failure.

Assume that $F(t)$ has a density function $f(t)$ and $h(t) \equiv f(t)/\overline{F}(t)$, which is continuous. The function $h(t)$ is also called the instantaneous failure rate or simply the failure rate and has the same monotone property as $[F(t+x) - F(t)]/\overline{F}(t)$ as shown in Section 1.1. Moreover, $H(t) \equiv \int_0^t h(u) du$ is called the cumulative hazard function and satisfies a relation $\overline{F}(t) = e^{-H(t)}$.

Theorem 4.1. Let $G_n(x) \equiv \Pr\{Y_n \leq x\}$ and $F_n(x) \equiv \Pr\{X_n \leq x\}$ ($n = 1, 2, \dots$). Then,

$$G_n(x) = 1 - \sum_{j=0}^{n-1} \frac{[H(x)]^j}{j!} e^{-H(x)} \quad (n = 1, 2, \dots) \quad (4.2)$$

$$F_n(x) = 1 - \int_0^\infty \overline{F}(t+x) \frac{[H(t)]^{n-2}}{(n-2)!} h(t) dt \quad (n = 2, 3, \dots). \quad (4.3)$$

Proof. By mathematical induction, we have

$$G_1(x) = F_1(x) = F(x)$$

$$G_{n+1}(x) = \int_0^\infty \Pr\{X_{n+1} \le x - t | Y_n = t\} \, \mathrm{d}G_n(t)$$

$$= \int_0^x \frac{F(x) - F(t)}{\overline{F}(t)} \frac{[H(t)]^{n-1}}{(n-1)!} f(t) \, \mathrm{d}t$$

$$= 1 - \sum_{j=0}^{n-1} \frac{[H(x)]^j}{j!} e^{-H(x)} - e^{-H(x)} \int_0^x \frac{[H(t)]^{n-1}}{(n-1)!} h(t) \, \mathrm{d}t$$

$$= 1 - \sum_{j=0}^{n} \frac{[H(x)]^j}{j!} e^{-H(x)} \qquad (n = 1, 2, \dots).$$

Similarly,

$$F_{n+1}(x) = \int_0^\infty \Pr\{X_{n+1} \le x | Y_n = t\} \, \mathrm{d}G_n(t)$$

$$= \int_0^\infty \frac{F(t+x) - F(t)}{\overline{F}(t)} \frac{[H(t)]^{n-1}}{(n-1)!} f(t) \, \mathrm{d}t$$

$$= 1 - \int_0^\infty \overline{F}(t+x) \frac{[H(t)]^{n-1}}{(n-1)!} h(t) \, \mathrm{d}t \qquad (n = 1, 2, \dots). \quad \blacksquare$$

It easily follows from Theorem 4.1 that

$$E\{Y_n\} \equiv \int_0^\infty \overline{G}_n(x) \mathrm{d}x = \sum_{j=0}^{n-1} \int_0^\infty \frac{[H(x)]^j}{j!} e^{-H(x)} \, \mathrm{d}x \qquad (n = 1, 2, \dots) \quad (4.4)$$

$$E\{X_n\} = E\{Y_n\} - E\{Y_{n-1}\} = \int_0^\infty \frac{[H(x)]^{n-1}}{(n-1)!} e^{-H(x)} \, \mathrm{d}x \qquad (n = 1, 2, \dots). \quad (4.5)$$

In particular, when $F(t) = 1 - e^{-\lambda t}$, i.e., $H(t) = \lambda t$,

$$F_n(x) = 1 - e^{-\lambda x}, \qquad G_n(x) = 1 - \sum_{j=0}^{n-1} \frac{(\lambda x)^j}{j!} e^{-\lambda x} \qquad (n = 1, 2, \dots)$$

$$E\{X_n\} = \frac{1}{\lambda}, \qquad E\{Y_n\} = \frac{n}{\lambda}.$$

Let $N(t)$ be the number of failures of a unit during $[0, t]$; i.e., $N(t) \equiv \max_n\{Y_n \le t\}$. Clearly,

$$p_n(t) \equiv \Pr\{N(t) = n\} = \Pr\{Y_n \le t < Y_{n+1}\} = G_n(t) - G_{n+1}(t)$$

$$= \frac{[H(t)]^n}{n!} e^{-H(t)} \qquad (n = 0, 1, 2, \dots) \quad (4.6)$$

and moreover,
$$E\{N(t)\} = V\{N(t)\} = H(t); \tag{4.7}$$
that is, failures occur at a non-homogeneous Poisson process with mean-value function $H(t)$ in Section 1.3 [21].

Next, assume that the failure rate $[F(t+x) - F(t)]/\overline{F}(t)$ or $h(t)$ is increasing in t for $x > 0$, $t \geq 0$. Then, there exists $\lim_{t \to \infty} h(t) \equiv h(\infty)$, which may possibly be infinity.

Theorem 4.2. If the failure rate is increasing then $E\{X_n\}$ is decreasing in n, and converges to $1/h(\infty)$ as $n \to \infty$, where $1/h(\infty) = 0$ whenever $h(\infty) = \infty$.

Proof. Let
$$\gamma(t) \equiv \int_0^\infty \left[1 - \frac{F(t+x) - F(t)}{\overline{F}(t)}\right] dx$$
which represents the mean residual lifetime of a unit with age t. Then, $\gamma(t)$ is decreasing in t from the assumption that $[F(t+x) - F(t)]/\overline{F}(t)$ is increasing, and
$$\lim_{t \to \infty} \gamma(t) = \lim_{t \to \infty} \frac{1}{\overline{F}(t)} \int_t^\infty \overline{F}(x) dx = \frac{1}{h(\infty)}.$$

Furthermore, noting from (4.1) that
$$E\{X_{n+1}\} = E\{\gamma(Y_n)\}$$
and using the relation $Y_{n+1} \geq Y_n$, we have the inequality
$$E\{X_{n+1}\} = E\{\gamma(Y_n)\} \leq E\{\gamma(Y_{n-1})\} = E\{X_n\} \quad (n = 1, 2, \dots).$$
Therefore, because $Y_n \to \infty$ as $n \to \infty$, we have, by monotone convergence,
$$\lim_{n \to \infty} E\{\gamma(Y_n)\} = \frac{1}{h(\infty)}$$
which completes the proof. ∎

Theorem 4.3. If failure rate $h(t)$ is increasing then
$$\frac{\int_0^T \{[H(t)]^n/n!\} f(t)\, dt}{\int_0^T \{[H(t)]^n/n!\} \overline{F}(t)\, dt} \tag{4.8}$$
is increasing in n and converges to $h(T)$ as $n \to \infty$ for any $T > 0$.

Proof. Letting

4.1 Definition of Minimal Repair

$$q(T) \equiv \int_0^T [H(t)]^{n+1} f(t)\, dt \int_0^T [H(t)]^n \overline{F}(t)\, dt$$
$$- \int_0^T [H(t)]^n f(t)\, dt \int_0^T [H(t)]^{n+1} \overline{F}(t)\, dt$$

we obviously have that $\lim_{T \to 0} q(T) = 0$, and because $h(t)$ is increasing,

$$\frac{dq(T)}{dT} = [H(T)]^n \overline{F}(T) \int_0^T [H(t)]^n \overline{F}(t)[H(T) - H(t)][h(T) - h(t)]\, dt \geq 0.$$

Thus, $q(T)$ is increasing in T from 0, and hence, $q(T) \geq 0$ for all $T > 0$, which implies that the function (4.8) is increasing in n.

Next, to prove that the function (4.8) converges to $h(T)$ as $n \to \infty$, we introduce the following result. If $\phi(t)$ and $\psi(t)$ are continuous, $\phi(b) \neq 0$ and $\psi(b) \neq 0$, then for $0 \leq a < b$,

$$\lim_{n \to \infty} \frac{\int_a^b t^n \phi(t)\, dt}{\int_a^b t^n \psi(t)\, dt} = \frac{\phi(b)}{\psi(b)}. \tag{4.9}$$

For, putting $t = bx$, $c = a/b$, $\phi(bx) = f(x)$, and $\psi(bx) = g(x)$, Equation (4.9) is rewritten as

$$\lim_{n \to \infty} \frac{\int_c^1 x^n f(x)\, dx}{\int_c^1 x^n g(x)\, dx} = \frac{f(1)}{g(1)}.$$

This is easily shown from the fact that

$$\lim_{n \to \infty} (n+1) \int_c^1 x^n f(x)\, dx = f(1)$$

for any c ($0 \leq c < 1$). Thus, letting $H(t) = x$ in (4.8) and using (4.9), it follows that

$$\lim_{n \to \infty} \frac{\int_0^T \{[H(t)]^n/n!\} f(t)\, dt}{\int_0^T \{[H(t)]^n/n!\} \overline{F}(t)\, dt} = \lim_{n \to \infty} \frac{\int_0^{H(T)} x^n e^{-x}\, dx}{\int_0^{H(T)} x^n e^{-x}/h(H^{-1}(x))\, dx} = h(T),$$

where $H^{-1}(x)$ is the inverse function of $x = H(t)$. ∎

In particular, when $F(t) = 1 - e^{-\lambda t}$,

$$\frac{\int_0^T \{[H(t)]^n/n!\} f(t)\, dt}{\int_0^T \{[H(t)]^n/n!\} \overline{F}(t)\, dt} = \lambda \qquad (n = 0, 1, 2, \ldots).$$

Let $G(t)$ represent any distribution with failure rate $r(t) \equiv g(t)/\overline{G}(t)$ and finite mean, where $g(t)$ is a density function of $G(t)$ and $\overline{G} \equiv 1 - G$.

Theorem 4.4. If both $h(t)$ and $r(t)$ are continuous and increasing then

$$\frac{\int_0^\infty \{[H(t)]^{n-1}/(n-1)!\}\overline{G}(t)f(t)\,dt}{\int_0^\infty \{[H(t)]^n/n!\}\overline{G}(t)\overline{F}(t)\,dt} \tag{4.10}$$

is increasing in n and converges to $h(\infty) + r(\infty)$ as $n \to \infty$.

Proof. Integrating by parts, we have

$$\int_0^\infty \frac{[H(t)]^{n-1}}{(n-1)!}\overline{G}(t)f(t)\,dt = \int_0^\infty \frac{[H(t)]^n}{n!}\overline{G}(t)f(t)\,dt + \int_0^\infty \frac{[H(t)]^n}{n!}\overline{F}(t)g(t)\,dt.$$

First, we show

$$\frac{\int_0^\infty [H(t)]^n \overline{G}(t)f(t)\,dt}{\int_0^\infty [H(t)]^n \overline{G}(t)\overline{F}(t)\,dt} \tag{4.11}$$

is increasing in n when $h(t)$ is increasing. By a similar method to that of proving Theorem 4.3, letting

$$q(T) \equiv \int_0^T [H(t)]^{n+1}\overline{G}(t)f(t)\,dt \int_0^T [H(t)]^n \overline{G}(t)\overline{F}(t)\,dt$$
$$- \int_0^T [H(t)]^n \overline{G}(t)f(t)\,dt \int_0^T [H(t)]^{n+1}\overline{G}(t)\overline{F}(t)\,dt$$

for any $T > 0$, we have $\lim_{T\to 0} q(T) = 0$ and $dq(T)/dT \geq 0$. Thus, $q(T) \geq 0$ for all $T > 0$, and hence, the function (4.11) is increasing in n. Similarly,

$$\frac{\int_0^\infty [H(t)]^n \overline{F}(t)g(t)\,dt}{\int_0^\infty [H(t)]^n \overline{F}(t)\overline{G}(t)\,dt} \tag{4.12}$$

is also increasing in n. Therefore, from (4.11) and (4.12), the function (4.10) is also increasing in n.

Next, we show that

$$\lim_{n\to\infty} \frac{\int_0^\infty [H(t)]^n \overline{G}(t)f(t)\,dt}{\int_0^\infty [H(t)]^n \overline{G}(t)\overline{F}(t)\,dt} = h(\infty). \tag{4.13}$$

Clearly,

$$\frac{\int_0^\infty [H(t)]^n \overline{G}(t)f(t)\,dt}{\int_0^\infty [H(t)]^n \overline{G}(t)\overline{F}(t)\,dt} \leq h(\infty).$$

On the other hand, we have, for any $T > 0$,

$$\frac{\int_0^\infty [H(t)]^n \overline{G}(t)f(t)\,dt}{\int_0^\infty [H(t)]^n \overline{G}(t)\overline{F}(t)\,dt} = \frac{\int_0^T [H(t)]^n \overline{G}(t)f(t)\,dt + \int_T^\infty [H(t)]^n \overline{G}(t)f(t)\,dt}{\int_0^T [H(t)]^n \overline{G}(t)\overline{F}(t)\,dt + \int_T^\infty [H(t)]^n \overline{G}(t)\overline{F}(t)\,dt}$$
$$\geq \frac{h(T)\int_T^\infty [H(t)]^n \overline{G}(t)\overline{F}(t)\,dt}{\int_0^T [H(t)]^n \overline{G}(t)\overline{F}(t)\,dt + \int_T^\infty [H(t)]^n \overline{G}(t)\overline{F}(t)\,dt}$$
$$= \frac{h(T)}{1 + \{\int_0^T [H(t)]^n \overline{G}(t)\overline{F}(t)\,dt / \int_T^\infty [H(t)]^n \overline{G}(t)\overline{F}(t)\,dt\}}.$$

Furthermore, the bracket of the denominator is, for $T < T_1$,

$$\frac{\int_0^T [H(t)]^n \overline{G}(t)\overline{F}(t)\,dt}{\int_T^\infty [H(t)]^n \overline{G}(t)\overline{F}(t)\,dt} \leq \frac{[H(T)]^n \int_0^T \overline{G}(t)\overline{F}(t)\,dt}{\int_{T_1}^\infty [H(t)]^n \overline{G}(t)\overline{F}(t)\,dt}$$

$$\leq \frac{[H(T)]^n \int_0^T \overline{G}(t)\overline{F}(t)\,dt}{[H(T_1)]^n \int_{T_1}^\infty \overline{G}(t)\overline{F}(t)\,dt} \to 0 \quad \text{as } n \to \infty.$$

Thus, we have

$$h(\infty) \geq \lim_{n\to\infty} \frac{\int_0^\infty [H(t)]^n \overline{G}(t) f(t)\,dt}{\int_0^\infty [H(t)]^n \overline{G}(t)\overline{F}(t)\,dt} \geq h(T)$$

which implies (4.13) because T is arbitrary. Similarly,

$$\lim_{n\to\infty} \frac{\int_0^\infty [H(t)]^n \overline{F}(t) g(t)\,dt}{\int_0^\infty [H(t)]^n \overline{F}(t)\overline{G}(t)\,dt} = r(\infty). \tag{4.14}$$

Therefore, combining (4.13) and (4.14), we complete the proof. ■

From Theorems 4.3 and 4.4, we easily have that for any function $\phi(t)$ that is continuous and $\phi(t) \neq 0$ for any $t > 0$, if the failure rate $h(t)$ is increasing then

$$\frac{\int_0^T \{[H(t)]^n/n!\}\phi(t) f(t)\,dt}{\int_0^T \{[H(t)]^n/n!\}\phi(t)\overline{F}(t)\,dt} \tag{4.15}$$

is increasing in n and converges to $h(T)$ as $n \to \infty$ for any $T > 0$.

In all results of Theorems 4.2 through 4.4 it can easily be seen that if the failure rates are strictly increasing then $E\{X_n\}$, the functions (4.8), (4.10), and (4.15) are also strictly increasing.

4.2 Periodic Replacement with Minimal Repair

A new unit begins to operate at time $t = 0$, and when it fails, only minimal repair is made. Also, a unit is replaced at periodic times kT ($k = 1, 2, \dots$) independent of its age, and any unit becomes as good as new after replacement (Figure 4.1). It is assumed that the repair and replacement times are negligible. Suppose that the failure times of a unit have a density function $f(t)$ and a distribution $F(t)$ with finite mean $\mu \equiv \int_0^\infty \overline{F}(t)dt < \infty$ and its failure rate $h(t) \equiv f(t)/\overline{F}(t)$.

Consider one cycle with constant time T ($0 < T \leq \infty$) from the planned replacement to the next one. Let c_1 be the cost of minimal repair and c_2 be the cost of the planned replacement. Then, the expected cost of one cycle is, from (3.2),

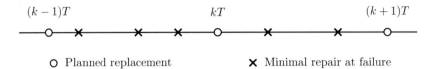

Fig. 4.1. Process of periodic replacement with minimal repair

$$c_1 E\{N_1(T)\} + c_2 E\{N_2(T)\} = c_1 H(T) + c_2$$

because the expected number of failures during one cycle is $E\{N_1(T)\} = \int_0^T h(t)dt \equiv H(T)$ from (4.7). Therefore, from (3.3), the expected cost rate is [1, p. 99],

$$C(T) = \frac{1}{T}[c_1 H(T) + c_2]. \tag{4.16}$$

If a unit is never replaced (*i.e.*, $T = \infty$) then $\lim_{T \to \infty} H(T)/T = h(\infty)$ if it exists, which may possibly be infinite, and $C(\infty) \equiv \lim_{T \to \infty} C(T) = c_1 h(\infty)$.

Furthermore, suppose that a unit is replaced when the total operating time is T. Then, the availability is given by

$$A(T) = \frac{T}{T + \beta_1 H(T) + \beta_2}, \tag{4.17}$$

where β_1 = time of minimal repair and β_2 = time of replacement. Thus, the policy maximizing $A(T)$ is the same as minimizing the expected cost rate $C(T)$ in (4.16) by replacing β_i with c_i.

We seek an optimum planned time T^* that minimizes the expected cost rate $C(T)$ in (4.16). Differentiating $C(T)$ with respect to T and setting it equal to zero, we have

$$Th(T) - H(T) = \frac{c_2}{c_1} \quad \text{or} \quad \int_0^T t\,dh(t) = \frac{c_2}{c_1}. \tag{4.18}$$

Suppose that the failure rate $h(t)$ is continuous and strictly increasing. Then, the left-hand side of (4.18) is also strictly increasing because

$$(T + \Delta T)h(T + \Delta T) - H(T + \Delta T) - Th(T) + H(T)$$
$$= T[h(T + \Delta T) - h(T)] + \int_T^{T+\Delta T} [h(T + \Delta T) - h(t)]\,dt > 0$$

for any $\Delta T > 0$. Thus, if a solution T^* to (4.18) exists then it is unique, and the resulting cost rate is

$$C(T^*) = c_1 h(T^*). \tag{4.19}$$

In addition, if $\int_0^\infty t\,dh(t) > c_2/c_1$ then there exists a finite solution to (4.18). Also, from (4.18),

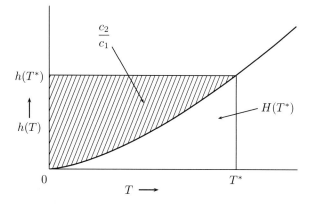

Fig. 4.2. Optimum T^* for failure rate $h(T)$

$$Th(T) - H(T) > T_1 h(T) - H(T_1)$$

for any $T > T_1$. Thus, if $h(t)$ is strictly increasing to infinity then there exists a finite and unique T^* that satisfies (4.18).

When $h(t)$ is strictly increasing, we have, from Theorem 3.3,

$$Th(T) - \int_0^T h(t)\,dt \geq h(T) \int_0^T \overline{F}(t)\,dt - F(T)$$

whose right-hand side agrees with (3.9). That is, an optimum T^* is not greater than that of an age replacement in Section 3.1. Thus, from Theorem 3.2, if $h(\infty) > (c_1 + c_2)/(\mu c_1)$ then a finite solution to (4.18) exists.

Figure 4.2 shows graphically an optimum time T^* given in (4.18) for the failure rate $h(T)$. If $h(T)$ were roughly drawn then T^* could be given by the time when the area covered with slash lines becomes equal to the ratio of c_2/c_1. So that, when $h(T)$ is a concave function, $H(T^*) > c_2/c_1$, and when $h(T)$ is a convex function, $H(T^*) < c_2/c_1$. For example, when the failure distribution is Weibull, i.e., $F(t) = 1 - \exp(-t^m)$ $(m > 1)$, $H(T^*) > c_2/c_1$ for $1 < m < 2$, $= c_2/c_1$ for $m = 2$ and $< c_2/c_1$ for $m > 2$. If the cumulative hazard function $H(t)$ were statistically estimated, the replacement time that satisfies $H(T) = c_2/c_1$ could be utilized as one indicator of replacement time [22] (see Example 3.1 in Chapter 3).

If the cost of minimal repair depends on the age t of a unit and is given by $c_1(t)$, the expected cost rate is

$$C(T) = \frac{1}{T}\left[\int_0^T c_1(t) h(t)\,dt + c_2\right]. \tag{4.20}$$

Finally, we consider a system consisting of n identical units that operate independently of each other. It is assumed that all are replaced together at times kT ($k = 1, 2, \ldots$) and each failed unit between replacements undergoes minimal repair. Then, the expected cost rate is

$$C(T; n) = \frac{1}{T}[nc_1 H(T) + c_2], \tag{4.21}$$

where c_1 = cost of minimal repair for one failed unit, and c_2 = cost of planned replacement for all units at time T.

4.3 Periodic Replacement with Nth Failure

A unit is replaced at time T or at the Nth ($N = 1, 2, \ldots$) failure after its installation, whichever occurs first, where T is a positive constant and previously specified. A unit undergoes only minimal repair at failures between replacements. This policy is called *Policy IV* [12].

From Theorem 4.1, the mean time to replacement is

$$T \Pr\{Y_N > T\} + \int_0^T t \, d\Pr\{Y_N \le t\} = \int_0^T \Pr\{Y_N > t\} \, dt$$

$$= \sum_{j=0}^{N-1} \int_0^T p_j(t) \, dt,$$

where $p_j(t)$ is given in (4.6), and the expected number of failures before replacement is

$$\sum_{j=0}^{N-1} j \Pr\{N(T) = j\} + (N-1) \Pr\{Y_N \le T\}$$

$$= N - 1 - \sum_{j=0}^{N-1} (N-1-j) p_j(T).$$

Therefore, from (3.3), the expected cost rate is

$$C(N; T) = \frac{c_1 \left[N - 1 - \sum_{j=0}^{N-1}(N-1-j) p_j(T) \right] + c_2}{\sum_{j=0}^{N-1} \int_0^T p_j(t) \, dt} \quad (N = 1, 2, \ldots), \tag{4.22}$$

where c_1 = cost of minimal repair and c_2 = cost of planned replacement at time T or at number N. It is evident that

$$C(\infty; T) \equiv \lim_{N \to \infty} C(N; T) = \frac{1}{T}[c_1 H(T) + c_2]$$

which agrees with (4.16) for the periodic replacement with planned time T.

4.3 Periodic Replacement with Nth Failure

Let T^* be the optimum time that minimizes $C(\infty; T)$ and is given by a unique solution to (4.18) if it exists, or $T^* = \infty$ if it does not. We seek an optimum number N^* such that $C(N^*; T) = \min_N C(N; T)$ for a fixed $0 < T \leq \infty$, when the failure rate $h(t)$ is continuous and strictly increasing.

Theorem 4.5. Suppose that $0 < T^* \leq \infty$.

(i) If $T > T^*$ then there exists a finite and unique minimum N^* that satisfies

$$L(N; T) \geq \frac{c_2}{c_1} \quad (N = 1, 2, \dots), \tag{4.23}$$

where

$$L(N; T) \equiv \frac{\sum_{j=N}^{\infty} p_j(T) \sum_{j=0}^{N-1} \int_0^T p_j(t)\,dt}{\int_0^T p_N(t)\,dt}$$

$$- \left[N - 1 - \sum_{j=0}^{N-1} (N - 1 - j) p_j(T) \right] \quad (N = 1, 2, \dots).$$

(ii) If $T \leq T^*$ or $T^* = \infty$ then no N^* satisfying (4.23) exists.

Proof. For simplicity of computation, we put $C(0; T) = \infty$. To find an N^* that minimizes $C(N; T)$ for a fixed T, we form the inequality $C(N + 1; T) \geq C(N; T)$, and have (4.23). Hence, we may seek a minimum N^* that satisfies (4.23).

Using the relation

$$\sum_{j=N+1}^{\infty} \frac{[H(T)]^j}{j!} e^{-H(T)} = \int_0^T \frac{[H(t)]^N}{N!}\,dF(t) \quad (N = 0, 1, 2, \dots)$$

we have, from Theorem 4.3,

$$L(N+1; T) - L(N; T)$$

$$= \sum_{j=0}^{N} \int_0^T p_j(t)\,dt \left[\frac{\sum_{j=N+1}^{\infty} p_j(T)}{\int_0^T p_{N+1}(t)\,dt} - \frac{\sum_{j=N}^{\infty} p_j(T)}{\int_0^T p_N(t)\,dt} \right] > 0$$

and

$$L(\infty; T) \equiv \lim_{N \to \infty} L(N; T) = Th(T) - H(T)$$

which is equal to the left-hand side of (4.18) and is strictly increasing in T.

Suppose that $0 < T^* < \infty$. If $L(\infty; T) > c_2/c_1$, i.e., $T > T^*$, then there exists a finite and unique minimum N^* that satisfies (4.23). On the other hand, if $L(\infty; T) \leq c_2/c_1$, i.e., $T \leq T^*$, then $C(N; T)$ is decreasing in N, and no solution satisfying (4.23) exists. Finally, if $T^* = \infty$ then no solution to (4.23) exists inasmuch as $L(\infty; T) < c_2/c_1$ for all T. ∎

This theorem describes that when a unit is planned to be replaced at time $T > T^*$ for some reason, it also should be replaced at the N^*th failure before time T.

If c_2 is the cost of planned replacement at the Nth failure and c_3 is the cost at time T, then the expected cost rate in (4.22) is rewritten as

$$C(N;T) = \frac{c_1\left[N-1-\sum_{j=0}^{N-1}(N-1-j)p_j(T)\right] + c_2\sum_{j=N}^{\infty}p_j(T) + c_3\sum_{j=0}^{N-1}p_j(T)}{\sum_{j=0}^{N-1}\int_0^T p_j(t)\,dt}. \qquad (4.24)$$

Similar replacement policies were discussed in [23–33].

Next, suppose that a unit is replaced only at the Nth failure. Then, the expected cost rate is, from (4.22),

$$C(N) \equiv \lim_{T\to\infty} C(N;T) = \frac{c_1(N-1) + c_2}{\sum_{j=0}^{N-1}\int_0^\infty p_j(t)\,dt} \qquad (N=1,2,\ldots). \qquad (4.25)$$

In a similar way to that of obtaining Theorem 4.5, we derive an optimum number N^* that minimizes $C(N)$.

Theorem 4.6. If $h(\infty) > c_2/(\mu c_1)$ then there exists a finite and unique minimum N^* that satisfies

$$L(N) \geq \frac{c_2}{c_1} \quad (N=1,2,\ldots) \qquad (4.26)$$

and the resulting cost rate is

$$\frac{c_1}{\int_0^\infty p_{N^*-1}(t)\,dt} < C(N^*) \leq \frac{c_1}{\int_0^\infty p_{N^*}(t)\,dt}, \qquad (4.27)$$

where

$$L(N) \equiv \lim_{T\to\infty} L(N;T) = \frac{\sum_{j=0}^{N-1}\int_0^\infty p_j(t)\,dt}{\int_0^\infty p_N(t)\,dt} - (N-1) \quad (N=1,2,\ldots).$$

Proof. The inequality $C(N+1) \geq C(N)$ implies (4.26). It is easily seen that $L(N+1) - L(N) > 0$ from Theorem 4.2. Thus, if a solution to (4.26) exists then it is unique.

Furthermore, we have the inequality

$$L(N) \geq \frac{\mu}{\int_0^\infty p_N(t)\,dt} \qquad (4.28)$$

because $\int_0^\infty p_N(t)\,dt$ is decreasing in N from Theorem 4.2. Therefore, if

$$\lim_{N\to\infty} \frac{\mu}{\int_0^\infty p_N(t)\,dt} > \frac{c_2}{c_1},$$

i.e., if $h(\infty) > c_2/(\mu c_1)$, then a solution to (4.26) exists, and it is unique. Also, we easily have (4.27) from the inequalities $L(N^* - 1) < c_2/c_1$ and $L(N^*) \geq c_2/c_1$. ∎

Suppose that $h(\infty) > c_2/(\mu c_1)$. Then, from (4.28), there exists a finite and unique minimum $\overline{N}$ that satisfies

$$\int_0^\infty p_N(t)\, dt \leq \frac{\mu c_1}{c_2} \quad (N = 1, 2, \ldots) \tag{4.29}$$

and $N^* \leq \overline{N}$.

Example 4.1. Suppose that the failure time of a unit has a Weibull distribution; i.e., $\overline{F}(t) = \exp(-t^m)$ for $m > 1$. Then, $h(t)$ is strictly increasing from 0 to infinity, and

$$\int_0^\infty \frac{[H(t)]^N}{N!} e^{-H(t)}\, dt = \frac{1}{m} \frac{\Gamma(N+1/m)}{\Gamma(N+1)}$$

$$\sum_{j=0}^{N-1} \int_0^\infty \frac{[H(t)]^j}{j!} e^{-H(t)}\, dt = \frac{\Gamma(N+1/m)}{\Gamma(N)}.$$

Thus, there exists a finite and unique minimum that satisfies (4.26), which is given by

$$N^* = \left[\frac{c_2 - c_1}{(m-1)c_1}\right] + 1,$$

where $[x]$ denotes the greatest integer contained in x. ∎

4.4 Modified Replacement Models

We show the following modified models of periodic replacement with minimal repair at failures: (1) replacement with discounting, (2) replacement in discrete time, (3) replacement of a used unit, (4) replacement with random and wearout failures, and (5) replacement with threshold level. The detailed derivations are omitted and optimum policies for each model are directly given.

(1) Replacement with Discounting

Suppose that all costs are discounted with rate α ($0 < \alpha < \infty$). In a similar way to that for obtaining (3.14) in **(1)** of Section 3.2, the total expected cost for an infinite time span is

$$C(T; \alpha) = \frac{c_1 \int_0^T e^{-\alpha t} h(t)\, dt + c_2 e^{-\alpha T}}{1 - e^{-\alpha T}}. \tag{4.30}$$

Differentiating $C(T; \alpha)$ with respect to T and setting it equal to zero

$$\frac{1 - e^{-\alpha T}}{\alpha} h(T) - \int_0^T e^{-\alpha t} h(t)\, dt = \frac{c_2}{c_1} \qquad (4.31)$$

and the resulting cost rate is

$$C(T^*; \alpha) = \frac{c_1}{\alpha} h(T^*) - c_2. \qquad (4.32)$$

Note that $\lim_{\alpha \to 0} \alpha C(T; \alpha) = C(T)$ in (4.16), and (4.31) agrees with (4.18) as $\alpha \to 0$.

(2) Replacement in Discrete Time

A unit is replaced at cycles kN ($k = 1, 2, \ldots$) and a failed unit between planned replacements undergoes only minimal repair. Then, using the same notation and methods in **(2)** of Section 3.2, the expected cost rate is

$$C(N) = \frac{1}{N}\left[c_1 \sum_{j=1}^{N} h_j + c_2 \right] \quad (N = 1, 2, \ldots) \qquad (4.33)$$

and an optimum number N^* is given by a minimum solution that satisfies

$$N h_{N+1} - \sum_{j=1}^{N} h_j \geq \frac{c_2}{c_1} \quad (N = 1, 2, \ldots). \qquad (4.34)$$

(3) Replacement of a Used Unit

Consider the periodic replacement with minimal repair at failures for a used unit. A unit is replaced at times kT ($k = 1, 2, \ldots$) by the same used unit with age x, where x ($0 \leq x < \infty$) is previously specified. Then, the expected cost rate is, from (4.16),

$$C(T; x) = \frac{1}{T}\left[c_1 \int_x^{T+x} h(t)\, dt + c_2(x) \right], \qquad (4.35)$$

where $c_1 = $ cost of minimal repair and $c_2(x) = $ acquisition cost of a used unit with age x which may be decreasing in x. In this case, (4.18) and (4.19) are rewritten as

$$Th(T + x) - \int_x^{T+x} h(t)\, dt = \frac{c_2(x)}{c_1} \qquad (4.36)$$

$$C(T^*; x) = c_1 h(T^* + x). \qquad (4.37)$$

Next, consider the problem that it is most economical to use a unit of a certain age. Suppose that x is a variable, and inversely, T is constant and $c_2(x)$

4.4 Modified Replacement Models   109

is differentiable. Then, differentiating $C(T;x)$ with respect to x and setting it equal to zero imply

$$h(T+x) - h(x) = -\frac{c_2'(x)}{c_1} \qquad (4.38)$$

which is a necessary condition that a finite x minimizes $C(T;x)$ for a fixed T.

(4) Replacement with Random and Wearout Failures

We consider a modified replacement policy for a unit with random and wearout failure periods, where an operating unit enters a wearout failure period at a fixed time T_0, after it has operated continuously in a random failure period. It is assumed that a unit is replaced at planned time $T+T_0$, where T_0 is constant and previously given, and it undergoes only minimal repair at failures between replacements [34, 35].

Suppose that a unit has a constant failure rate λ for $0 < t \leq T_0$ in a random failure period and $\lambda + h(t - T_0)$ for $t > T_0$ in a wearout failure period. Then, the expected cost rate is

$$C(T; T_0) = c_1 \lambda + \frac{c_1 H(T) + c_2}{T + T_0}. \qquad (4.39)$$

Thus, if $h(t)$ is strictly increasing and there exists a solution T^* that satisfies

$$(T + T_0) h(T) - H(T) = \frac{c_2}{c_1} \qquad (4.40)$$

then it is unique and the resulting cost rate is

$$C(T^*; T_0) = c_1 [\lambda + h(T^*)]. \qquad (4.41)$$

Furthermore, it is easy to see that T^* is a decreasing function of T_0 because the left-hand side of (4.40) is increasing in T_0 for a fixed T. Thus, an optimum time T^* is less than the optimum one given in (4.18) as we have expected.

(5) Replacement with Threshold Level

Suppose that if more failures have occurred between periodic replacements then the total cost would be higher than expected. For example, if more than K failures have occurred and the number of K parts is needed for providing against $K-1$ spares during a planned interval, an extra cost would result from the downtime, the ordering and delivery of spares, and repair. Let $N(T)$ be the total number of failures during $(0, T]$ and K be its threshold number. Then, from (4.16) and (4.6), the expected cost rate is

$$C(T;K) = \frac{1}{T}[c_1 H(T) + c_2 + c_3 \Pr\{N(T) \geq K\}]$$

$$= \frac{1}{T}\left[c_1 H(T) + c_2 + c_3 \sum_{j=K}^{\infty} p_j(T)\right], \qquad (4.42)$$

where c_3 = additional cost when the number of failures has exceeded a threshold level K.

4.5 Replacements with Two Different Types

Periodic replacement with minimal repair is modified and extended in several ways. We show typical models of periodic replacement with (1) two types of failures and (2) two types of units.

(1) Two Types of Failures

We may generally classify failure into failure modes: partial and total failures, slight and serious failures, minor and major failures, or simply faults and failures. Generalized replacement models of two types of failures were proposed in [36–40].

Consider a unit with two types of failures. When a unit fails, type 1 failure occurs with probability p $(0 \leq p \leq 1)$ and is removed by minimal repair, and type 2 failure occurs with probability $1-p$ and is removed by replacement. Type 1 failure is a minor failure that is easily restored to the same operating state by minimal repair, and type 2 failure incurs a total breakdown and needs replacement or repair.

A unit is replaced at the time of type 2 failure or Nth type 1 failure, whichever occurs first. Then, the expected number of minimal repairs, *i.e.*, type 1 failures before replacement, is

$$(N-1)p^N + \sum_{j=1}^{N}(j-1)p^{j-1}(1-p) = \begin{cases} \dfrac{p - p^N}{1-p} & \text{for } 0 \leq p < 1 \\ N-1 & \text{for } p = 1. \end{cases}$$

Thus, the expected cost rate is, from (4.25),

$$C(N;p) = \frac{c_1[(p-p^N)/(1-p)] + c_2}{\sum_{j=0}^{N-1} p^j \int_0^\infty p_j(t)\, dt} \qquad (N = 1, 2, \dots) \qquad (4.43)$$

for $0 \leq p < 1$, where c_1 = cost of minimal repair for type 1 failure and c_2 = cost of replacement at the Nth type 1 or type 2 failure. When $p \to 1$, $C(N;1) \equiv \lim_{p \to 1} C(N;p)$ is equal to (4.25) and the optimum policy is given in Theorem 4.6. When $p = 0$, $C(N;0) = c_2/\mu$, which is constant for all N, and a unit is replaced only at type 2 failure. Therefore, we need only discuss an optimum policy in the case of $0 < p < 1$ when the failure rate $h(t)$ is strictly increasing. To simplify equations, we denote $\mu_p \equiv \int_0^\infty [\overline{F}(t)]^p \, dt = \int_0^\infty e^{-pH(t)} \, dt$. When $p = 1$, $\mu_1 = \mu$ which is the mean time to failure of a unit.

Theorem 4.7. (i) If $h(\infty) > [c_1 p + c_2(1-p)]/[c_1(1-p)\mu_{1-p}]$ then there exists a finite and unique minimum $N^*(p)$ that satisfies

$$L(N;p) \geq \frac{c_2}{c_1} \quad (N = 1, 2, \ldots), \tag{4.44}$$

where

$$L(N;p) \equiv \frac{\sum_{j=0}^{N-1} p^j \int_0^\infty p_j(t)\, dt}{\int_0^\infty p_N(t)\, dt} - \frac{p - p^N}{1-p} \quad (N = 1, 2, \ldots).$$

(ii) If $h(\infty) \leq [c_1 p + c_2(1-p)]/[c_1(1-p)\mu_{1-p}]$ then $N^*(p) = \infty$, and the resulting cost rate is

$$C(\infty;p) \equiv \lim_{N \to \infty} C(N;p) = \frac{c_1[p/(1-p)] + c_2}{\mu_{1-p}}. \tag{4.45}$$

Proof. The inequality $C(N+1;p) \geq C(N;p)$ implies (4.44). Furthermore, it is easily seen from Theorem 4.2 that $L(N;p)$ is an increasing function of N, and hence, $\lim_{N \to \infty} L(N;p) = \mu_{1-p} h(\infty) - [p/(1-p)]$. Thus, in a similar way to that of obtaining Theorem 4.6, if $h(\infty) > [c_1 p + c_2(1-p)]/[c_1(1-p)\mu_{1-p}]$ then there exists a finite and unique minimum $N^*(p)$ that satisfies (4.44). On the other hand, if $h(\infty) \leq [c_1 p + c_2(1-p)]/[c_1(1-p)\mu_{1-p}]$ then $L(N;p) < c_2/c_1$ for all N, and hence, $N^*(p) = \infty$, and we have (4.45). ∎

It is easily noted that $\partial L(N;p)/\partial p > 0$ for all N. Thus, if $h(\infty) > [c_1 p + c_2(1-p)]/[c_1(1-p)\mu_{1-p}]$ for $0 < p < 1$ then $N^*(p)$ is decreasing in p, and $\overline{N} \geq N^*(p) \geq N^*$, where both N^* and $\overline{N}$ exist and are given in (4.26) and (4.29), respectively.

Until now, it has been assumed that the replacement costs for both the Nth type 1 failure and type 2 failure are the same. In reality, they may be different from each other. It is supposed that c_2 is the replacement cost of the Nth type 1 failure and c_3 is the replacement cost of the type 2 failure. Then, the expected cost rate in (4.43) is rewritten as

$$C(N;p) = \frac{c_1[(p-p^N)/(1-p)] + c_2 p^N + c_3(1-p^N)}{\sum_{j=0}^{N-1} p^j \int_0^\infty p_j(t)\, dt} \quad (N = 1, 2, \ldots).$$

$$\tag{4.46}$$

Example 4.2. We compute an optimum number $N^*(p)$ that minimizes the expected cost rate $C(N;p)$ in (4.46) when $\overline{F}(t) = \exp(-t^m)$ for $m > 1$. When

$c_2 = c_3$, it is shown from Theorem 4.7 that $N^*(p)$ exists uniquely and is decreasing in p for $0 < p < 1$. Furthermore, when $p = 1$, $N^*(p)$ is given in Example 4.1. If $c_1 + (c_3 - c_2)(1-p) > 0$ then $N^*(p)$ is given by a minimum value such that

$$\frac{(1-p)\Gamma(N+1)}{\Gamma(N+1/m)} \sum_{j=0}^{N-1} \frac{p^j \Gamma(j+1/m)}{\Gamma(j+1)} + p^N \geq \frac{c_1 p + c_3(1-p)}{c_1 + (c_3 - c_2)(1-p)}.$$

It is easily seen that $N^*(p)$ is small when c_1/c_2 or c_3/c_2 for $c_2 > c_1$ is large. Conversely, if $c_1 + (c_3 - c_2)(1-p) \leq 0$ then $N^*(p) = \infty$.

Table 4.1. Variation in the optimum number $N^*(p)$ for probability p of type 1 failure and ratio of c_3 to c_2 when $m = 2$ and $c_1/c_2 = 0.1$

p	\multicolumn{7}{c}{c_3/c_2}
	0.8	0.9	1.0	1.2	1.5	2.0	3.0
0.1	∞	∞	30	6	2	1	1
0.2	∞	∞	27	6	3	1	1
0.3	∞	220	24	6	3	2	1
0.4	∞	112	22	7	3	2	1
0.6	288	39	17	7	4	2	1
0.7	64	25	15	8	5	3	2
0.8	26	17	13	8	6	4	2
0.9	14	12	11	9	7	5	4
1.0	10	10	10	10	10	10	10

Table 4.1 gives the optimum number $N^*(p)$ for probability p of type 1 failure and the ratio of cost c_3 to cost c_2 when $m = 2$ and $c_1/c_2 = 0.1$. It is of great interest that $N^*(p)$ is increasing in p for $c_3 > c_2$, however, it is decreasing for $c_3 \leq c_2$. We can explain the reason why $N^*(p)$ is increasing in p for c_3/c_2. If $c_3 > c_2$ then the replacement cost for type 1 failure is cheaper than that for type 2 failure and the number of its failures increases with p, and so, $N^*(p)$ is large when p is large. This situation reflects a real situation. On the other hand, if $c_3 \leq c_2$ then it is not useful to replace the unit frequently before type 2 failure, however, the total cost of minimal repairs for type 1 increases as the number of its failures does with p. Thus, it may be better to replace the unit preventively at some number N when p is large. Evidently, $N^*(p)$ is rapidly increasing when c_1 is small enough. ∎

(2) Two Types of Units

Most systems consist of vital and nonvital parts or essential and nonessential units. If vital parts fail then a system becomes dangerous or incurs high cost. It would be wise to make replacements or overhauls before failure at

4.5 Replacements with Two Different Types

periodic times. The optimum replacement policies for systems with two units were derived in [42–48]. Furthermore, the optimum inspection schedule of a production system [49] and a storage system [50] with two types of units was studied.

Consider a system with two types of units that operate statistically independently. When unit 1 fails, it undergoes minimal repair instantaneously and begins to operate again. When unit 2 fails, the system is replaced without repairing unit 2. Unit 1 has a failure distribution $F_1(t)$, the failure rate $h_1(t)$ and $H_1(t) \equiv \int_0^t h_1(u) du$, which have the same assumptions as those in Section 4.2, whereas unit 2 has a failure distribution $F_2(t)$ with finite mean μ_2 and the failure rate $h_2(t)$, where $\overline{F}_i \equiv 1 - F_i$ $(i = 1, 2)$.

Suppose that the system is replaced at the time of unit 2 failure or Nth unit 1 failure, whichever occurs first. Then, the mean time to replacement is

$$\sum_{j=0}^{N-1} \int_0^\infty t p_j(t) \, dF_2(t) + \int_0^\infty t \overline{F}_2(t) p_{N-1}(t) h_1(t) \, dt = \sum_{j=0}^{N-1} \int_0^\infty \overline{F}_2(t) p_j(t) \, dt,$$

where $p_j(t) = \{[H_1(t)]^j / j!\} e^{-H_1(t)}$ $(j = 0, 1, 2, \dots)$, and the expected number of minimal repairs before replacement is

$$\sum_{j=0}^{N-1} j \int_0^\infty p_j(t) \, dF_2(t) + (N-1) \int_0^\infty \overline{F}_2(t) p_{N-1}(t) h_1(t) \, dt$$

$$= \sum_{j=0}^{N-2} \int_0^\infty \overline{F}_2(t) p_j(t) h_1(t) \, dt,$$

where $\sum_{j=0}^{-1} \equiv 0$. Thus, the expected cost rate is

$$C(N) = \frac{c_1 \sum_{j=0}^{N-2} \int_0^\infty \overline{F}_2(t) p_j(t) h_1(t) \, dt + c_2}{\sum_{j=0}^{N-1} \int_0^\infty \overline{F}_2(t) p_j(t) \, dt} \quad (N = 1, 2, \dots). \tag{4.47}$$

When $\overline{F}_2(t) \equiv 1$ for $t \geq 0$, $C(N)$ is equal to (4.25), and when $\overline{F}_2(t) \equiv 1$ for $t \leq T$ and 0 for $t > T$, this is equal to (4.22).

We have the following optimum number N^* that minimizes $C(N)$.

Theorem 4.8. Suppose that $h_1(t)$ is continuous and increasing. If there exists a minimum N^* that satisfies

$$L(N) \geq \frac{c_2}{c_1} \quad (N = 1, 2, \dots) \tag{4.48}$$

then it is unique and it minimizes $C(N)$, where

$$L(N) \equiv \frac{\int_0^\infty \overline{F}_2(t)p_{N-1}(t)h_1(t)\,dt}{\int_0^\infty \overline{F}_2(t)p_N(t)\,dt} \sum_{j=0}^{N-1}\int_0^\infty \overline{F}_2(t)p_j(t)\,dt$$

$$-\sum_{j=0}^{N-2}\int_0^\infty \overline{F}_2(t)p_j(t)h_1(t)\,dt \quad (N=1,2,\ldots).$$

Proof. The inequality $C(N+1) \geq C(N)$ implies (4.48). In addition,

$$L(N+1) - L(N) = \sum_{j=0}^{N}\int_0^\infty \overline{F}_2(t)p_j(t)\,dt$$

$$\times \left[\frac{\int_0^\infty \overline{F}_2(t)p_N(t)h_1(t)\,dt}{\int_0^\infty \overline{F}_2(t)p_{N+1}(t)\,dt} - \frac{\int_0^\infty \overline{F}_2(t)p_{N-1}(t)h_1(t)\,dt}{\int_0^\infty \overline{F}_2(t)p_N(t)\,dt}\right] \geq 0$$

because $\int_0^\infty \overline{F}_2(t)p_N(t)h_1(t)dt / \int_0^\infty \overline{F}_2(t)p_{N+1}(t)dt$ is increasing in N from Theorem 4.4, when $h_1(t)$ is increasing. Thus, if a minimum solution to (4.48) exists then it is unique. ∎

Furthermore, we also have, from Theorem 4.4,

$$L(\infty) \equiv \lim_{N \to \infty} L(N) = \mu_2[h_1(\infty) + h_2(\infty)] - \int_0^\infty \overline{F}_2(t)h_1(t)\,dt.$$

Thus, if $h_1(t)+h_2(t)$ is strictly increasing and $h_1(\infty)+h_2(\infty) > (1/\mu_2)[(c_2/c_1) + \int_0^\infty \overline{F}_2(t)h_1(t)\,dt]$ then there exists a finite and unique minimum N^* that satisfies (4.48). For example, suppose that $h_2(t)$ is strictly increasing and $h_1(t)$ is increasing. Then, because $L(\infty) \geq \mu_2 h_2(\infty)$, if $h_2(\infty) > c_2/(\mu_2 c_1)$ then a finite minimum to (4.48) exists uniquely.

If c_2 is the replacement cost of the Nth failure of unit 1 and c_3 is the replacement cost of unit 2 failure, then the expected cost rate $C(N)$ in (4.47) is rewritten as

$$C(N) = \frac{c_1 \sum_{j=0}^{N-2}\int_0^\infty \overline{F}_2(t)p_j(t)h_1(t)\,dt + c_2 \int_0^\infty \overline{F}_2(t)p_{N-1}(t)h_1(t)\,dt + c_3\left[1 - \int_0^\infty \overline{F}_2(t)p_{N-1}(t)h_1(t)\,dt\right]}{\sum_{j=0}^{N-1}\int_0^\infty \overline{F}_2(t)p_j(t)\,dt}$$

$$(N=1,2,\ldots). \qquad (4.49)$$

References

1. Barlow RE and Proschan F (1965) Mathematical Theory of Reliability. J Wiley & Sons, New York.
2. Holland CW, McLean RA (1975) Applications of replacement theory. AIIE Trans 7:42–47.

3. Tilquin C, Cléroux R (1975) Periodic replacement with minimal repair at failure and adjustment costs. Nav Res Logist Q 22:243–254.
4. Boland PJ (1982) Periodic replacement when minimal repair costs vary with time. Nav Res Logist Q 29:541–546.
5. Boland PJ, Proschan F (1982) Periodic replacement with increasing minimal repair costs at failure. Oper Res 30:1183–1189.
6. Chen M, Feldman RM (1997) Optimal replacement policies with minimal repair and age-dependent costs. Eur J Oper Res 98:75–84.
7. Aven T (1983) Optimal replacement under a minimal repair strategy–A general failure model. Adv Appl Prob 15:198–211.
8. Makis V, Jardine AKS (1991) Optimal replacement in the proportional hazard model. INFOR 30:172–183.
9. Makis V, Jardine AKS (1992) Optimal replacement policy for a general model with imperfect repair. J Oper Res Soc 43:111–120.
10. Makis V, Jardine AKS (1993) A note on optimal replacement policy under general repair. Eur J Oper Res 69:75–82.
11. Bagai I, Jain K (1994) Improvement, deterioration, and optimal replacement under age-replacement with minimal repair. IEEE Trans Reliab 43:156–162.
12. Morimura H (1970) On some preventive maintenance policies for IFR. J Oper Res Soc Jpn 12:94–124.
13. Pulcini G (2003) Mechanical reliability and maintenance models. In: Pham H (ed) Handbook of Reliability Engineering. Springer, London:317–348.
14. Nakagawa T, Kowada M (1983) Analysis of a system with minimal repair and its application to replacement policy. Eur J Oper Res 12:176–182.
15. Nakagawa T (1981) A summary of periodic replacement with minimal repair at failure. J Oper Res Soc Jpn 24:213–227.
16. Nakagawa T (1981) Generalized models for determining optimal number of minimal repairs before replacement. J Oper Res Soc Jpn 24:325–337.
17. Nakagawa T (1983) Optimal number of failures before replacement time. IEEE Trans Reliab R-32:115–116.
18. Nakagawa T (1984) Optimal policy of continuous and discrete replacement with minimal repair at failure. Nav Res Logist Q 31:543–550.
19. Nakagawa T, Yasui K (1991) Periodic-replacement models with threshold levels. IEEE Trans Reliab 40:395–397.
20. Nakagawa T (1987) Optimum replacement policies for systems with two types of units. In: Osaki S, Cao JH (eds) Reliability Theory and Applications Proceedings of the China-Japan Reliability Symposium, Shanghai, China.
21. Murthy DNP (1991) A note on minimal repair. IEEE Trans Reliab 40:245–246.
22. Kumar D (2000) Maintenance scheduling using monitored parameter values. In: Ben-Daya M, Duffuaa SO, Raouf A (eds) Maintenance, Modeling and Optimization. Kluwer Academic, Boston:345–374.
23. Park KS (1979) Optimal number of minimal repairs before replacement. IEEE Trans Reliab R-28:137–140.
24. Tapiero CS, Ritcken P (1985) Note on the (N,T) replacement rule. IEEE Trans Reliab R-34:374–376.
25. Stadje W, Zuckerman D (1990) Optimal strategies for some repair replacement models. Adv Appl Prob 22:641–656.
26. Lam Y (1988) A note on the optimal replacement problem. Adv Appl Prob 20:479–482.

27. Lam Y (1990) A repair replacement model. Adv Appl Prob 22:494–497.
28. Lam Y (1991) An optimal repairable replacement model for deteriorating system. J Appl Prob 28:843–851.
29. Lam Y (1999) An optimal maintenance model for a combination of secondhand-new or outdated-updated system. Eur J Oper Res 119:739–752.
30. Lam Y, Zhang YL (2003) A geometric-process maintenance model for a deteriorating system under a random environment. IEEE Trans Reliab 52:83–89.
31. Zhang YL (2002) A geometric-process repair-model with good-as-new preventive repair. IEEE Trans Reliab 51:223–228.
32. Ritchken P, Wilson JG (1990) (m, T) group maintenance policies. Manage Sci 36:632–639.
33. Sheu SH (1993) A generalized model for determining optimal number of minimal repairs before replacement. Eur J Oper Res 69:38–49.
34. Mine H, Kawai H (1974) Preventive replacement of a 1-unit system with a wearout state. IEEE Trans Reliab R23:24–29.
35. Maillart LM, Pollock SM (2002) Cost-optimal condition-monitoring for predictive maintenance of 2-phase systems. IEEE Trans Reliab 51:322–330.
36. Beichelt F, Fischer K (1980) General failure model applied to preventive maintenance policies. IEEE Trans Reliab R-29:39–41.
37. Beichelt F (1981) A generalized block-replacement policy. IEEE Trans Reliab R-30:171–172.
38. Murthy DNP, Maxwell MR (1981) Optimal age replacement policies for items from a mixture. IEEE Trans Reliab R-30:169–170.
39. Block HW, Borges WS, Savitz TH (1985) Age-dependent minimal repair. J Appl Prob 22:370–385.
40. Block HW, Borges WS, Savitz TH (1988) A general age replacement model with minimal repair. Nav Res Logist Q 35:365–372.
41. Berg M (1995) The marginal cost analysis and its application to repair and replacement policies. Eur J Oper 82:214–224.
42. Scheaffer RL (1975) Optimum age replacement in the bivariate exponential case. IEEE Trans Reliab R-24: 214–215.
43. Berg M (1976) Optimal replacement policies for two-unit machines with increasing running costs I. Stoch Process Appl 4:80–106.
44. Berg M (1978) General trigger-off replacement procedures for two-unit systems. Nav Res Logist Q 25:15–29.
45. Yamada S, Osaki S (1981) Optimum replacement policies for a system composed of components. IEEE Trans Reliab R-30:278–283.
46. Bai DS, Jang JS, Kwon YI (1983) Generalized preventive maintenance policies for a system subject to deterioration. IEEE Trans Reliab R-32:512–514.
47. Murthy DNP, Nguyen DG (1985) Study of two-component system with failure interaction. Nav Res Logist Q 32:239–247.
48. Pullen KW, Thomas MU (1986) Evaluation of an opportunistic replacement policy for a 2-unit system. IEEE Trans Reliab R-35:320–324.
49. Makis V, Jiang X (2001) Optimal control policy for a general EMQ model with random machine failure. In: Rahim MA, Ben-Daya M (eds) Integrated Models in Production, Planning, Inventory, Quality, and Maintenance. Kluwer Academic, Boston:67–78.
50. Ito K, Nakagawa T (2000) Optimal inspection policies for a storage system with degradation at periodic tests. Math Comput Model 31:191–195.

5

Block Replacement

If a system consists of a block or group of units, their ages are not observed and only their failures are known, all units may be replaced periodically independently of their ages in use. The policy is called *block replacement* and is commonly used with complex electronic systems and many electrical parts. Such block replacement was studied and compared with other replacements in [1, 2]. Furthermore, the n-stage block replacement was proposed in [3, 4]. The adjustment costs, which are increasing with the age of a unit, were introduced in [5]. More general replacement policies were considered and summarized in [6–10]. The block replacement of a two-unit system with failure dependence was considered in [11]. The optimum problem of provisioning spare parts for block replacement was discussed in [12] as an example of railways. The question, "Which is better, age or block replacement?", was answered in [13].

This chapter summarizes the block replacement from the book [1] based mainly on our original work: In Sections 5.1 and 5.2, we consider two periodic replacement policies with planned time T in which failed units are always replaced at each failure and a failed unit remains failed until time T. We obtain the expected cost rates for each policy and analytically discuss optimum replacement times that minimize them [14]. In Section 5.3, we propose the combined model of block replacement and no replacement at failure in Sections 5.1 and 5.2, and discuss the optimization problem with two variables [15]. In Section 5.4, we first summarize the periodic replacements in Section 4.1 and Sections 5.1 and 5.2, and show that they are written theoretically on general forms [14]. Next, we introduce four combined models of age, periodic, and block replacements [16, 17].

5.1 Replacement Policy

A new unit begins to operate at time $t = 0$, and a failed unit is instantly detected and is replaced with a new one. Furthermore, a unit is replaced

5 Block Replacement

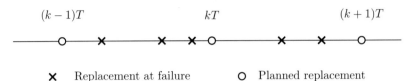

Fig. 5.1. Process of block replacement

at periodic times kT $(k = 1, 2, \ldots)$ independent of its age. Suppose that each unit has an identical failure distribution $F(t)$ with finite mean μ, and $F^{(n)}(t)$ $(n = 1, 2, \ldots)$ is the n-fold Stieltjes convolution of $F(t)$ with itself; i.e., $F^{(n)}(t) \equiv \int_0^t F^{(n-1)}(t-u) \mathrm{d}F(u)$ $(n = 1, 2, \ldots)$ and $F^{(0)}(t) \equiv 1$ for $t \geq 0$.

Consider one cycle with constant time T from the planned replacement to the next one (see Figure 5.1). Let c_1 be the cost of replacement for a failed unit and c_2 be the cost of the planned replacement. Then, because the expected number of failed units during one cycle is $M(T) \equiv \sum_{n=1}^{\infty} F^{(n)}(T)$ from (1.19), the expected cost in one cycle is, from (3.2) in Chapter 3,

$$c_1 E\{N_1(T)\} + c_2 E\{N_2(T)\} = c_1 M(T) + c_2.$$

Therefore, from (3.3), the expected cost rate is

$$C(T) = \frac{1}{T}[c_1 M(T) + c_2]. \tag{5.1}$$

If a unit is replaced only at failures, i.e., $T = \infty$, then $\lim_{T \to \infty} M(T)/T = 1/\mu$ from Theorem 1.2, and the expected cost rate is

$$C(\infty) \equiv \lim_{T \to \infty} C(T) = \frac{c_1}{\mu}.$$

Next, compare the expected costs between age replacement and block replacement. Letting $A(T) \equiv c_1 F(T) + c_2 \overline{F}(T)$ and $B(T) \equiv c_1 M(T) + c_2$, we have the renewal equations [18],

$$B(T) = A(T) + \int_0^T B(T-t) \, \mathrm{d}F(t) \quad \text{or} \quad B(T) = A(T) + \int_0^T A(T-t) \, \mathrm{d}M(t);$$

i.e., $A(T)$ and $B(T)$ determine each other.

We seek an optimum planned replacement time T^* that minimizes $C(T)$ in (5.1). It is assumed that $M(t)$ is differentiable and define $m(t) \equiv \mathrm{d}M(t)/\mathrm{d}t$, where $M(t)$ is called the *renewal function* and $m(t)$ is called the *renewal density* in Section 1.3. Then, differentiating $C(T)$ with respect to T and setting it equal to zero, we have

$$Tm(T) - M(T) = \frac{c_2}{c_1}. \tag{5.2}$$

5.1 Replacement Policy

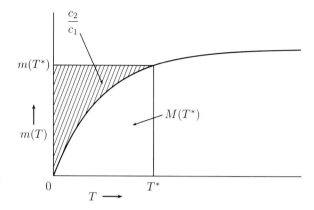

Fig. 5.2. Optimum T^* for renewal density $m(T)$

This equation is a necessary condition that there exists a finite T^*, and the resulting cost rate is

$$C(T^*) = c_1 m(T^*). \tag{5.3}$$

Figure 5.2 shows graphically an optimum time T^* on the horizontal axis given in (5.2) for the renewal density $m(T)$, and the expected cost rate $m(T^*) = C(T^*)/c_1$ on the vertical axis.

Let σ^2 be the variance of $F(t)$. Then, from (1.25), there exists a large T such that $C(T) < C(\infty)$ if $c_2/c_1 < [1 - (\sigma^2/\mu^2)]/2$ [19].

In general, it might be difficult to compute explicitly a renewal function $M(t)$. In this case, because $F^{(n)}(t) \leq [F(t)]^n$, we may use the following upper and lower bounds [13].

$$\frac{1}{T}\{c_1[F^{(1)}(T) + F^{(2)}(T)] + c_2\} < C(T) \leq \frac{1}{T}\{c_1[F^{(1)}(T) + F^{(2)}(T) + F(T)^3/\overline{F}(T)] + c_2\}. \tag{5.4}$$

Next, consider a system with n identical units that operate independently of each other. It is assumed that all units together are replaced immediately upon failure. Then, the expected cost rate is

$$C(T;n) = \frac{1}{T}[c_1 n M(T) + c_2], \tag{5.5}$$

where c_1 = cost of replacement at each failure and c_2 = cost of planned replacement for all n units at time T.

Suppose that all costs are discounted with rate α $(0 < \alpha < \infty)$. In similar ways to those of obtaining (3.15) in **(1)** of Section 3.2 and (4.30) in **(1)** of Section 4.4, the total expected cost for an infinite time span is

$$C(T;\alpha) = \frac{c_1 \int_0^T e^{-\alpha t} m(t)\, dt + c_2 e^{-\alpha T}}{1 - e^{-\alpha T}}. \tag{5.6}$$

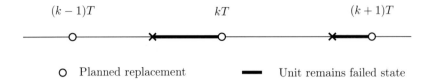

Fig. 5.3. Process of no replacement at failure

Differentiating $C(T; \alpha)$ with respect to T and setting it equal to zero,

$$\frac{1 - e^{-\alpha T}}{\alpha} m(T) - \int_0^T e^{-\alpha t} m(t)\, dt = \frac{c_2}{c_1} \tag{5.7}$$

and the resulting cost rate is

$$C(T^*; \alpha) = \frac{c_1}{\alpha} m(T^*) - c_2. \tag{5.8}$$

5.2 No Replacement at Failure

A unit is always replaced at times kT ($k = 1, 2, \ldots$), but it is not replaced at failure, and hence, it remains in failed state for the time interval from a failure to its detection (see Figure 5.3). This can be applied to the maintenance model where a unit is not monitored continuously, and its failures can be detected only at times kT and some maintenance is done [20].

Let c_1 be the downtime cost per unit of time elapsed between a failure and its replacement, and c_2 be the cost of planned replacement. Then, the mean time from a failure to its detection is

$$\int_0^T (T - t)\, dF(t) = \int_0^T F(t)\, dt \tag{5.9}$$

and the expected cost rate is

$$C(T) = \frac{1}{T}\left[c_1 \int_0^T F(t)\, dt + c_2 \right]. \tag{5.10}$$

Differentiating $C(T)$ with respect to T and setting it equal to zero,

$$TF(T) - \int_0^T F(t)\, dt = \frac{c_2}{c_1} \quad \text{or} \quad \int_0^T t\, dF(t) = \frac{c_2}{c_1}. \tag{5.11}$$

Thus, if $\mu > c_2/c_1$ then there exists an optimum time T^* that uniquely satisfies (5.11), and the resulting cost rate is

$$C(T^*) = c_1 F(T^*). \tag{5.12}$$

Figure 5.4 graphically shows an optimum time T^* on the horizontal axis given in (5.11) for the distribution $F(T)$, and the expected cost rate $F(T^*) = C(T^*)/c_1$ on the vertical axis.

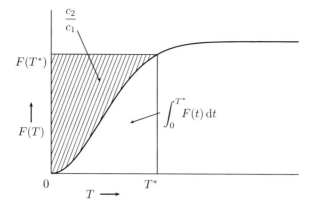

Fig. 5.4. Optimum T^* for distribution $F(T)$

5.3 Replacement with Two Variables

In the block replacement model, it may be wasteful to replace a failed unit with a new one just before the planned replacement. Three modifications of the model from this viewpoint have been suggested. When a failure occurs just before the planned replacement, it remains failed until the replacement time [19,21,22] or it is replaced with a used one [23–26]. An operating unit with young age is not replaced at planned replacement and remains in service [27, 28].

We consider the combined model of block replacement in Section 5.1 and no replacement at failure in Section 5.2. Failed units are replaced with a new one during $(0, T_0]$, and after T_0, if a failure occurs in an interval (T_0, T), then the replacement is not made in this interval and the unit remains failed until the planned time T. Using the results of a renewal theory in Section 1.3, the expected cost rate is obtained, and the optimum T_0^* and T^* to minimize it are analytically derived. This is a problem of minimizing an objective function with two dependent variables, which extends the standard replacement problem. This is transformed into a problem with one variable and is solved by the usual calculus method.

A unit is replaced at planned time T. If a unit fails during $(0, T_0]$ for $0 \le T_0 \le T$ then it is replaced with a new one, whereas if it fails in an interval (T_0, T) then it remains failed for the time interval from its failure to time T. Let $\gamma(x)$ denote the residual life of a unit at time x in a renewal process. Then, from (1.29) in Section 1.3, the distribution of $\gamma(x)$ is given by

$$G(t;x) \equiv \Pr\{\gamma(x) \le t\} = F(x+t) - \int_0^x \overline{F}(x+t-u)\,\mathrm{d}M(u). \quad (5.13)$$

Thus, the mean time from a failure to replacement time T in an interval (T_0, T) is

5 Block Replacement

$$\int_0^{T-T_0} (T - T_0 - t)\, dG(t; T_0) = \int_0^{T-T_0} G(t; T_0)\, dt.$$

Therefore, in similar ways to those of obtaining (5.1) and (5.10), the expected cost rate is

$$C(T_0, T) = \frac{1}{T}\left[c_1 M(T_0) + c_2 + c_3 \int_0^{T-T_0} G(t; T_0)\, dt\right], \qquad (5.14)$$

where c_1 = cost of replacement at failure, c_2 = cost of planned replacement, and c_3 = downtime cost from a failure to its detection. This is equal to (5.1) when $T = T_0$, and to (5.10) when $T_0 = 0$ by replacing c_3 with c_1.

We seek optimum times T_0^* and T^* that minimize $C(T_0, T)$. Differentiating $C(T_0, T)$ with respect to T_0 for a fixed T and setting it equal to zero,

$$\int_0^{T-T_0} \overline{F}(t)\, dt = \frac{c_1}{c_3}. \qquad (5.15)$$

We consider the following three cases.

Case 1. If $c_3 \leq c_1/\mu$ then $C(T_0, T)$ is increasing in T_0, and hence, $T_0^* = 0$ and the expected cost rate is

$$C(0, T) = \frac{1}{T}\left[c_2 + c_3 \int_0^T F(t)\, dt\right]. \qquad (5.16)$$

Replacing c_1 with c_3 in Section 5.2, we can obtain an optimum policy.

Case 2. If $c_3 > c_1/\mu$ then there exists a unique a $(0 < a < \infty)$ that satisfies $\int_0^a \overline{F}(t) dt = c_1/c_3$. Thus, $T_0^* = T - a$ $(a \leq T < \infty)$ and the problem of minimizing $C(T_0, T)$ for both T_0 and T corresponds to the problem of minimizing $C(T - a, T)$ as follows.

$$C(T - a, T) = \frac{1}{T}\left[c_1 M(T - a) + c_2 + c_3 \int_0^a G(t; T - a)\, dt\right]. \qquad (5.17)$$

Differentiating $C(T - a, T)$ with respect to T and setting it equal to zero,

$$TG(a; T - a) - \int_0^a G(t; T - a)\, dt - M(T - a)\int_0^a \overline{F}(t)\, dt = \frac{c_2}{c_3}, \qquad (5.18)$$

which is a necessary condition that a finite T^* minimizes $C(T - a, T)$. In general, it is very difficult to discuss whether a solution T to (5.18) exists.

Case 3. Suppose that $c_3 > c_1/\mu$ and $m(t)$ is strictly increasing. Let $Q(T; a)$ be the left-hand side of (5.18). Then, using the renewal equation of $m(t)$:

5.3 Replacement with Two Variables

$$m(t) = f(t) + \int_0^t f(t-u)m(u)\,du$$

we have the inequality

$$\frac{dQ(T;a)}{dT} = T\left[f(T) - \overline{F}(a)m(T-a) + \int_0^{T-a} f(T-u)m(u)\,du\right]$$

$$= T\left[m(T) - \int_0^T f(T-u)m(u)\,du - \overline{F}(a)m(T-a)\right.$$

$$\left. + \int_0^{T-a} f(T-u)m(u)\,du\right]$$

$$> T\overline{F}(a)[m(T) - m(T-a)] > 0.$$

Furthermore, from (1.30) in a renewal process,

$$G(t;T) \to \frac{1}{\mu}\int_0^t \overline{F}(u)\,du \quad \text{as } T \to \infty$$

and from (1.25),

$$M(T) = \frac{T}{\mu} + \frac{\mu_2}{2\mu^2} - 1 + o(1) \quad \text{as } T \to \infty,$$

where μ_2 is the second moment of $F(t)$; i.e., $\mu_2 \equiv \int_0^\infty t^2 dF(t)$. Thus,

$$Q(a;a) = \int_0^a [F(a) - F(t)]\,dt \geq 0$$

$$Q(\infty;a) \equiv \lim_{T \to \infty} Q(T;a)$$

$$= \int_0^a \overline{F}(t)\,dt\left(\frac{a}{\mu} - \frac{\mu_2}{2\mu^2} + 1\right) - \frac{1}{\mu}\int_0^a \int_0^t \overline{F}(u)\,du\,dt.$$

From the above discussion, we can obtain the following optimum policy.

(i) If $Q(a;a) \geq c_2/c_3$ then a solution to (5.18) does not exist, and $C(T-a;T)$ is increasing in T. Hence, $T_0^* = 0$ by putting $T = a$, and T^* is given by a solution of equation

$$TF(T) - \int_0^T F(t)\,dt = \frac{c_2}{c_3}$$

and

$$C(0;T^*) = c_3 F(T^*).$$

(ii) If $Q(a;a) < c_2/c_3 < Q(\infty;a)$ then there exists a unique $T^*(a < T^* < \infty)$ that satisfies (5.18), and hence, $T_0^* = T^* - a$ and

$$C(T_0^*, T^*) = c_3 G(a; T_0^*).$$

(iii) If $Q(\infty, a) \leq c_2/c_3$ then $T_0^* = \infty$; i.e., a unit is replaced only at failure and $C(\infty, \infty) = c_1/\mu$.

Example 5.1. Suppose that $f(t) = te^{-t}$. Then, the renewal density is $m(t) = (1 - e^{-2t})/2$ which is strictly increasing, and hence, $Q(T; a)$ is also strictly increasing from $2 - (2 + 2a + a^2)e^{-a}$ to $2 - [2 + (7a/4) + (a^2/2)]e^{-a}$. Thus, we have the following optimum policy.

(i) If $c_3 \leq c_2/2$ then $T_0^* = 0$ and $T^* = \infty$, and $C(0, \infty) = c_3$.
(ii) If $c_2/2 < c_3 \leq c_1/2$, or $c_3 > c_1/2$ and $(c_1 - c_2)/c_3 \geq (1+a)ae^{-a}$ then $T_0^* = 0$ and T^* is given by a solution of the equation

$$(2 + 2T + T^2)e^{-T} = 2 - \frac{c_2}{c_3}$$

and

$$C(0, T^*) = c_3[1 - (1 + T^*)e^{-T^*}],$$

where a satisfies uniquely the equation

$$2 - (2 + a)e^{-a} = \frac{c_1}{c_3}.$$

(iii) If $c_3 > c_1/2$ and $[(3/4) + (a/2)]ae^{-a} < (c_1 - c_2)/c_3 < (1+a)ae^{-a}$, then $T_0^* = T^* - a$ and T^* $(a < T^* < \infty)$ is given by

$$2(1 - e^{-a}) + \frac{a}{2}e^{-a}\left[T - a - 2 - (1+T)(1 + e^{-2(T-a)}) - \frac{1}{2}(1 - e^{-2(T-a)})\right] = \frac{c_2}{c_3}$$

and

$$C(T_0^*, T^*) = c_3 \left\{ 1 - e^{-a}\left[1 + \frac{a}{2}(1 + e^{-2T_0^*})\right]\right\}.$$

(iv) If $c_3 > c_1/2$ and $(c_1 - c_2)/c_3 \leq [(3/4) + (a/2)]ae^{-a}$ then $T_0^* = T^* = \infty$, and $C(\infty, \infty) = c_1/2$.

Table 5.1 gives the optimum times T_0^* and T^*, and the expected cost rate $C(T_0^*, T^*)$ for $c_1 = 5, c_2 = 1, c_3 = 1, 2, \ldots, 10$. For the standard replacement policy in Section 5.1, when $c_1 > 4c_2$, the optimum time uniquely satisfies

$$1 - (1 + 2T)e^{-2T} = \frac{4c_2}{c_1}$$

and the resulting cost rate is

$$C(T^*) = \frac{c_1}{2}(1 - e^{-2T^*}).$$

In this case, $T^* = 1.50$ and $C(T^*) = 2.38$. This indicates that the maintenance with two variables becomes more effective as cost c_3 is smaller. ∎

Table 5.1. Variation in optimum times T_0^* and T^*, expected cost rate $C(T_0^*, T^*)$ for c_3 when $f(t) = te^{-t}$, $c_1 = 5$, and $c_2 = 1$

c_3	T_0^*	T^*	$C(T_0^*, T^*)$
1	0	2.67	0.75
2	0	1.73	1.03
3	0	1.40	1.23
4	0	1.22	1.38
5	0	1.10	1.51
6	0.11	1.03	1.62
7	0.22	1.00	1.71
8	0.33	0.99	1.80
9	0.41	0.99	1.86
10	0.48	0.99	1.92

5.4 Combined Replacement Models

This section represents the results of periodic replacements in Sections 4.2, 5.1, and 5.2 on the general forms. It is theoretically shown that these replacement models come to the same one essentially. Furthermore, we propose the combined replacement models of age, periodic, and block replacements. These modified and extended replacements would be more realistic than the usual ones, and moreover, offer interesting topics to reliability theoreticians.

5.4.1 Summary of Periodic Replacement

In general, the results of periodic replacements in Sections 4.2, 5.1, and 5.2 are summarized as follows. The expected cost rate is

$$C(T) = \frac{1}{T}\left[c_1 \int_0^T \varphi(t)\,dt + c_2\right], \tag{5.19}$$

where $\varphi(t)$ is $h(t)$, $m(t)$, and $F(t)$, respectively. Differentiating $C(T)$ with respect to T and setting it equal to zero,

$$T\varphi(T) - \int_0^T \varphi(t)\,dt = \frac{c_2}{c_1} \quad \text{or} \quad \int_0^T t\,d\varphi(t) = \frac{c_2}{c_1}. \tag{5.20}$$

If there exists T^* that satisfies (5.20) then the expected cost rate is

$$C(T^*) = c_1 \varphi(T^*). \tag{5.21}$$

For the periodic replacement with discounting rate $\alpha > 0$,

$$C(T; \alpha) = \frac{c_1 \int_0^T e^{-\alpha t} \varphi(t)\,dt + c_2 e^{-\alpha T}}{1 - e^{-\alpha T}} \tag{5.22}$$

126   5 Block Replacement

$$\frac{1-e^{-\alpha T}}{\alpha}\varphi(T) - \int_0^T e^{-\alpha t}\varphi(t)\,dt = \frac{c_2}{c_1} \qquad (5.23)$$

$$C(T^*;\alpha) = \frac{c_1}{\alpha}\varphi(T^*) - c_2. \qquad (5.24)$$

Moreover, the expected cost rate is rewritten on a general form

$$C(T) = \frac{1}{T}[\Phi(T) + c_2],$$

where $\Phi(T)$ is the total expected cost during $(0,T]$, and the optimum policies were discussed under several conditions in [29–31]. Furthermore, if the maintenance cost depends on time t and is given by $c(t)$, the expected cost rate is [32]

$$C(T) = \frac{1}{T}\left[\int_0^T c(t)\varphi(t)\,dt + c_2\right].$$

Finally, we consider a system consisting of n units that operate independently of each other and have parameter function $\varphi_i(t)$ $(i = 1, 2, \ldots, n)$. It is assumed that all units are replaced together at times kT $(k = 1, 2, \ldots)$. Then, the expected cost rate is

$$C(T) = \frac{1}{T}\left[\sum_{i=1}^n c_i \int_0^T \varphi_i(t)\,dt + c_2\right],$$

where c_i = cost of maintenance for each failed unit. Such group maintenance policies for multiunit systems were analyzed in [33–36], and their overviews were presented in [37, 38].

If the failure rate $h(t)$, renewal density $m(t)$, and failure distribution $F(t)$ are statistically estimated and graphically drawn, we could derive roughly optimum replacement T^* on the horizontal axis and the expected cost rate $C(T^*)/c_1$ on the vertical axis from Figures 4.2, 5.2, and 5.4.

5.4.2 Combined Replacement

This section summarizes the combined replacement models of age, periodic, and block replacements.

(1) Periodic and No Replacement at Failure

We propose the combined model of periodic replacement with minimal repair at failures in Section 4.2 and no replacement at failure in Section 5.2: A unit is replaced at planned time T, where T is given by a solution to (4.18) and minimizes $C(T)$ in (4.16). If a unit fails during $(0, T_0]$ $(0 \leq T_0 \leq T)$ then it undergoes only minimal repair at failures, whereas if it fails in an interval

(T_0, T) then the minimal repair is not made and it remains failed until the planned time T.

Consider one cycle from time $t = 0$ to the time that a unit is replaced at planned time T. Then, the total expected cost in one cycle is given by the sum of the minimal repair cost during $(0, T_0]$, the planned replacement cost, and the downtime cost when a unit fails in an interval (T_0, T). The mean downtime from a failure to the replacement is

$$\frac{1}{\overline{F}(T_0)} \int_{T_0}^{T} (T - t) \, \mathrm{d}F(t) = T - T_0 - \frac{1}{\overline{F}(T_0)} \int_{T_0}^{T} \overline{F}(t) \, \mathrm{d}t.$$

Thus, from (4.16) in Section 4.2, the expected cost rate is

$$C(T_0; T) = \frac{1}{T} \left\{ c_1 H(T_0) + c_2 + c_3 \left[T - T_0 - \frac{1}{\overline{F}(T_0)} \int_{T_0}^{T} \overline{F}(t) \, \mathrm{d}t \right] \right\}, \quad (5.25)$$

where c_1 = cost of minimal repair at failure, c_2 = cost of planned replacement at time T, and c_3 = downtime cost per unit of time from a failure to its replacement. This is equal to (4.16) when $T = T_0$, and (5.10) when $T_0 = 0$ by replacing c_3 with c_1.

We seek an optimum T_0^* that minimizes $C(T_0; T)$ for a fixed T when the failure rate $h(t)$ is strictly increasing. Differentiating $C(T_0; T)$ with respect to T_0 and setting it equal to zero, we have

$$\frac{1}{\overline{F}(T_0)} \int_{T_0}^{T} \overline{F}(t) \, \mathrm{d}t = \frac{c_1}{c_3}. \quad (5.26)$$

It is easy to see that the left-hand side of (5.26) is strictly decreasing in T_0 from $\int_0^T \overline{F}(t) \mathrm{d}t$ to 0, because $h(t)$ is strictly increasing. Thus, we have the following optimum policy.

(i) If $\int_0^T \overline{F}(t) \mathrm{d}t > c_1/c_3$ then there exists a finite and unique T_0^* that satisfies (5.26). In this case, optimum T_0^* is an increasing function of T because the left-hand side of (5.26) is increasing in T.
(ii) If $\int_0^T \overline{F}(t) \mathrm{d}t \le c_1/c_3$ then $T_0^* = 0$; i.e., no minimal repair is made.

(2) Periodic and Age Replacements

We consider two combined models of periodic and age replacements and obtain optimum replacement policies. First, suppose that if a unit fails during $(0, T_0]$ then it undergoes minimal repair at failures. However, if a unit fails in an interval (T_0, T) then it is replaced with a new one before time T, whereas if it does not fail in an interval (T_0, T) then it is replaced at time T.

Because the probability that a unit fails in an interval (T_0, T) is $[F(T) - F(T_0)]/\overline{F}(T_0)$, the mean time from T_0 to replacement is

$$\frac{1}{\overline{F}(T_0)}\left[(T-T_0)\overline{F}(T) + \int_{T_0}^T (t-T_0)\,\mathrm{d}F(t)\right] = \frac{1}{\overline{F}(T_0)}\int_{T_0}^T \overline{F}(t)\,\mathrm{d}t$$

and the expected cost rate is

$$C(T_0, T) = \frac{c_1 H(T_0) + c_2 + c_3[F(T) - F(T_0)]/\overline{F}(T_0)}{T_0 + \int_{T_0}^T \overline{F}(t)\,\mathrm{d}t/\overline{F}(T_0)}, \tag{5.27}$$

where c_1 and c_2 are given in (5.25) and $c_3 = $ additional cost of no planned replacement at failure. This corresponds to periodic replacement in Section 4.2 when $T_0 = T$, and age replacement in Section 3.1 when $T_0 = 0$.

We seek an optimum T_0^* that minimizes $C(T_0; T)$, where a finite T satisfies (4.18) when $h(t)$ is strictly increasing. Differentiating $C(T_0, T)$ in (5.27) with respect to T_0 and setting it equal to zero,

$$Q_1(T_0; T) = c_2 + c_3 - c_1, \tag{5.28}$$

where

$$Q_1(T_0; T) \equiv \frac{T_0}{\int_{T_0}^T \overline{F}(t)\,\mathrm{d}t}[c_1\overline{F}(T_0) - c_3\overline{F}(T)] - c_1 H(T_0).$$

Also, we have $Q_1(0; T) = 0$, and differentiating $Q_1(T_0; T)$ with respect to T_0,

$$\frac{\mathrm{d}Q_1(T_0; T)}{\mathrm{d}T_0} = \left[1 + \frac{T_0 \overline{F}(T_0)}{\int_{T_0}^T \overline{F}(t)\,\mathrm{d}t}\right]\left[\frac{c_1\overline{F}(T_0) - c_3\overline{F}(T)}{\int_{T_0}^T \overline{F}(t)\,\mathrm{d}t} - c_1 h(T_0)\right].$$

First, suppose that $c_3 \geq c_1$. Then, $Q_1(0; T) < c_2 + c_3 - c_1$, $\lim_{T_0 \to T} Q_1(T_0; T) = -\infty$ for $c_3 > c_1$ and $\lim_{T_0 \to T} Q_1(T_0; T) = c_2$ for $c_3 = c_1$. Furthermore, putting $\mathrm{d}Q_1(T_0; T)/\mathrm{d}T_0 = 0$ and arranging it, we have

$$h(T_0)\int_{T_0}^T \overline{F}(t)\,\mathrm{d}t - \overline{F}(T_0) = -\frac{c_3}{c_1}\overline{F}(T). \tag{5.29}$$

Thus, if $c_3\overline{F}(T) \geq c_1[1 - h(0)\int_0^T \overline{F}(t)\mathrm{d}t]$ then $\mathrm{d}Q_1(T_0; T)/\mathrm{d}T_0 \leq 0$. Conversely, if $c_3\overline{F}(T) < c_1[1 - h(0)\int_0^T \overline{F}(t)\mathrm{d}t]$ then (5.29) has one solution in $0 < T_0 < T$, and its extreme value is

$$Q_1(T_0; T) = c_1[T_0 h(T_0) - H(T_0)] < c_2$$

inasmuch as $th(t) - H(t)$ is an increasing function of t and $Th(T) - H(T) = c_2/c_1$. In both cases, $Q_1(T_0; T) \leq c_2 + c_3 - c_1$ for all $T_0 (0 \leq T_0 \leq T)$; i.e., $C(T_0; T)$ is decreasing in T_0, and hence, $T_0^* = T$.

Next, suppose that $c_3 < c_1$. Then, $Q_1(T_0; T)$ is strictly increasing in T_0 because $\mathrm{d}Q_1(T_0; T)/\mathrm{d}T_0 > 0$ from (ii) of Theorem 1.1 and $\lim_{T_0 \to T} Q_1(T_0; T) = \infty$. If $c_2 + c_3 > c_1$ then $Q_1(0; T) < c_2 + c_3 - c_1$, and hence, there exists a unique T_0^* $(0 < T_0^* < T)$ that satisfies (5.28), and it minimizes $C(T_0; T)$. On the other hand, if $c_2 + c_3 \leq c_1$ then $Q_1(0; T) \geq c_2 + c_3 - c_1$. Thus, $C(T_0; T)$ is increasing in T_0, and hence, $T_0^* = 0$.

From the above discussion, we have the following optimum policy.

(i) If $c_3 \geq c_1$ then $T_0^* = T$; i.e., a unit undergoes only minimal repair until the replacement time comes.

(ii) If $c_2 + c_3 > c_1 > c_3$ then there exists a unique T_0^* $(0 < T_0^* < T)$ that satisfies (5.28), and the resulting cost rate is

$$C(T_0^*; T) = \frac{c_1 \overline{F}(T_0^*) - c_3 \overline{F}(T)}{\int_{T_0^*}^T \overline{F}(t)\,dt} \quad (5.30)$$

and the expected cost rate is between two costs:

$$c_1 h(T_0^*) < C(T_0^*; T) < c_1 h(T).$$

(iii) If $c_1 \geq c_2 + c_3$ then $T_0^* = 0$; i.e., a unit is replaced at failure or at time T, whichever occurs first.

This policy was called the (T_0, T) policy, and it was proved that if $c_2 + c_3 > c_1 > c_3$ and $h(t)$ is strictly increasing to infinity then there exist finite and unique T_0^* and T^* $(0 < T_0^* < T^* < \infty)$ that minimize $C(T_0; T)$ in (5.27) [39]. Some modified models of this policy were proposed in [40–44].

Next, suppose that if a unit fails during $(0, T]$ then it undergoes minimal repair at failures. However, a unit is not replaced at time T and is replaced at the first failure after time T or at time T_1 $(T_1 > T)$, whichever occurs first, where T satisfies (4.18).

Changing T_0 and T into T and T_1 in (5.27), the expected cost rate is

$$C(T_1; T) = \frac{c_1 H(T) + c_2 + c_3 [F(T_1) - F(T)]/\overline{F}(T)}{T + \int_T^{T_1} \overline{F}(t)\,dt/\overline{F}(T)}. \quad (5.31)$$

This corresponds to periodic replacement when $T = T_1$, and age replacement when $T = 0$. We seek an optimum T_1^* that minimizes $C(T_1; T)$ for a fixed T given in (4.18) when $h(t)$ is strictly increasing. Differentiating $C(T_1; T)$ with respect to T_1 and putting it to zero,

$$Q_2(T_1; T) = \frac{c_1}{c_3} Th(T), \quad (5.32)$$

where

$$Q_2(T_1; T) \equiv h(T_1)\left[T + \frac{\int_T^{T_1} \overline{F}(t)\,dt}{\overline{F}(T)}\right] - \frac{F(T_1) - F(T)}{\overline{F}(T)}.$$

From the assumption that $h(t)$ is strictly increasing, $Q_2(T_1; T)$ is also strictly increasing with $Q_2(T; T) = Th(T)$ and

$$Q_2(\infty; T) \equiv \lim_{T_1 \to \infty} Q_2(T_1; T) = h(\infty)\left[T + \frac{\int_T^\infty \overline{F}(t)\,dt}{\overline{F}(T)}\right] - 1.$$

Thus, if $c_3 \geq c_1$ then $Q_2(T; T) \geq (c_1/c_3)Th(T)$, and $T_1^* = T$. Conversely, if $c_3 < c_1$ and $h(\infty) > K(T)$ then $Q_2(T; T) < (c_1/c_3)Th(T) < Q_2(\infty; T)$, where

$$K(T) \equiv \frac{(c_1/c_3)Th(T) + 1}{T + \int_T^\infty \overline{F}(t)\,dt/\overline{F}(T)}.$$

Hence, there exists a finite and unique T_1^* that satisfies (5.32), and it minimizes $C(T_1;T)$. Finally, if $c_3 < c_1$ and $h(\infty) \leq K(T)$ then $Q_2(\infty;T) \leq (c_1/c_3)Th(T)$, and $T_1^* = \infty$.

Therefore, we have the following optimum policy.

(i) If $c_1 \leq c_3$ then $T_1^* = T$; i.e., a unit is replaced only at time T.
(ii) If $c_1 > c_3$ and $h(\infty) > K(T)$ then there exists a finite and unique T_1^* ($T < T_1^* < \infty$) that satisfies (5.32) and the resulting cost rate is

$$C(T_1^*;T) = c_3 h(T_1^*). \tag{5.33}$$

(iii) If $c_1 > c_3$ and $h(\infty) \leq K(T)$ then $T_1^* = \infty$; i.e., a unit is replaced at the first failure after time T, and the expected cost rate is $C(\infty;T) = c_3 K(T)$.

We compare $C(T_0;T)$ and $C(T_1;T)$ when $c_2 + c_3 > c_1 > c_3$ and $h(\infty) > K(T)$. From (5.30) and (5.33), if

$$\frac{c_1}{c_3} > \frac{1}{\overline{F}(T_0^*)}\left[\overline{F}(T) + h(T_1^*)\int_{T_0^*}^T \overline{F}(t)\,dt\right]$$

then the replacement after time T is better than the replacement before time T; i.e., a unit should be replaced late rather than early, and *vice versa*.

We consider the cases of $T_1 = \infty$ and $c_3 = 0$; i.e., a unit undergoes minimal repair at failures until time T, and after that, it is replaced at the first failure [45]. In this case, the expected cost rate is, from (5.31),

$$C(T) \equiv \lim_{T_1 \to \infty} C(T_1;T) = \frac{c_1 H(T) + c_2}{T + \int_T^\infty \overline{F}(t)\,dt/\overline{F}(T)}. \tag{5.34}$$

By a similar method to the previous models, we have the following results.

(i) If $c_1 \geq c_2$ then $T^* = 0$; i.e., a unit is replaced at each failure.
(ii) If $c_1 < c_2$ and $Q_3(\infty) > (c_2 - c_1)/c_1$ then there exists a finite and unique T^* that satisfies

$$Q_3(T) = \frac{c_2 - c_1}{c_1}, \tag{5.35}$$

where

$$Q_3(T) \equiv \frac{T\overline{F}(T)}{\int_T^\infty \overline{F}(t)\,dt} - H(T)$$

and the expected cost rate is

$$C(T^*) = \frac{c_1 \overline{F}(T^*)}{\int_{T^*}^\infty \overline{F}(t)\,dt}. \tag{5.36}$$

(iii) If $c_1 < c_2$ and $Q_3(\infty) \le (c_2 - c_1)/c_1$ then $T^* = \infty$; i.e., a unit undergoes only minimal repair at any failure.

Note that when the failure rate $h(t)$ is strictly increasing, $Q_3(T)$ is also strictly increasing and

$$\frac{T\overline{F}(T)}{\int_T^\infty \overline{F}(t)\,dt} - H(T) > Th(T) - H(T) \ge T_1 h(T) - H(T_1)$$

for any $T \ge T_1$. Thus, if $h(t)$ is strictly increasing to infinity then $Q_3(\infty) = \infty$ and there exists a finite and unique T^* that satisfies (5.35).

(3) Block and Age Replacement

We consider two combined models of block and age replacements. First, suppose that if a unit fails during $(0, T_0]$ then it is replaced at each failure. However, if a unit fails in an interval (T_0, T) then it is replaced with a new one before time T, whereas if it does not fail in (T_0, T) then it is replaced at time T.

From (5.13) in Section 5.3, the probability that a unit fails in an interval (T_0, T) is

$$G(T - T_0; T_0) = F(T) - \int_0^{T_0} \overline{F}(T - t)\,dM(t)$$

and the mean time to replacement after time T_0 is

$$\int_0^{T-T_0} (t + T_0)\,dG(t; T_0) + T\overline{G}(T - T_0; T_0) = T_0 + \int_0^{T-T_0} \overline{G}(t; T_0)\,dt.$$

Thus, the expected cost rate is

$$C(T_0; T) = \frac{c_1 M(T_0) + c_2 + c_3 G(T - T_0; T_0)}{T_0 + \int_0^{T-T_0} \overline{G}(t; T_0)\,dt}. \tag{5.37}$$

This corresponds to age replacement when $T_0 = 0$ and block replacement when $T = T_0$.

Next, suppose that if a unit fails during $(0, T]$ then it is replaced at each failure. However, a unit is not replaced at time T, and is replaced at the first failure after time T or at time T_1 ($T_1 \ge T$), whichever occurs first. Then, changing T_0 and T into T and T_1 in (5.37), the expected cost rate is

$$C(T_1; T) = \frac{c_1 M(T) + c_2 + c_3 G(T_1 - T; T)}{T + \int_0^{T_1 - T} \overline{G}(t; T)\,dt}. \tag{5.38}$$

This corresponds to age replacement when $T = 0$ and block replacement when $T_1 = T$.

Moreover, if a unit is replaced at the first failure after time T and $c_3 = 0$, the expected cost rate is

$$C(T) \equiv \lim_{T_1 \to \infty} C(T_1; T) = \frac{c_1 M(T) + c_2}{T + \int_T^\infty \overline{F}(t)\,dt + \int_0^T [\int_{T-t}^\infty \overline{F}(u)\,du]\,dM(t)}. \tag{5.39}$$

(4) Block and Periodic Replacements

A unit is replaced at each failure during $(0, T_0]$ and at planned time T ($T_0 \le T$). However, if a unit fails in an interval (T_0, T) then it undergoes minimal repair. Then, from (1.28) in Section 1.3, the expected number of failures in (T_0, T) is

$$\int_0^{T_0} [H(T-t) - H(T_0-t)] \, \mathrm{d}\Pr\{\delta(T_0) \le T_0 - t\}$$
$$= \overline{F}(T_0)[H(T) - H(T_0)] + \int_0^{T_0} [H(T-t) - H(T_0-t)]\overline{F}(T_0-t) \, \mathrm{d}M(t),$$

where $\delta(t)$ = age of a unit at time t in a renewal process. Thus, the expected cost rate is

$$C(T_0; T) = \frac{1}{T} \left[\begin{array}{l} c_1 M(T_0) + c_2 + c_3\{\overline{F}(T_0)[H(T) - H(T_0)] \\ + \int_0^{T_0} [H(T-t) - H(T_0-t)]\overline{F}(T_0-t) \, \mathrm{d}M(t)\} \end{array} \right], \quad (5.40)$$

where c_1 = cost of replacement at failure, c_2 = cost of planned replacement at time T, and c_3 = cost of minimal repair. This corresponds to periodic replacement when $T_0 = 0$ and block replacement when $T = T_0$.

References

1. Barlow RE and Proschan F (1965) Mathematical Theory of Reliability. J Wiley & Sons, New York.
2. Schweitzer PJ (1967) Optimal replacement policies for hyperexponentially and uniformly distributed lifetimes. Oper Res 15:360–362.
3. Marathe VP, Nair KPK (1966) Multistage planned replacement strategies. Oper Res 14:874–887.
4. Jain A, Nair KPK (1974) Comparison of replacement strategies for items that fail. IEEE Trans Reliab R-23:247–251.
5. Tilquin C, Cléroux R (1975) Block replacement policies with general cost structures. Technometrics 17:291–298.
6. Archibald TW, Dekker R (1996) Modified block-replacement for multi-component systems. IEEE Trans Reliab R-45:75–83.
7. Sheu SH (1991) A generalized block replacement policy with minimal repair and general random repair costs for a multi-unit system. J Oper Res Soc 42:331–341.
8. Sheu SH (1994) Extended block replacement policy with used item and general random minimal repair cost. Eur J Oper Res 79:405–416.
9. Sheu SH (1996) A modified block replacement policy with two variables and general random repair cost. J Appl Prob 33:557–572.
10. Sheu SH (1999) Extended optimal replacement model for deteriorating systems. Eur J Oper Res 112:503–516.
11. Scarf PA, Deara M (2003) Block replacement policies for a two-component system with failure dependence. Nav Res Logist 50:70–87.

12. Brezavšcek A, Hudoklin A (2003) Joint optimization of block-replacement and periodic-review spare-provisioning policy. IEEE Trans Reliab 52:112–117.
13. Gertsbakh I (2000) Reliability Theory with Applications to Preventive Maintenance. Springer, New York.
14. Nakagawa T (1979) A summary of block replacement policies. RAIRO Oper Res 13:351–361.
15. Nakagawa T (1982) A modified block replacement with two variables. IEEE Trans Reliab R-31:398–400.
16. Nakagawa T (1981) A summary of periodic replacement with minimal repair at failure. J Oper Res Soc Jpn 24:213–227.
17. Nakagawa T (1983) Combined replacement models. RAIRO Oper Res 17:193–203.
18. Savits TH (1988) A cost relationship between age and block replacement policies. J Appl Prob 25:789–796.
19. Cox DR (1962) Renewal Theory. Methuen, London.
20. Sandoh H, Nakagawa T (2003) How much should we reweigh? J Oper Res Soc 54:318–321.
21. Crookes PCI (1963) Replacement strategies. Oper Res Q 14:167–184.
22. Blanning RW (1965) Replacement strategies. Oper Res Q 16:253-254.
23. Bhat BR (1969) Used item replacement policy. J Appl Prob 6:309–318.
24. Tango T (1979) A modified block replacement policy using less reliable items. IEEE Trans Reliab R-28:400–401.
25. Murthy DNP, Nguyen DG (1982) A note on extended block replacement policy with used items. J Appl Prob 19:885–889.
26. Ait Kadi D, Cléroux R (1988) Optimal block replacement policies with multiple choice at failure. Nav Res Logist 35:99–110.
27. Berg M, Epstein B (1976) A modified block replacement policy. Nav Res Logist Q 23:15–24.
28. Berg M, Epstein B (1979) A note on a modified block replacement policy for units with increasing marginal running costs. Nav Res Logist Q 26:157–160.
29. Dekker R (1996) A framework for single-parameter maintenance activities and its use in optimisation, priority setting and combining. In: Özekici S (ed) Reliability and Maintenance of Complex Systems. Springer, New York:170–188.
30. Dekker R (1995) A general framework for optimisation priority setting, planning and combining of maintenance activities. Eur J Oper Res 82:225–240.
31. Aven T, Dekker R (1997) A useful framework for optimal replacement models. Reliab Eng Syst Saf 58:61–67.
32. Boland PJ (1982) Periodic replacement when minimal repair costs vary with time. Nav Res Logist Q 29:541–546.
33. Sivazlian BD, Mahoney JF (1978) Group replacement of a multicomponent system which is subject to deterioration only. Adv Appl Prob 10:867–885.
34. Okumoto K, Elsayed EA (1983) An optimum group maintenance policy. Nav Res Logist Q 30:667–674.
35. Assaf D, Shanthikumar G (1987) Optimal group maintenance policies with continuous and periodic inspections. Manage Sci 33:1440–1452.
36. Dekker R, Roelvink IFK (1995) Marginal cost criteria for preventive replacement of a group of components. Eur J Oper Res 84:467–480.
37. Van der Duyn Schouten FA (1996) Maintenance policies for multicomponent systems: An overview. In: Özekici S (ed) Reliability and Maintenance of Complex Systems. Springer, New York:117–136.

38. Van Dijkhuizen GC (2000) Maintenance grouping in multi-step multi-component production systems. In: Ben-Daya M, Duffuaa SO, Raouf A (eds) Maintenance, Modelling and Optimization. Kluwer Academic, Boston:283–306.
39. Tahara A, Nishida T (1975) Optimal replacement policy for minimal repair model. J Oper Res Soc Jpn 18:113–124.
40. Phelps RI (1981) Replacement policies under minimal repair. J Oper Res Soc 32:549–554.
41. Phelps RI (1983) Optimal policy for minimal repair. J Oper Res Soc 34:425–427.
42. Park KS, Yoo YK (1993) (τ, k) block replacement policy with idle count. IEEE Trans Reliab 42:561–565.
43. Bai DS, Yun WY (1986) An age replacement policy with minimal repair cost limit. IEEE Trans Reliab R-35:452–454.
44. Pham H, Wang H (2000) Optimal (τ, T) opportunistic maintenance of a k-out-of-n: G system with imperfect PM and partial failure. Nav Res Logist 47:223–239.
45. Muth E (1977) An optimal decision rule for repair vs. replacement. IEEE Trans Reliab 26:179–181.

6

Preventive Maintenance

An operating unit is repaired or replaced when it fails. If a failed unit undergoes repair, it needs a repair time which may not be negligible. After the repair completion, a unit begins to operate again. If a failed unit cannot be repaired and spare units are not on hand, it takes a replacement time that might not be negligible. A unit forms an alternating renewal process that repeats up and down states alternately in Section 1.3.2. Some reliability quantities such as availabilities, expected number of failures, and repair limit times have already been derived in Chapter 2.

When a unit is repaired after failure, *i.e.*, *corrective maintenance* is done, it may require much time and high cost. In particular, the downtime of such systems as computers, plants, and radar should be made as short as possible by decreasing the number of system failures. In this case, to maintain a unit to prevent failures, we need to do *preventive maintenance* (PM), but not to do it too often from the viewpoints of reliability and cost.

The optimum PM policy that maximizes the availability was first derived in [1]. Optimum PM policies for more general systems were discussed in [2–6]. The PM policies for series systems by modifying the opportunistic replacement [7] and for a system with spare units [8, 9] were studied. The PM model where the failure distribution is uncertain was considered in [10]. Furthermore, several maintenance models in Europe were presented and a good survey of applied PM models was given in [11]. The PM programs of plants and aircraft were given in [12–15]. Several imperfect PM policies where a unit may not be new at PM are discussed in Chapter 7.

In this chapter, we summarize appropriate PM policies that are suitable for some systems: In Section 6.1, we consider the PM of a one-unit system and obtain the reliability quantities such as renewal functions and transition probabilities [2]. Using these results, we derive optimum PM policies that maximize the availabilities, the expected earning rate, and the interval reliability [16]. In Section 6.2, we consider the PM of a two-unit standby system and analytically derive optimum policies that maximize the mean time to system failure and the availability [17–19]. In Section 6.3, we propose the modified PM policy which is planned at periodic times, and when the total number of failures has exceeded a specified number, the PM is done at the next planned time [20]. This is applied to the analysis of restarts for a computer system, the number of uses, the number of shocks for a cumulative damage model, and the number of unit failures for a parallel system.

6.1 One-Unit System with Repair

Consider a one-unit system. When a unit fails, it undergoes repair immediately, and once repaired, it is returned to the operating state. It is assumed that the failure distribution of a unit is a general distribution $F(t)$ with finite mean $\mu \equiv \int_0^\infty \overline{F}(t)dt$, where $\overline{F} \equiv 1 - F$, and the repair distribution $G_1(t)$ is also a general distribution with finite mean β_1.

We discuss a preventive maintenance policy for a one-unit system with repair. When a unit operates for a planned time T $(0 < T \leq \infty)$ without failure, we stop its operation for PM. The distribution of time to the PM completion is assumed to be a general distribution $G_2(t)$ with finite mean β_2, which may be different from the repair distribution $G_1(t)$.

It was pointed out [2] that the optimum PM policy maximizing the availability of the system is reduced to the standard age replacement problem as described in Section 3.1 if the mean time β_1 to repair is replaced with the replacement cost c_1 of a failed unit and the mean time β_2 to PM with the cost c_2 of exchanging a nonfailed unit.

6.1.1 Reliability Quantities

We derive renewal functions and transition probabilities of a one-unit system with repair and PM, using the same regeneration-point techniques found in Section 1.3.3 on Markov renewal processes. The expected number of system failures and the availability are easily given by these functions and probabilities, respectively.

To analyze the above system, we define the following system states.

State 0: Unit is operating.
State 1: Unit is under repair.
State 2: Unit is under PM.

These system states represent the continuous states of the system and the system makes a Markov renewal process (see Figure 6.1). We can obtain renewal functions and transition probabilities by using the same techniques as those in Section 1.3.

Let $M_{ij}(t)$ $(i, j = 0, 1, 2)$ be the expected number of visits to state j during $(0, t]$, starting from state i. For instance, $M_{02}(t)$ represents the expected number of exchanges of nonfailed units during $(0, t]$, given that a unit began to operate at time 0.

6.1 One-Unit System with Repair   137

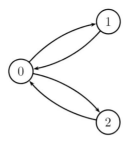

Fig. 6.1. Process of one-unit system

For convenience, we define $D(t)$ as the distribution of a degenerate random variable placing unit mass at T $(0 < T \leq \infty)$; i.e., $D(t) \equiv 0$ for $t < T$ and 1 for $t \geq T$. Then, by considering the transitions between the system states, we have the following renewal-type equations of $M_{ij}(t)$.

$$M_{00}(t) = \int_0^t \overline{D}(u)\,\mathrm{d}F(u) * M_{10}(t) + \int_0^t \overline{F}(u)\,\mathrm{d}D(u) * M_{20}(t)$$

$$M_{01}(t) = \int_0^t \overline{D}(u)\,\mathrm{d}F(u) * [1 + M_{11}(t)] + \int_0^t \overline{F}(u)\,\mathrm{d}D(u) * M_{21}(t)$$

$$M_{02}(t) = \int_0^t \overline{D}(u)\,\mathrm{d}F(u) * M_{12}(t) + \int_0^t \overline{F}(u)\,\mathrm{d}D(u) * [1 + M_{22}(t)]$$

$$M_{i0}(t) = G_i(t) * [1 + M_{00}(t)] \quad (i = 1, 2)$$

$$M_{1j}(t) = G_1(t) * M_{0j}(t), \qquad M_{2j}(t) = G_2(t) * M_{0j}(t) \quad (j = 1, 2),$$

where the asterisk represents the Stieltjes convolution; i.e., $a(t) * b(t) \equiv \int_0^t b(t-u)\,\mathrm{d}a(u)$ for any $a(t)$ and $b(t)$, and $\overline{\Psi} \equiv 1 - \Psi$ for any distribution Ψ.

Let $\Psi^*(s)$ be the Laplace–Stieltjes (LS) transform of any function $\Psi(t)$; i.e., $\Psi^*(s) \equiv \int_0^\infty e^{-st}\mathrm{d}\Psi(t)$ for $s > 0$. Forming the LS transforms of the above equations, we have

$$M_{00}^*(s) = \int_0^T e^{-st}\,\mathrm{d}F(t)M_{10}^*(s) + e^{-sT}\overline{F}(T)M_{20}^*(s)$$

$$M_{01}^*(s) = \int_0^T e^{-st}\,\mathrm{d}F(t)[1 + M_{11}^*(s)] + e^{-sT}\overline{F}(T)M_{21}^*(s)$$

$$M_{02}^*(s) = \int_0^T e^{-st}\,\mathrm{d}F(t)M_{12}^*(s) + e^{-sT}\overline{F}(T)[1 + M_{22}^*(s)]$$

$$M_{i0}^*(s) = G_i^*(s)[1 + M_{00}^*(s)] \quad (i = 1, 2)$$

$$M_{1j}^*(s) = G_1^*(s)M_{0j}^*(s), \qquad M_{2j}^*(s) = G_2^*(s)M_{0j}^*(s) \quad (j = 1, 2).$$

Thus, solving the equations for $M_{0j}^*(s)$ $(j = 0, 1, 2)$, we have

$$M_{00}^*(s) = \frac{G_1^*(s)\int_0^T e^{-st}\,\mathrm{d}F(t) + G_2^*(s)e^{-sT}\overline{F}(T)}{1 - G_1^*(s)\int_0^T e^{-st}\,\mathrm{d}F(t) - G_2^*(s)e^{-sT}\overline{F}(T)} \tag{6.1}$$

$$M_{01}^*(s) = \frac{\int_0^T e^{-st}\,\mathrm{d}F(t)}{1 - G_1^*(s)\int_0^T e^{-st}\,\mathrm{d}F(t) - G_2^*(s)e^{-sT}\overline{F}(T)} \tag{6.2}$$

$$M_{02}^*(s) = \frac{e^{-sT}\overline{F}(T)}{1 - G_1^*(s)\int_0^T e^{-st}\,\mathrm{d}F(t) - G_2^*(s)e^{-sT}\overline{F}(T)}. \tag{6.3}$$

Furthermore, from (1.63), the limiting values $M_j \equiv \lim_{t\to\infty} M_{0j}(t)/t = \lim_{s\to 0} sM_{0j}^*(s)$; i.e., the expected numbers of visits to state j per unit of time in the steady-state are

$$M_0 = \frac{1}{\int_0^T \overline{F}(t)\,\mathrm{d}t + \beta_1 F(T) + \beta_2 \overline{F}(T)} \tag{6.4}$$

$$M_1 = \frac{F(T)}{\int_0^T \overline{F}(t)\,\mathrm{d}t + \beta_1 F(T) + \beta_2 \overline{F}(T)} \tag{6.5}$$

$$M_2 = \frac{\overline{F}(T)}{\int_0^T \overline{F}(t)\,\mathrm{d}t + \beta_1 F(T) + \beta_2 \overline{F}(T)}. \tag{6.6}$$

Next, let $P_{ij}(t)$ $(i,j = 0,1,2)$ be the transition probability that the system is in state j at time t, starting from state i at time 0. Then, in a similar way, we have the following renewal equations of the transition probabilities.

$$P_{00}(t) = \overline{F}(t)\overline{D}(t) + \int_0^t \overline{D}(u)\,\mathrm{d}F(u) * P_{10}(t) + \int_0^t \overline{F}(u)\,\mathrm{d}D(u) * P_{20}(t)$$

$$P_{0j}(t) = \int_0^t \overline{D}(u)\,\mathrm{d}F(u) * P_{1j}(t) + \int_0^t \overline{F}(u)\,\mathrm{d}D(u) * P_{2j}(t) \quad (j = 1,2)$$

$$P_{i0}(t) = G_i(t) * P_{00}(t) \quad (i = 1,2)$$

$$P_{jj}(t) = \overline{G}_j(t) + G_j(t) * P_{0j}(t) \quad (j = 1,2)$$

$$P_{12}(t) = G_1(t) * P_{02}(t), \quad P_{21}(t) = G_2(t) * P_{01}(t).$$

Thus, forming the LS transforms and solving them for $P_{0j}^*(s)$ $(j = 0,1,2)$,

$$P_{00}^*(s) = \frac{1 - \int_0^T e^{-st}\,\mathrm{d}F(t) - e^{-sT}\overline{F}(T)}{1 - G_1^*(s)\int_0^T e^{-st}\,\mathrm{d}F(t) - G_2^*(s)e^{-sT}\overline{F}(T)} \tag{6.7}$$

$$P_{01}^*(s) = \frac{[1 - G_1^*(s)]\int_0^T e^{-st}\,\mathrm{d}F(t)}{1 - G_1^*(s)\int_0^T e^{-st}\,\mathrm{d}F(t) - G_2^*(s)e^{-sT}\overline{F}(T)} \tag{6.8}$$

$$P_{02}^*(s) = \frac{[1 - G_2^*(s)]e^{-sT}\overline{F}(T)}{1 - G_1^*(s)\int_0^T e^{-st}\,\mathrm{d}F(t) - G_2^*(s)e^{-sT}\overline{F}(T)}. \tag{6.9}$$

Furthermore, the limiting probabilities $P_j \equiv \lim_{t\to\infty} P_{ij}(t) = \lim_{s\to 0} P_{ij}^*(s)$ are

6.1 One-Unit System with Repair

$$P_0 = \frac{\int_0^T \overline{F}(t)\,dt}{\int_0^T \overline{F}(t)\,dt + \beta_1 F(T) + \beta_2 \overline{F}(T)} \tag{6.10}$$

$$P_1 = \frac{\beta_1 F(T)}{\int_0^T \overline{F}(t)\,dt + \beta_1 F(T) + \beta_2 \overline{F}(T)} \tag{6.11}$$

$$P_2 = \frac{\beta_2 \overline{F}(T)}{\int_0^T \overline{F}(t)\,dt + \beta_1 F(T) + \beta_2 \overline{F}(T)}, \tag{6.12}$$

where $P_0 + P_1 + P_2 = 1$. It is of great interest to have the relation that $P_j = \beta_j M_j$ ($j = 1, 2$).

Also, note that the probability $P_{00}(t)$ represents the pointwise availability of the system at time t, given that a unit began to operate at time 0, and $P_{01}(t) + P_{02}(t)$ is the pointwise unavailability at time t. It is also noted that the limiting probability P_0 represents the steady-state availability, and $P_1 + P_2$ is the steady-state unavailability.

6.1.2 Optimum Policies

(1) Availability

We derive an optimum PM time T^* maximizing the availability P_0 that is a function of T. From (6.10), P_0 is rewritten as

$$P_0 = 1 \Big/ \left[1 + \frac{\beta_1 F(T) + \beta_2 \overline{F}(T)}{\int_0^T \overline{F}(t)\,dt}\right]. \tag{6.13}$$

Thus, the policy maximizing P_0 is the same as minimizing the expected cost rate $C(T)$ in (3.4) by replacing β_i with c_i ($i = 1, 2$). We have the same theorems as those in Section 3.1 under the assumption that $\beta_1 > \beta_2$.

(2) Expected Earning Rate

Introduce the following earnings in specifying the PM policy. Let e_0 be a net earning per unit of time made by the production of an operating unit. Furthermore, let e_1 be an earning rate per unit of time while a unit is under repair and e_2 be an earning rate per unit of time while a unit is under PM. Both e_1 and e_2 are usually negative, and may be $e_0 > e_2 > e_1$. Then, from (6.10) to (6.12), the expected earning rate is

$$E(T) \equiv e_0 P_0 + e_1 P_1 + e_2 P_2 = \frac{e_0 \int_0^T \overline{F}(t)\,dt + e_1 \beta_1 F(T) + e_2 \beta_2 \overline{F}(T)}{\int_0^T \overline{F}(t)\,dt + \beta_1 F(T) + \beta_2 \overline{F}(T)}. \tag{6.14}$$

We can also obtain an optimum policy that maximizes $E(T)$ by a similar method. If $e_0 = 0$, i.e., we consider no earning of the operating unit, then $E(T)$ agrees with that of [22, p. 42].

(3) Emergency Event

Suppose that a unit is required for operation when an emergency event occurs. A typical example of such a model is standby generators in hospitals or buildings whenever the electric power stops. In any case, it is catastrophic and dangerous that the unit has failed when an emergency event occurs. We wish to lessen the probability of such an event by adopting the PM policy.

It is assumed that an emergency event occurs randomly in time; i.e., it occurs according to an exponential distribution $(1 - e^{-\alpha t})$ $(0 < \alpha < \infty)$ [23]. Then, the probability $1 - A(T)$ that the unit has failed when an emergency event occurs is

$$1 - A(T) = \int_0^\infty [P_{01}(t) + P_{02}(t)] \, d(1 - e^{-\alpha t})$$
$$= P_{01}^*(\alpha) + P_{02}^*(\alpha).$$

Thus, from (6.8) and (6.9), we have

$$A(T) = \frac{\int_0^T \alpha e^{-\alpha t} \overline{F}(t) \, dt}{1 - G_1^*(\alpha) \int_0^T e^{-\alpha t} \, dF(t) - G_2^*(\alpha) e^{-\alpha T} \overline{F}(T)}. \qquad (6.15)$$

We can derive an optimum policy that maximizes $A(T)$ by a similar method, under the assumption that $G_2^*(\alpha) > G_1^*(\alpha)$; i.e., the PM rate of a nonfailed unit is greater than the repair rate of a failed unit.

6.1.3 Interval Reliability

Interval reliability $R(x, T_0)$ is defined in Chapter 1 as the probability that at a specified time T_0, a unit is operating and will continue to operate for an interval of time x. In this section, we consider the case where T_0 is distributed exponentially. A typical model is a standby generator, in which T_0 is the time until the electric power stops and x is the required time until the electric power recovers. In this case, the interval reliability represents the probability that a standby generator will be able to operate while the electric power is interrupted.

Consider a one-unit system that is repaired upon failure and brought back to operation after the repair completion. The failure time has a general distribution $F(t)$ with finite mean μ and the repair time has a general distribution $G(t)$ with finite mean β. We set the PM time T $(0 < T \le \infty)$ for the operating unit. However, the PM of the operating unit is not done during the interval $[T_0, T_0 + x]$ even if it is time for PM. It is assumed that the distribution of time to the PM completion is the same as the repair distribution $G(t)$.

Similar to (2.28) in Section 2.1, we obtain the interval reliability $R(T; x, T_0)$:

$$R(T; x, T_0) = \overline{F}(T_0 + x)\overline{D}(T_0) + \int_0^{T_0} \overline{F}(T_0 + x - u)\overline{D}(T_0 - u) \, dM_{00}(u),$$

6.1 One-Unit System with Repair

where $M_{00}(t)$ represents the expected number of occurrences of the recovery of operating state during $(0, t]$, and its LS transform can be given by putting $G_1 = G_2 = G$ in (6.1). Thus, forming the Laplace transform of the above equation, we have

$$R^*(T; x, s) \equiv \int_0^\infty e^{-sT_0} R(T; x, T_0)\, dT_0$$

$$= \frac{e^{sx} \int_x^{T+x} e^{-st} \overline{F}(t)\, dt}{1 - G^*(s) + sG^*(s) \int_0^T e^{-st} \overline{F}(t)\, dt}. \tag{6.16}$$

Thus, the limiting interval reliability is

$$R(T; x) \equiv \lim_{T_0 \to \infty} R(T; x, T_0) = \lim_{s \to 0} s R^*(T; x, s)$$

$$= \frac{\int_x^{T+x} \overline{F}(t)\, dt}{\int_0^T \overline{F}(t)\, dt + \beta} \tag{6.17}$$

and the interval reliability when T_0 is a random variable with an exponential distribution $(1 - e^{-\alpha t})$ $(0 < \alpha < \infty)$ is

$$R(T; x, \alpha) \equiv \int_0^\infty R(T; x, T_0)\, d(1 - e^{-\alpha T_0}) = \alpha R^*(T; x, \alpha). \tag{6.18}$$

It is noted that $R(T; x)$ and $R(T; x, \alpha)/\alpha$ agree with (2.30) and (2.29), respectively, in the case of no PM; i.e., $T = \infty$.

First, we seek an optimum PM time that maximizes the interval reliability $R(T; x)$ in (6.17) for a fixed $x > 0$. Let $\lambda(t; x) \equiv [F(t + x) - F(t)]/\overline{F}(t)$ for $t \geq 0$. Then, both $\lambda(t; x)$ and $h(t) \equiv f(t)/\overline{F}(t)$ are called the *failure rate* and have the same properties as mentioned in Section 1.1. It is noted that $h(t)$ has already played an important role in analyzing the replacement models in Chapters 3 and 4. Let

$$K_1 \equiv \frac{\int_0^x \overline{F}(t)\, dt + \beta}{\mu + \beta} = 1 - R(\infty; x).$$

Then, we have the following optimum policy.

Theorem 6.1. Suppose that the failure rate $\lambda(t; x)$ is continuous and strictly increasing in t for $x > 0$.

(i) If $\lambda(\infty; x) > K_1$ then there exists a finite and unique T^* $(0 < T^* < \infty)$ that satisfies

$$\lambda(T; x) \left[\int_0^T \overline{F}(t)\, dt + \beta \right] - \int_0^T [\overline{F}(t) - \overline{F}(t + x)]\, dt = \beta \tag{6.19}$$

and the resulting interval reliability is

$$R(T^*; x) = 1 - \lambda(T^*; x). \tag{6.20}$$

(ii) If $\lambda(\infty; x) \le K_1$ then $T^* = \infty$; i.e., no PM is done.

Proof. Differentiating $R(T; x)$ in (6.17) with respect to T and putting it equal to zero, we have (6.19). Letting $Q_1(T)$ be the left-hand side of (6.19), it is easy to prove that $Q_1(T)$ is strictly increasing,

$$Q_1(0) \equiv \lim_{T \to 0} Q_1(T) = \beta F(x)$$

$$Q_1(\infty) \equiv \lim_{T \to \infty} Q_1(T) = \lambda(\infty; x)(\mu + \beta) - \int_0^x \overline{F}(t)\,dt.$$

If $\lambda(\infty; x) > K_1$ then $Q_1(\infty) > \beta > Q_1(0)$. Thus, from the monotonicity and the continuity of $Q_1(T)$, there exists a finite and unique T^* that satisfies (6.19) and maximizes $R(T; x)$. Furthermore, from (6.19), we clearly have (6.20).

If $\lambda(\infty; x) \le K_1$ then $Q_1(\infty) \le \beta$; i.e., $R(T; x)$ is strictly increasing. Thus, the optimum PM time is $T^* = \infty$. ∎

It is of interest that $1 - \lambda(T^*; x)$ in (6.20) represents the probability that a unit with age T^* does not fail in a finite interval $(T^*, T^* + x]$.

In the case (i) of Theorem 6.1, we can get the following upper limit of the optimum PM time T^*.

Theorem 6.2. *Suppose that the failure rate $\lambda(t; x)$ is continuous and strictly increasing, $\lambda(0; x) < K_1 < \lambda(\infty; x)$. Then, there exists a finite and unique $\overline{T}$ that satisfies $\lambda(T; x) = K_1$ and $T^* < \overline{T}$.*

Proof. From the assumption that $\lambda(t; x)$ is strictly increasing, we have

$$\lambda(T; x) < \frac{\int_T^\infty [\overline{F}(t) - \overline{F}(t+x)]\,dt}{\int_T^\infty \overline{F}(t)\,dt} \qquad \text{for } 0 \le T < \infty.$$

Thus, we have the inequality

$$Q_1(T) > \lambda(T; x)(\mu + \beta) - \int_0^x \overline{F}(t)\,dt \qquad \text{for } 0 \le T < \infty.$$

Therefore, if there exists $\overline{T}$ that satisfies

$$\lambda(T; x)(\mu + \beta) - \int_0^x \overline{F}(t)\,dt = \beta,$$

i.e., $\lambda(\overline{T}; x) = K_1$, then $T^* < \overline{T}$. It can be seen that $\overline{T}$ is finite and unique inasmuch as $\lambda(T; x)$ is strictly increasing, and $\lambda(0; x) < K_1 < \lambda(\infty; x)$. If $\lambda(0; x) \ge K_1$ then we may take $\overline{T} = \infty$. ∎

If the time for PM has a distribution $G_2(t)$ with mean β_2 and the time for repair has a distribution $G_1(t)$ with mean β_1, then the limiting interval reliability is given by

6.1 One-Unit System with Repair

$$R(T;x) = \frac{\int_x^{T+x} \overline{F}(t)\,dt}{\int_0^T \overline{F}(t)\,dt + \beta_1 F(T) + \beta_2 \overline{F}(T)}. \qquad (6.21)$$

Next, we consider the optimum PM policy that maximizes the interval reliability when T_0 is distributed exponentially. From (6.16) and (6.18),

$$R(T;x,\alpha) = \frac{\alpha e^{\alpha x} \int_x^{T+x} e^{-\alpha t} \overline{F}(t)\,dt}{\alpha G^*(\alpha) \int_0^T e^{-\alpha t} \overline{F}(t)\,dt + 1 - G^*(\alpha)}. \qquad (6.22)$$

Let

$$K_2(\alpha) \equiv \frac{1 - F^*(\alpha)G^*(\alpha) - \alpha G^*(\alpha) e^{\alpha x} \int_x^{\infty} e^{-\alpha t} \overline{F}(t)\,dt}{1 - F^*(\alpha)G^*(\alpha)} = 1 - G^*(\alpha)R(\infty;x,\alpha).$$

Then, in similar ways to those of obtaining Theorems 6.1 and 6.2, we can get the following theorems without proof.

Theorem 6.3. Suppose that the failure rate $\lambda(t;x)$ is continuous and strictly increasing.

(i) If $\lambda(\infty;x) > K_2(\alpha)$ then there exists a finite and unique T^* ($0 < T^* < \infty$) that satisfies

$$\lambda(T;x)\left[1 - G^*(\alpha) + \alpha G^*(\alpha)\int_0^T e^{-\alpha t}\overline{F}(t)\,dt\right]$$

$$- \alpha G^*(\alpha)\int_0^T e^{-\alpha t}[\overline{F}(t) - \overline{F}(t+x)]\,dt = 1 - G^*(\alpha) \qquad (6.23)$$

and the resulting interval reliability is

$$R(T^*;x,\alpha) = \frac{1}{G^*(\alpha)}[1 - \lambda(T^*;x)]. \qquad (6.24)$$

(ii) If $\lambda(\infty;x) \le K_2(\alpha)$ then $T^* = \infty$.

Theorem 6.4. Suppose that the failure rate $\lambda(t;x)$ is continuous and strictly increasing, $\lambda(0;x) < K_2(\alpha) < \lambda(\infty;x)$. Then, there exists a finite and unique $\overline{T}$ that satisfies $\lambda(T;x) = K_2(\alpha)$, and $T^* < \overline{T}$.

Example 6.1. We compute the PM time T^* that maximizes the limiting interval reliability $R(T;x)$ in (6.17) numerically when $F(t) = 1 - (1+\lambda t)e^{-\lambda t}$; that is,

$$\lambda(t;x) = 1 - \left(1 + \frac{\lambda x}{1+\lambda t}\right)e^{-\lambda x}$$

$$K_1 = \frac{(2/\lambda)(1 - e^{-\lambda x}) - xe^{-\lambda x} + \beta}{2/\lambda + \beta}.$$

Table 6.1. Dependence of interval of time x in the optimum time T^*, the upper bound $\overline{T}$, the interval reliabilities $R(T^*; x)$ and $R(\infty; x)$ when $\beta = 1$ and $2/\lambda = 10$

x	T^*	$\overline{T}$	$R(T^*; x)$	$R(\infty; x)$
1	∞	∞	0.819	0.819
2	16.64	17.00	0.732	0.731
3	10.60	11.50	0.654	0.649
4	8.43	9.66	0.583	0.572
5	7.30	8.75	0.517	0.502
6	6.60	8.20	0.457	0.438
7	6.12	7.83	0.402	0.381
8	5.77	7.57	0.352	0.330
9	5.50	7.38	0.307	0.286
10	5.30	7.22	0.267	0.246
11	5.13	7.10	0.231	0.211

The failure rate $\lambda(t; x)$ is strictly increasing with $\lambda(0; x) \equiv F(x)$ and $\lambda(\infty; x) = 1 - e^{-\lambda x}$. From (i) of Theorem 6.1, if $x \leq \beta$ then we should do no PM of the operating unit. Otherwise, the optimum time T^* is a unique solution of the equation

$$\lambda T(x - \beta) - x(1 - e^{-\lambda T}) = (1 + \lambda x)\beta$$

and

$$R(T^*; x) = \left(1 + \frac{\lambda x}{1 + \lambda T^*}\right) e^{-\lambda x}.$$

From Theorem 6.2, we have the upper limit $\overline{T}$ of T^*:

$$\lambda T^* < \frac{x + (\lambda x + 1)\beta}{x - \beta}.$$

Table 6.1 presents T^*, $R(T^*; x)$, $R(\infty; x)$, and $\overline{T}$ for $x = 1, 2, \ldots, 11$ when $\mu = 2/\lambda = 10$ and $\beta = 1$. ∎

6.2 Two-Unit System with Repair

We discuss a system that consists of one operating unit with one standby unit, where the repair and PM are considered. Such a system is called a *two-unit standby redundant system with repair and PM*. Two PM policies, strict PM and slide PM were considered and the method of approximating the optimum PM policy that maximizes the mean time to system failure (MTTF) was obtained in [24]. It was shown in [25] that the MTTF of the same system with PM is greater than that of a similar system with only repair maintenance.

The replacement of a two-unit system was considered in [26–28], where at the failure points of one unit, the other unit is replaced if its age exceeds a control limit. The maintenance strategy for a two-unit standby system with waiting time for repair was obtained in [29]. The opportunistic replacement policies for two-unit systems [30, 31] and imperfectly monitored two-unit parallel systems [32] were examined. The early results of two-unit systems were extensively surveyed in [33].

We adopt a slide PM policy which is explained below, and derive optimum PM policies that maximize the MTTF and the availability. It is noted that in this section, the availability refers to the steady-state availability.

6.2.1 Reliability Quantities

Consider a standby redundant system that consists of two identical units. The failure time of an operating unit has a general distribution $F(t)$ with finite mean μ, and its repair time also has a general distribution $G_1(t)$ with finite mean β_1. When an operating unit fails and the other unit is in standby state, a failed unit undergoes repair immediately and a standby unit takes over its operation. However, when an operating unit fails while the other unit is still under repair, a new failed unit has to wait until a failed unit under repair is completed by a repairperson. This situation means system failure.

It is assumed that a failed unit recovers its functioning as good as new upon the repair completion, and that a standby unit neither deteriorates nor fails during the standby interval. Furthermore, each switchover is perfect and instantaneous, where each switchover is made when an operating unit fails, a failed unit undergoes repair, and the other standby unit takes over its operation. Two units are alternately used for its operation as described above. Even if system failure occurs, the system can operate again upon the repair completion. Thus, the system assumes up and down states repeatedly.

Next, we adopt the same PM policy as in Section 6.1. When a unit operates for a planned time T $(0 \leq T \leq \infty)$ without failure, we stop the operation of a working unit for PM. It is assumed that the time to the PM completion has a general distribution $G_2(t)$ with finite mean β_2, and after PM, a unit recovers its operation as good as new.

Also, the following two assumptions are made.

(1) The PM of an operating unit is done only if the other unit is on standby.
(2) An operating unit, which forfeited the PM due to assumption (1), undergoes PM just upon the repair completion of a failed unit or the PM completion.

Assumption (1) can be thought of as avoiding the situation when the PM of an operating unit causes system failure. Another reason is that the PM of an operating unit can be done only when a repairperson is free. However, if we make only assumption (1), an operating unit, which forfeited the PM because of assumption (1), undergoes no PM forever. This contradicts the concept of

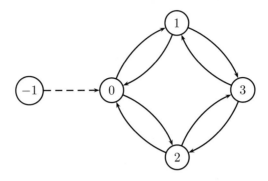

Fig. 6.2. State-transition diagram of a two-unit standby system with PM

the PM policy. Thus, we make assumption (2) which seems to be reasonable in practice.

To analyze the above system, we define the following system states.

State -1: One unit begins to operate and the other unit is on standby.
State 0: One unit is operating and the other unit is on standby.
State 1: One unit is operating and the other unit is under repair.
State 2: One unit is operating and the other unit is under PM.
State 3: One unit is under repair or PM and the other unit waits for repair.

We consider the epochs or time instants at which the system makes a transition into the states. State -1 represents an initial state and transits to State 0 in the time zero. The epochs for States 1 and 2 are regeneration points. However, the epoch for State 0 is not a regeneration point except for the transition from State -1, and represents that the repair or PM of one unit is completed while the other unit is operating. The epoch for State 3 is also not a regeneration point and represents that an operating unit fails while the other unit is under repair or PM, and means that system failure occurs.

The system states defined above are regarded as those of Markov renewal processes with finite-state space as described in Section 1.3 (see Figure 6.2). Let us define the mass function $Q_{ij}(t)$ from state i ($i = -1, 1, 2$) to state j ($j = 0, 1, 2, 3$) as the probability that after making a transition into state i, the system next makes a transition into j in an amount of time less than or equal to time t. Then, from the method similar to that of Section 1.3.4, we have the following mass functions.

$$Q_{-11}(t) = \int_0^t \overline{D}(u)\,dF(u) \qquad (6.25)$$

$$Q_{-12}(t) = \int_0^t \overline{F}(u)\,dD(u) \qquad (6.26)$$

6.2 Two-Unit System with Repair

$$Q_{i0}(t) = \int_0^t \overline{F}(u)\,dG_i(u), \quad Q_{i3}(t) = \int_0^t \overline{G}_i(u)\,dF(u) \quad (i=1,2), \quad (6.27)$$

where $D(t)$ is the distribution of a degenerate random variable placing unit mass at T, and $\overline{\Phi} \equiv 1 - \Phi$ for any function.

Noting that epochs for States 0 and 3 are not regeneration points, we define new mass functions $Q_{ij}^{(k)}(t)$ from state i to state j ($i,j = 1,2$) via state k ($k = 0,3$). That is, $Q_{ij}^{(k)}(t)$ represents the probability that after making a transition into state i, the system next makes a transition into state k and finally into state j, in an amount of time less than or equal to time t.

The following cases in State 1 are considered.

(1) After the repair completion of a failed unit, an operating unit fails.
(2) An operating unit is stopped operating for PM.
(3) After the failure of an operating unit, the repair of the other failed unit is completed.

In the case (1), the system comes back to state 1 via state 0:

$$Q_{11}^{(0)}(t) = \int_0^t \overline{D}(u) G_1(u)\,dF(u). \quad (6.28)$$

In the case (2), two exclusive cases are further considered. (a) After the repair completion of a failed unit, the PM of an operating unit begins or (b) the PM of an operating unit comes before the repair completion of a failed unit. In this case, the PM is not done from assumption (1). After that, when the repair of a failed unit is completed and an operating unit has not yet failed, the PM of an operating unit is done from assumption (2). In either case, the system goes to State 2 via 0:

$$Q_{12}^{(0)}(t) = \int_0^t \overline{F}(u) G_1(u)\,dD(u) + \int_0^t \overline{F}(u) D(u)\,dG_1(u). \quad (6.29)$$

In the case (3), the system comes back to State 1 via State 3, irrespective of the PM:

$$Q_{11}^{(3)}(t) = \int_0^t F(u)\,dG_1(u). \quad (6.30)$$

In a similar fashion, substituting $G_1(t)$ into $G_2(t)$ in (6.28), (6.29), and (6.30), respectively, we have

$$Q_{21}^{(0)}(t) = \int_0^t \overline{D}(u) G_2(u)\,dF(u) \quad (6.31)$$

$$Q_{22}^{(0)}(t) = \int_0^t \overline{F}(u) G_2(u)\,dD(u) + \int_0^t \overline{F}(u) D(u)\,dG_2(u) \quad (6.32)$$

$$Q_{21}^{(3)}(t) = \int_0^t F(u)\,dG_2(u). \quad (6.33)$$

Note that we do not need to consider $Q_{12}^{(3)}(t)$ from Assumption (1).

(1) First-Passage Time Distribution

We derive the first-passage time distribution to system failure and its MTTF, using the above mass functions. Let $F_{ij}(t)$ ($i = -1, 1, 2; j = 1, 2, 3$) denote the first-passage time distribution from the epoch for state i to the epoch for state j.

First, consider the first-passage time distribution $F_{ij}(t)$ ($i, j = 1, 2$). Recall that the epochs for States 1 and 2 are the regeneration points, but the epochs for States 0 and 3 are not. Then, from the renewal-theoretic arguments,

$$F_{i1}(t) = Q_{i1}^{(0)}(t) + Q_{i1}^{(3)}(t) + Q_{i2}^{(0)}(t) * F_{21}(t) \qquad (i = 1, 2)$$
$$F_{i2}(t) = Q_{i2}^{(0)}(t) + [Q_{i1}^{(0)}(t) + Q_{i1}^{(3)}(t)] * F_{12}(t) \qquad (i = 1, 2),$$

where the asterisk denotes the Stieltjes convolution.

Next, consider the first-passage time distribution $F_{i3}(t)$ ($i = 1, 2$). Then, we have the following renewal-type equation.

$$F_{i3}(t) = Q_{i3}(t) + Q_{i1}^{(0)}(t) * F_{13}(t) + Q_{i2}^{(0)}(t) * F_{23}(t) \qquad (i = 1, 2).$$

If the system starts from State -1 then

$$F_{-13}(t) = Q_{-11}(t) * F_{13}(t) + Q_{-12}(t) * F_{23}(t).$$

Taking the LS transform of the above equations and solving them for $F_{-13}^*(s)$, we have

$$F_{-13}^*(s) = \frac{\begin{array}{c} Q_{-11}^*(s)\{Q_{13}^*(s)[1 - Q_{22}^{*(0)}(s)] + Q_{12}^{*(0)}(s)Q_{23}^*(s)\} \\ + Q_{-12}^*(s)\{Q_{23}^*(s)[1 - Q_{11}^{*(0)}(s)] + Q_{21}^{*(0)}Q_{13}^*(s)\} \end{array}}{[1 - Q_{22}^{*(0)}(s)][1 - Q_{11}^{*(0)}(s)] - Q_{12}^{*(0)}(s)Q_{21}^{*(0)}(s)} \qquad (6.34)$$

which is the LS transform of the first-passage time distribution to system failure, starting from State -1.

Therefore, the mean time to system failure is

$$l_{-13}(T) \equiv \int_0^\infty t\, dF_{-13}(t) = -\lim_{s \to 0} \frac{dF_{-13}^*(s)}{ds} = \lim_{s \to 0} \frac{1 - F_{-13}^*(s)}{s}$$

$$= \frac{\begin{array}{c} \{\gamma_1 + \int_0^T [q_1 + G_1(t)]\overline{F}(t)\, dt\} \int_0^T [q_2 + G_2(t)]\, dF(t) \\ + \{\gamma_2 + \int_0^T [q_2 + G_2(t)]\overline{F}(t)\, dt\}\{1 - \int_0^T [q_1 + G_1(t)]\, dF(t)\} \end{array}}{q_1 \int_0^T [q_2 + G_2(t)]\, dF(t) + q_2\{1 - \int_0^T [q_1 + G_1(t)]\, dF(t)\}} \qquad (6.35)$$

or

$$l_{-13}(T) = \frac{\begin{array}{c} \{\mu(1 + q_1) - \int_T^\infty [q_1 + G_1(t)]\overline{F}(t)\, dt\}\{1 - \int_T^\infty [q_2 + G_2(t)]\, dF(t)\} \\ + \{\mu(1 + q_2) - \int_T^\infty [q_2 + G_2(t)]\overline{F}(t)\, dt\} \int_T^\infty [q_1 + G_1(t)]\, dF(t) \end{array}}{q_1\{1 - \int_T^\infty [q_2 + G_2(t)]\, dF(t)\} + q_2 \int_T^\infty [q_1 + G_1(t)]\, dF(t)}, \qquad (6.36)$$

where

$$q_i \equiv \int_0^\infty \overline{G}_i(t)\,\mathrm{d}F(t), \qquad \gamma_i \equiv \int_0^\infty \overline{G}_i(t)\overline{F}(t)\,\mathrm{d}t \qquad (i=1,2).$$

Consider some special cases of PM policies. In the first case, an operating unit undergoes the PM immediately upon the repair or PM completion. In this case, setting $T=0$ in (6.35), we have

$$l_{-13}(0) = \frac{1}{q_2}\int_0^\infty \overline{G}_2(t)\overline{F}(t)\,\mathrm{d}t. \tag{6.37}$$

In the second case where no PM is done, setting $T=\infty$ in (6.36),

$$l_{-13}(\infty) = \mu\left(1 + \frac{1}{q_1}\right). \tag{6.38}$$

(2) Transition Probabilities

We derive transition probability from state i to state j by the method similar to that of obtaining $F_{ij}(t)$. Let $P_{ij}(t)$ denote the transition probability that the system is in state j $(j=0,1,2,3)$ at time t, starting from the epoch for state i $(i=-1,1,2)$ at time 0. From the renewal-theoretic arguments,

$$P_{i0}(t) = Q_{i0}(t) - Q_{i1}^{(0)}(t) - Q_{i2}^{(0)}(t) + F_{i1}(t)*P_{10}(t) + F_{i2}(t)*P_{20}(t)$$
$$P_{ii}(t) = 1 - Q_{i0}(t) - Q_{i3}(t) + F_{ii}(t)*P_{ii}(t)$$
$$P_{i3}(t) = Q_{i3} - Q_{11}^{(3)}(t) + F_{i1}(t)*P_{13}(t) + F_{i2}(t)*P_{23}(t) \qquad (i=1,2)$$
$$P_{12}(t) = F_{12}(t)*P_{22}(t), \qquad P_{21}(t) = F_{21}(t)*P_{11}(t).$$

If the system starts from State -1 then

$$P_{-10}(t) = 1 - Q_{-11}(t) - Q_{-12}(t) + Q_{-11}(t)*P_{10}(t) + Q_{-12}(t)*P_{20}(t)$$
$$P_{-1j}(t) = Q_{-11}(t)*P_{1j}(t) + Q_{-12}(t)*P_{2j}(t) \qquad (j=1,2,3).$$

We can explicitly obtain the LS transforms $P_{ij}^*(s)$ of transition probabilities $P_{ij}(t)$.

In particular, the limiting probabilities $P_j \equiv \lim_{t\to\infty} P_{ij}(t) = \lim_{s\to 0} P_{ij}^*(s)$ are:

$$P_0 = \frac{1}{\gamma_{11}}\int_0^T G_1(t)\overline{F}(t)\,\mathrm{d}t + \frac{1}{\gamma_{22}}\int_0^T G_2(t)\overline{F}(t)\,\mathrm{d}t \tag{6.39}$$

$$P_j = \frac{1}{\gamma_{jj}}\int_0^\infty \overline{G}_j(t)\overline{F}(t)\,\mathrm{d}t \qquad (j=1,2) \tag{6.40}$$

$$P_3 = \frac{1}{\gamma_{11}}\int_0^\infty \overline{G}_1(t)F(t)\,\mathrm{d}t + \frac{1}{\gamma_{22}}\int_0^\infty \overline{G}_2(t)F(t)\,\mathrm{d}t, \tag{6.41}$$

where

$$\gamma_{11} \equiv \beta_1 + \int_0^T G_1(t)\overline{F}(t)\,dt + \left[\beta_2 + \int_0^T G_2(t)\overline{F}(t)\,dt\right]\frac{\int_T^\infty G_1(t)\,dF(t)}{1 - \int_T^\infty G_2(t)\,dF(t)}$$

$$\gamma_{22} \equiv \beta_2 + \int_0^T G_2(t)\overline{F}(t)\,dt + \left[\beta_1 + \int_0^T G_1(t)\overline{F}(t)\,dt\right]\frac{1 - \int_T^\infty G_2(t)\,dF(t)}{\int_T^\infty G_1(t)\,dF(t)}.$$

It is evident that $P_0 + P_1 + P_2 + P_3 = 1$, $P_0 + P_1 + P_2$ represents the steady-state availability and P_3 represents the steady-state unavailability.

6.2.2 Optimum Policies

It is interesting to derive optimum PM times T^* that maximize the mean time $l(T) \equiv l_{-13}(T)$ in (6.35) and the availability $A(T) \equiv P_0 + P_1 + P_2$ in (6.39) and (6.40). It is assumed that $G_1(t) < G_2(t)$ for $0 < t < \infty$; i.e., the probability that the repair is completed up to time t is less than the probability that the PM is completed up to time t. Let $h(t) \equiv f(t)/\overline{F}(t)$ be the failure rate with $h(\infty) \equiv \lim_{t\to\infty} h(t)$.

First, we give the optimum PM policy that maximizes $l(T)$ in (6.35). Let

$$L_1(t) \equiv \frac{q_1 G_2(t) - q_2 G_1(t)}{q_1 - q_2}$$

$$k_1 \equiv \frac{q_2}{q_1\gamma_2 - q_2\gamma_1}, \quad K_1 \equiv \frac{q_1}{\mu(q_1 - q_2)}.$$

Theorem 6.5. Suppose that $G_1(t) < G_2(t)$ for $0 < t < \infty$, and the failure rate $h(t)$ is continuous and strictly increasing.

(i) If $h(\infty) > K_1$, $q_1\gamma_2 > q_2\gamma_1$, and $h(0) < k_1$, or $h(\infty) > K_1$ and $q_1\gamma_2 \le q_2\gamma_1$, then there exists a finite and unique T^* $(0 < T^* < \infty)$ that satisfies

$$h(T)\left[\int_0^T L_1(t)\overline{F}(t)\,dt + \int_0^\infty \overline{L}_1(t)\overline{F}(t)\,dt\right] - \int_0^T L_1(t)\,dF(t)$$
$$= \frac{q_1 \int_0^\infty \overline{L}_1(t)\,dF(t) + q_2 \int_0^\infty L_1(t)\,dF(t)}{q_1 - q_2}. \qquad (6.42)$$

(ii) If $h(\infty) < K_1$ then $T^* = \infty$; i.e., no PM is done, and the mean time is given in (6.38).
(iii) If $q_1\gamma_2 > q_2\gamma_1$ and $h(0) > k_1$ then $T^* = 0$; i.e., the PM is done just upon the repair or PM completion.

Proof. First note that $q_1 > q_2$, $\gamma_1 > \gamma_2$, and $L_1(t) > 0$ from the assumption $G_1(t) < G_2(t)$ for $0 < t < \infty$. Further note that

6.2 Two-Unit System with Repair

$$\int_0^\infty \overline{L}_1(t)\overline{F}(t)\,dt = \frac{q_1\gamma_2 - q_2\gamma_1}{q_1 - q_2}.$$

Differentiating $l(T)$ in (6.35) with respect to T and putting it equal to zero imply (6.42). Letting the left-hand side of (6.42) be denoted by $Q_1(T)$, we have

$$\frac{dQ_1(T)}{dT} = \frac{dh(T)}{dT}\left[\int_0^T L_1(t)\overline{F}(t)\,dt + \int_0^\infty \overline{L}_1(t)\overline{F}(t)\,dt\right]$$

$$Q_1(0) \equiv \lim_{T \to 0} Q_1(T) = h(0)\int_0^\infty \overline{L}_1(t)\overline{F}(t)\,dt$$

$$Q_1(\infty) \equiv \lim_{T \to \infty} Q_1(T) = \mu h(\infty) - \int_0^\infty L_1(t)\,dF(t).$$

Thus, if $q_1\gamma_2 > q_2\gamma_1$ then $Q_1(T)$ is continuous and positive for $T > 0$, and is strictly increasing. Furthermore, let

$$K_0 \equiv \frac{q_1\int_0^\infty \overline{L}_1(t)\,dF(t) + q_2\int_0^\infty L_1(t)\,dF(t)}{q_1 - q_2} > 0.$$

If $h(0) < k_1$ and $h(\infty) > K_1$ then $Q_1(0) < K_0 < Q_1(\infty)$. Therefore, there exists a finite and unique T^* $(0 < T^* < \infty)$ that satisfies (6.42), and it maximizes $l(T)$. If $h(0) \geq k_1$ then $Q_1(0) \geq 0$. Thus, $l(T)$ is strictly decreasing in T, and hence, $T^* = 0$. If $h(\infty) \leq K_1$ then $Q_1(\infty) \leq 0$. Thus, $l(T)$ is strictly increasing in T, and hence, $T^* = \infty$; i.e., no PM is done.

On the other hand, if $q_1\gamma_2 < q_2\gamma_1$ then $Q_1(0) < 0$, $Q_1'(0) \leq 0$, and $Q_1'(\infty) > 0$. Furthermore, it is easy to see that there exists a unique solution T_1 to $dQ_1(T)/dT = 0$ for $0 < T < \infty$, Thus, $Q_1(T)$ is a unimodal function, and hence, $Q_1(T)$ is strictly increasing during the interval $[T_1, \infty)$. If $q_1\gamma_2 = q_2\gamma_1$ then $Q_1(0) = 0$ and $Q_1(T)$ is strictly increasing. In both cases, i.e., $q_1\gamma_2 \leq q_2\gamma_1$, if $h(\infty) > K_1$ then there exists a finite and unique T^* $(0 < T^* < \infty)$ that satisfies (6.42). Conversely, if $h(\infty) \leq K_1$ then $Q_1(T) \leq K_0$ for any finite T. Thus, the optimum PM time is $T^* = \infty$. ∎

By a similar method to that of Theorem 6.4, if there exists $\overline{T}$ that satisfies $h(T) = K_1$ then $T^* < \overline{T}$.

Next, we derive the optimum PM policy that maximizes the availability. From (6.39) and (6.40), the availability is given by

$$A(T) \equiv P_0 + P_1 + P_2$$

$$= \frac{[\gamma_1 + \int_0^T G_1(t)\overline{F}(t)\,dt][1 - \int_T^\infty G_2(t)\,dF(t)]}{[\beta_1 + \int_0^T G_1(t)\overline{F}(t)\,dt][1 - \int_T^\infty G_2(t)\,dF(t)]} \\ + [\gamma_2 + \int_0^T G_2(t)\overline{F}(t)\,dt]\int_T^\infty G_1(t)\,dF(t)}{+ [\beta_2 + \int_0^T G_2(t)\overline{F}(t)\,dt]\int_T^\infty G_1(t)\,dF(t)}. \quad (6.43)$$

6 Preventive Maintenance

When an operating unit undergoes PM immediately upon the repair or PM completion, the availability is

$$A(0) = \frac{q_2 \gamma_1 + (1-q_1)\gamma_2}{q_2 \beta_1 + (1-q_1)\beta_2}. \tag{6.44}$$

When no PM is done, the availability is

$$A(\infty) = \frac{\mu}{\mu + \beta_1 - \gamma_1}. \tag{6.45}$$

Let

$$\rho_i \equiv \int_0^\infty \overline{G}_i(t) F(t)\,dt = \beta_i - \gamma_i \quad (i=1,2)$$

$$L_2(t) \equiv \frac{\rho_1 G_2(t) - \rho_2 G_1(t)}{\rho_1 - \rho_2}$$

$$k_2 \equiv \frac{\rho_1 q_2 + \rho_2(1-q_1)}{\rho_1 \beta_2 - \rho_2 \beta_1}, \qquad K_2 \equiv \frac{\rho_1}{\mu(\rho_1 - \rho_2)}.$$

Theorem 6.6. Suppose that $G_1(t) < G_2(t)$ for $0 < t < \infty$, and the failure rate $h(t)$ is continuous and strictly increasing.

(i) If $h(\infty) > K_2$, $\rho_1 \beta_2 > \rho_2 \beta_1$, and $h(0) < k_2$, or $h(\infty) > K_2$ and $\rho_1 \beta_2 \le \rho_2 \beta_1$, then there exists a finite and unique T^* $(0 < T^* < \infty)$ that satisfies

$$h(T)\left[\int_0^T L_2(t)\overline{F}(t)\,dt + \int_0^\infty \overline{L}_2(t)\,dt\right] - \int_0^T L_2(t)\,dF(t)$$

$$= \frac{\rho_1 \int_0^\infty \overline{L}_2(t)\,dF(t) + \rho_2 \int_0^\infty L_2(t)\,dF(t)}{\rho_1 - \rho_2}. \tag{6.46}$$

(ii) If $h(\infty) \le K_2$ then $T^* = \infty$; i.e., no PM is done, and the availability is given in (6.45).
(iii) If $\rho_1 \beta_2 > \rho_2 \beta_1$ and $h(0) \ge k_2$ then $T^* = 0$, and the availability is given in (6.44).

Proof. Differentiating $A(T)$ in (6.43) with respect to T and putting it equal to zero imply (6.46). In a similar method to that of proving Theorem 6.5, we can prove this theorem. ∎

By a similar method to that of Theorem 6.4, if there exists $\overline{T}$ that satisfies $h(T) = K_2$ then $T^* < \overline{T}$.

It has been shown that the problem of maximizing the availability is formally coincident with that of minimizing the expected cost [19].

Example 6.2. We give two numerical problems: to maximize the mean time to system failure, and to maximize the availability. For the two problems, we assume that $G_i(t) = 1 - \exp(-\theta_i t)$ $(\theta_2 > \theta_1)$ and $F(t) = 1 - (1 + \lambda t)e^{-\lambda t}$. It is noted that the failure distribution is a gamma distribution with a shape parameter 2, and the failure rate $h(t) = \lambda^2 t/(1+\lambda t)$ which is strictly increasing from 0 to λ.

Consider the first problem of maximizing the mean time to system failure. Then, we have

$$q_i = \left(\frac{\lambda}{\lambda + \theta_i}\right)^2 \quad (i = 1, 2)$$

$$\overline{L}_1(t) = \frac{(\lambda + \theta_2)^2 e^{-\theta_2 t} - (\lambda + \theta_1)^2 e^{-\theta_1 t}}{(\lambda + \theta_2)^2 - (\lambda + \theta_1)^2}.$$

From Theorem 6.5, if $\lambda \leq K_1$; i.e., $(\lambda + \theta_2)^2 \leq 2(\lambda + \theta_1)^2$, we should do no PM of the operating unit. If $(\lambda + \theta_2)^2 > 2(\lambda + \theta_1)^2$, we should adopt a finite and unique PM time T^* that satisfies

$$\frac{1}{1 + \lambda T}\{(2\lambda T + e^{-\lambda T})[(\lambda + \theta_2)^2 - (\lambda + \theta_1)^2]$$
$$- \lambda^2(e^{-(\lambda + \theta_2)T} - e^{-(\lambda + \theta_1)T})\} = (\lambda + \theta_2)^2.$$

The above equation is derived from (6.42). In this case, the mean time to system failure is

$$l(T^*) = \frac{1}{\lambda^4 T^*}\{(2\lambda T^* + e^{-\lambda T^*})[(\lambda + \theta_1)^2 + \lambda^2] - \lambda^2 e^{-(\lambda + \theta_1)T^*}\}.$$

Furthermore, from the inequality $T^* < \overline{T}$, we have

$$\lambda T^* < \frac{(\lambda + \theta_2)^2}{(\lambda + \theta_2)^2 - 2(\lambda + \theta_1)^2}$$

which is useful in computing T^*.

Table 6.2 shows the numerical examples of the dependence of the mean PM time $1/\theta_2$ in the optimum PM time T^* and the other quantities when $\lambda = 2$ and $\theta_1 = 1$. For instance, if $\theta_2 = 8$ then $T^* = 0.292$, and its associated mean time is $l(T^*) = 4.447$. If we do no PM, i.e., $T = \infty$, we have $l(\infty) = 3.250$. Thus, we have $[l(T^*) - l(\infty)]/l(\infty) \times 100\% = 36.8\%$ gain in time by adopting the optimum PM policy.

Let us now consider the second problem of maximizing the availability. Then, we have

$$p_i = \frac{\lambda^2}{\theta_i(\lambda + \theta_i)^2} \quad (i = 1, 2)$$

$$\overline{L}_2(t) = \frac{\theta_2(\lambda + \theta_2)^2 e^{-\theta_2 t} - \theta_1(\lambda + \theta_1)^2 e^{-\theta_1 t}}{\theta_2(\lambda + \theta_2)^2 - \theta_1(\lambda + \theta_1)^2}.$$

Table 6.2. Dependence of the mean PM time $1/\theta_2$ in the optimum time T^* when $2/\lambda = 1$ hr, $1/\theta_1 = 1$ hr, and $l(\infty) = 3.250$ hr

θ_2	T^*	$l(T^*)$	$\frac{l(T^*)-l(\infty)}{l(\infty)} \times 100\%$
3.0	1.750	3.263	0.4
4.0	0.857	3.398	4.6
5.0	0.557	3.618	11.3
6.0	0.435	3.876	19.2
7.0	0.349	4.156	27.9
8.0	0.292	4.447	36.8
9.0	0.251	4.745	46.0
10.0	0.219	5.047	55.3

From Theorem 6.6, if $\lambda \leq K_2$, i.e., $\theta_2(\lambda + \theta_2)^2 \leq 2\theta_1(\lambda + \theta_1)^2$, we should do no PM of the operating unit. If $\theta_2(\lambda + \theta_2)^2 > 2\theta_1(\lambda + \theta_1)^2$, we should adopt PM with finite and unique time T^* that satisfies

$$\frac{1}{1+\lambda T}\{(2\lambda T + e^{-\lambda T})[\theta_2(\lambda + \theta_2)^2 - \theta_1(\lambda + \theta_1)^2] \\ - \lambda^2(\theta_2 e^{-(\lambda+\theta_2)T} - \theta_1 e^{-(\lambda+\theta_1)T})\} = \theta_2(\lambda + \theta_2)^2$$

which is derived from (6.46). In this case, the availability is

$$A(T^*) = \frac{\theta_1(\lambda + \theta_1)^2(2\lambda T^* + e^{-\lambda T^*}) - \lambda^2 \theta_1 e^{-(\lambda+\theta_1)T^*}}{\lambda^4 T^* + \theta_1(\lambda + \theta_1)^2(2\lambda T^* + e^{-\lambda T^*}) - \lambda^2 \theta_1 e^{-(\lambda+\theta_1)T^*}}.$$

Furthermore, from the inequality $T^* < \overline{T}$, we have

$$\lambda T^* < \frac{\theta_2(\lambda + \theta_2)^2}{\theta_2(\lambda + \theta_2)^2 - 2\theta_1(\lambda + \theta_1)^2}.$$

Table 6.3 shows the numerical examples of the dependence of the mean PM time $1/\theta_2$ in the optimum time T^* and the other quantities when $\lambda = 2$ and $\theta_1 = 1$. ∎

6.3 Modified Discrete Preventive Maintenance Policies

This section proposes a new preventive maintenance policy which is more practical in the real world than the usual PM policies [20]. Failures of a unit occur at a nonhomogeneous Poisson process and the PM is planned at periodic times kT ($k = 1, 2, \dots$) to prevent failures, where T meaning a day, a week, a month, and so on, is given. If the total number of failures equals or exceeds a specified number N, the PM should be done only at the next planned time, and otherwise, no PM would be done; i.e., we postpone PM until more than N failures have occurred. Of course, we may consider PM as replacement.

6.3 Modified Discrete Preventive Maintenance Policies

Table 6.3. Dependence of the mean PM time $1/\theta_2$ in the optimum time T^* when $2/\lambda = 1$ hr, $1/\theta_1 = 1$ hr, and $A(\infty) = 0.643$

θ_2	T^*	$A(T^*)$	$\frac{A(T^*)-A(\infty)}{A(\infty)} \times 100\%$
2.0	1.038	0.698	8.5
3.0	0.402	0.729	13.4
4.0	0.257	0.757	17.7
5.0	0.191	0.779	21.2
6.0	0.153	0.797	24.0
7.0	0.128	0.812	26.4
8.0	0.110	0.825	28.4
9.0	0.096	0.837	30.2
10.0	0.086	0.847	31.7

This is a modification of the policy where a unit undergoes replacement at the number of N failures in Section 4.3, and would be practical in many cases because the PM is done only when an operating unit is idle. For example, when $T = 6$ days, the PM is done at the next Sunday if more than N failures have occurred from the previous PM to the end of this week. Moreover, this would be more economical than the usual periodic policy in Section 4.2, inasmuch as we employ PM only if some failures have occurred.

We derive the expected cost rate of a unit, using the theory of Poisson processes, and determine an optimum number N^* of failures before PM that minimizes the expected cost rate. It is given by a unique solution of the equation when the intensity function $h(t)$ of a Poisson process is strictly increasing to infinity. Furthermore, we consider the model where a unit fails with a certain probability due to faults and the PM is done if more than N faults have occurred, and discuss an optimum policy. Finally, we suggest three PM models where the PM is done by counting the number of occurrences of use, shock, and unit failure.

These models are actually applied to the PM of a computer system: the system stops because of intermittent faults of transient failures [34] due to noise, temperature, power supply variations, and poor electric contacts. The PM should be done on weekends if more than N faults or failures have occurred. For similar reasons, these could be applied to equipment with microcomputers such as numerical control motors and machines, and more extensively to some machines in the factories of manufacturing companies. This policy was applied to the PM of hard disks [35] and a phased array radar [36].

6.3.1 Number of Failures

Consider a unit that should operate for an infinite time span. It is assumed that:

(1) Failures occur at a nonhomogeneous Poisson process with an intensity function $h(t)$, and a mean-value function $H(t) \equiv \int_0^t h(u)\,\mathrm{d}u$ rep-

resents the expected number of failures during $(0,t]$, and $p_j[H(t)] \equiv \{[H(t)]^j/j!\}e^{-H(t)}$ $(j=0,1,2,\ldots)$ is the probability that j failures exactly occur during $(0,t]$.

(2) The PM is planned at times kT $(k=1,2,\ldots)$ where a positive T $(0 < T < \infty)$ is given. A unit becomes like new by PM; *i.e.*, the time returns to zero after PM.

(3) If the total number of failures exceeds a specified number N ($N = 1, 2, \ldots$), the PM is done at the next planned time. Otherwise, a unit is left as it is.

(4) A unit undergoes minimal repair at each failure in Section 4.1. The repair and PM times are negligible; *i.e.*, the time considered here is measured only by the total operating time of a unit.

(5) An intensity function $h(t)$ is continuous and strictly increasing.

The probability that the PM is done at time $(k+1)T$ $(k=0,1,2,\ldots)$, because more than N failures have occurred during $(0,(k+1)T]$ when the number of failures was less than N until time kT, is

$$\sum_{j=0}^{N-1} p_j[H(kT)] \sum_{i=N-j}^{\infty} p_i[H((k+1)T) - H(kT)]$$

$$= \sum_{j=0}^{N-1} p_j[H(kT)] - \sum_{j=0}^{N-1} p_j[H(kT)] \sum_{i=0}^{N-j-1} p_i[H((k+1)T) - H(kT)]$$

$$= \sum_{j=0}^{N-1} \{p_j[H(kT)] - p_j[H((k+1)T)]\}$$

because $\sum_{i=N-j}^{\infty} p_i[\cdot] = 1 - \sum_{i=0}^{N-j-1} p_i[\cdot]$, and $\sum_{j=0}^{N-1} p_j(a) \sum_{i=0}^{N-j-1} p_i(b) = \sum_{j=0}^{N-1} p_j(a+b)$ by the property of a Poisson process in Section 1.3. Thus, the mean time to PM is

$$\sum_{k=0}^{\infty}[(k+1)T]\sum_{j=0}^{N-1}\{p_j[H(kT)] - p_j[H((k+1)T)]\}$$

$$= T\left\{\sum_{k=0}^{\infty}(k+1)\sum_{j=0}^{N-1} p_j[H(kT)] - \sum_{k=0}^{\infty}(k+1)\sum_{j=0}^{N-1} p_j[H((k+1)T)]\right\}$$

$$= T\sum_{k=0}^{\infty}\sum_{j=0}^{N-1} p_j[H(kT)]. \tag{6.47}$$

Furthermore, the expected number of failures when the PM is done at time $(k+1)T$ is

6.3 Modified Discrete Preventive Maintenance Policies

$$\sum_{j=0}^{N-1} p_j[H(kT)] \sum_{i=N-j}^{\infty} (i+j) p_i[H((k+1)T) - H(kT)]$$

$$= \sum_{j=0}^{N-1} p_j[H(kT)] \sum_{i=N}^{\infty} i p_{i-j}[H((k+1)T) - H(kT)].$$

Exchanging the summations and noting that $\binom{i}{j} \equiv 0$ for $i < j$, we can show that this is equal to

$$\sum_{i=N}^{\infty} \sum_{j=0}^{N-1} \frac{i}{i!} \binom{i}{j} [H(kT)]^j [H((k+1)T) - H(kT)]^{i-j} e^{-H((k+1)T)}$$

$$= \sum_{i=N}^{\infty} \frac{i}{i!} e^{-H((k+1)T)} \Bigg\{ [H((k+1)T)]^i$$

$$- \sum_{j=N}^{\infty} \binom{i}{j} [H(kT)]^j [H((k+1)T) - H(kT)]^{i-j} \Bigg\}$$

$$= H((k+1)T) \sum_{j=N-1}^{\infty} p_j[H((k+1)T)] - H(kT) \sum_{j=N-1}^{\infty} p_j[H(kT)]$$

$$- [H((k+1)T) - H(kT)] \sum_{j=N}^{\infty} p_j[H(kT)].$$

Using the relation $\sum_{j=N}^{\infty} p_j[\cdot] = 1 - \sum_{j=0}^{N-1} p_j[\cdot]$ and summing it over k from 0 to ∞, the expected number of failures, i.e., minimal repairs before PM, is

$$\sum_{k=0}^{\infty} [H((k+1)T) - H(kT)] \sum_{j=0}^{N-1} p_j[H(kT)]. \qquad (6.48)$$

Therefore, from Theorem 1.6 in Chapter 1 and (3.3), the expected cost rate is

$$C_1(N) = \frac{c_1 \sum_{k=0}^{\infty} [H((k+1)T) - H(kT)] \sum_{j=0}^{N-1} p_j[H(kT)] + c_2}{T \sum_{k=0}^{\infty} \sum_{j=0}^{N-1} p_j[H(kT)]}$$

$$(N = 1, 2, \dots), \qquad (6.49)$$

where $c_1 = $ cost of minimal repair and $c_2 = $ cost of planned PM.

We find an optimum number N^* that minimizes $C_1(N)$. Let

$$q_1(N) = \frac{\sum_{k=0}^{\infty} [H((k+1)T) - H(kT)] p_N[H(kT)]}{\sum_{k=0}^{\infty} p_N[H(kT)]} \qquad (N = 1, 2, \dots).$$

Then, we have the following theorem.

Theorem 6.7. When $h(t)$ is strictly increasing, $q_1(N)$ is also strictly increasing and $\lim_{N\to\infty} q_1(N) = h(\infty)$.

Proof. From the notation of $q_1(N)$, we have

$$q_1(N+1) - q_1(N)$$
$$= \frac{\sum_{k=0}^{\infty}[H((k+1)T) - H(kT)]p_{N+1}[H(kT)]\sum_{j=0}^{\infty}p_N[H(jT)] - \sum_{k=0}^{\infty}[H((k+1)T) - H(kT)]p_N[H(kT)]\sum_{j=0}^{\infty}p_{N+1}[H(jT)]}{\sum_{k=0}^{\infty}p_{N+1}[H(kT)]\sum_{j=0}^{\infty}p_N[H(jT)]}.$$

The numerator on the right-hand side is

$$\frac{1}{N+1}\left\{\sum_{k=0}^{\infty}[H((k+1)T) - H(kT)]p_N[H(kT)]\sum_{j=0}^{\infty}p_N[H(jT)][H(kT) - H(jT)]\right\}$$

$$= \frac{1}{N+1}\left\{\sum_{k=0}^{\infty}[H((k+1)T) - H(kT)]p_N[H(kT)]\sum_{j=0}^{k}p_N[H(jT)][H(kT) - H(jT)]\right.$$

$$\left. - \sum_{k=0}^{\infty}p_N[H(kT)]\sum_{j=0}^{k}[H((j+1)T) - H(jT)]p_N[H(jT)][H(kT) - H(jT)]\right\}$$

$$= \frac{1}{N+1}\left\{\sum_{k=0}^{\infty}p_N[H(kT)]\sum_{j=0}^{k}p_N[H(jT)][H(kT) - H(jT)]\right.$$

$$\left. \times [H((k+1)T) - H(kT) - H((j+1)T) + H(jT)]\right\} > 0$$

because $h(t)$ is strictly increasing.

Next, prove that $\lim_{N\to\infty} q_1(N) = h(\infty)$. We easily obtain $q_1(N) \leq h(\infty)$ for any finite N, and hence, we need only to show that $\lim_{N\to\infty} q_1(N) \geq h(\infty)$. For any positive number n, $q_1(N)$ is rewritten as

$$q_1(N) = \frac{\sum_{k=0}^{n}[H((k+1)T) - H(kT)]p_N[H(kT)] + \sum_{k=n+1}^{\infty}[H((k+1)T) - H(kT)]p_N[H(kT)]}{\sum_{k=0}^{n}p_N[H(kT)] + \sum_{k=n+1}^{\infty}p_N[H(kT)]}$$

$$\geq \frac{h(nT)}{1 + \{\sum_{k=0}^{n}p_N[H(kT)]/\sum_{k=n+1}^{\infty}p_N[H(kT)]\}}$$

and

$$\lim_{N\to\infty}\frac{\sum_{k=0}^{n}p_N[H(kT)]}{\sum_{k=n+1}^{\infty}p_N[H(kT)]} \leq \lim_{N\to\infty}\sum_{k=0}^{n}\left[\frac{H(kT)}{H((n+1)T)}\right]^N e^{[H((n+1)T) - H(nT)]}$$

$$= 0.$$

6.3 Modified Discrete Preventive Maintenance Policies

Therefore,
$$\lim_{N \to \infty} q_1(N) \geq h(n)$$

which completes the proof, because n is arbitrary. ∎

From Theorem 6.7, it is easy to see that

$$C_1(\infty) \equiv \lim_{N \to \infty} C_1(N) = \frac{c_1 h(\infty)}{T}. \tag{6.50}$$

Thus, if an intensity function $h(t)$ tends to infinity as $t \to \infty$, there exists a finite N^* to minimize $C_1(N)$.

We derive an optimum number N^* that minimizes the expected cost rate $C_1(N)$ in (6.49) when $h(t)$ is strictly increasing.

Theorem 6.8. Suppose that $h(t)$ is continuous and strictly increasing.

(i) If $L_1(\infty) > c_2/c_1$ then there exists a finite and unique minimum that satisfies

$$L_1(N) \geq \frac{c_2}{c_1} \quad (N = 1, 2, \dots) \tag{6.51}$$

and the resulting expected cost rate is given by

$$c_1 q_1(N^* - 1) < TC_1(N^*) \leq c_1 q_1(N^*), \tag{6.52}$$

where

$$L_1(N) \equiv q_1(N) \sum_{k=0}^{\infty} \sum_{j=0}^{N-1} p_j[H(kT)]$$

$$- \sum_{k=0}^{\infty} [H((k+1)T) - H(kT)] \sum_{j=0}^{N-1} p_j[H(kT)] \quad (N = 1, 2, \dots).$$

(ii) If $L_1(\infty) \leq c_2/c_1$ then $N^* = \infty$; i.e., the PM is not done and the expected cost rate is given in (6.50).

Proof. Forming the inequality $C_1(N+1) \geq C_1(N)$, we have

$$c_1 \left\{ \sum_{k=0}^{\infty} [H((k+1)T) - H(kT)] \sum_{j=0}^{N} p_j[H(kT)] \sum_{k=0}^{\infty} \sum_{j=0}^{N-1} p_j[H(kT)] \right.$$

$$\left. - \sum_{k=0}^{\infty} [H((k+1)T) - H(kT)] \sum_{j=0}^{N-1} p_j[H(kT)] \sum_{k=0}^{\infty} \sum_{j=0}^{N} p_j[H(kT)] \right\}$$

$$\geq c_2 \sum_{k=0}^{\infty} p_N[H(kT)].$$

160   6 Preventive Maintenance

Dividing both sides by $c_1 \sum_{k=0}^{\infty} p_N[H(kT)]$ and arranging them,

$$q_1(N) \sum_{k=0}^{\infty} \sum_{j=0}^{N-1} p_j[H(kT)] - \sum_{k=0}^{\infty} [H((k+1)T) - H(kT)] \sum_{j=0}^{N-1} p_j[H(kT)] \geq \frac{c_2}{c_1}$$

which implies (6.51). Furthermore, from Theorem 6.7,

$$L_1(N+1) - L_1(N) = [q_1(N+1) - q_1(N)] \sum_{k=0}^{\infty} \sum_{j=0}^{N-1} p_j[H(kT)] > 0$$

and hence, $L_1(N)$ is also strictly increasing. Therefore, if $L_1(\infty) > c_2/c_1$ then there exists a finite and unique minimum N^* that satisfies (6.51), and from $L_1(N^* - 1) < c_2/c_1$ and $L_1(N^*) \geq c_2/c_1$, we have (6.52).

Next, we investigate the limit of $L_1(N)$. In a similar way to that of proving Theorem 6.7, we can easily have

$$q_1(N) \geq \frac{\sum_{k=0}^{\infty} [H((k+1)T) - H(kT)] \sum_{j=1}^{N-1} p_j[H(kT)]}{\sum_{k=0}^{\infty} \sum_{j=1}^{N-1} p_j[H(kT)]}.$$

Thus,

$$L_1(N) > q_1(N) \sum_{k=0}^{\infty} p_0[H(kT)]$$
$$- \sum_{k=0}^{\infty} [H((k+1)T) - H(kT)] p_0[H(kT)] \qquad (N = 2, 3, \dots)$$

which implies

$$\lim_{N \to \infty} L_1(N) \geq h(\infty) \sum_{k=0}^{\infty} p_0[H(kT)]$$
$$- \sum_{k=0}^{\infty} [H((k+1)T) - H(kT)] p_0[H(kT)].$$

Therefore, if $h(\infty) > TC_1(1)/c_1$ then (6.51) has a finite solution in N.

From the above discussion, we can conclude that if $h(t)$ is strictly increasing to infinity, then there exists a unique minimum N^* such that $L_1(N) \geq c_2/c_1$ and it minimizes $C_1(N)$ in (6.49). ∎

6.3.2 Number of Faults

Faults of a unit occur at a nonhomogeneous Poisson process with an intensity function $h(t)$. A unit stops its operation due to faults and the restart is made

6.3 Modified Discrete Preventive Maintenance Policies

instantaneously by the detection of these faults. The restart succeeds with probability α $(0 < \alpha < 1)$ and the unit returns to a normal condition. On the other hand, the restart fails with probability $\beta \equiv 1 - \alpha$, and the unit needs repair. Then, if the total number of successes of restart is more than N, the PM can be done at the next scheduled time. A unit becomes like new by PM or repair, and the times for faults, restarts, PMs and repairs are negligible. The other assumptions are the same as those in Section 6.3.1.

Let $\overline{F}_\beta(t)$ be the probability that a unit survives because all restarts are successful during $(0,t]$. Then,

$$\overline{F}_\beta(t) = \sum_{j=0}^{\infty} \Pr\{\text{unit survives to time } t \mid j \text{ faults}\} \times \Pr\{j \text{ faults in } (0,t]\}$$

$$= \sum_{j=0}^{\infty} \alpha^j p_j[H(t)] = \mathrm{e}^{-\beta H(t)}. \tag{6.53}$$

Furthermore, the probability that the PM is done at time $(k+1)T$ $(k = 0, 1, 2, \dots)$, because more than N restarts have succeeded until $(k+1)T$ when j $(j = 0, 1, 2, \dots, N-1)$ successful restarts were made during $(0, kT]$, is

$$\sum_{j=0}^{N-1} \alpha^j p_j[H(kT)] \sum_{i=N-j}^{\infty} \alpha^i p_i[H((k+1)T) - H(kT)]$$

$$= \sum_{j=0}^{N-1} \alpha^j \{p_j[H(kT)] \mathrm{e}^{-\beta[H((k+1)T) - H(kT)]} - p_j[H((k+1)T)]\}.$$

Thus, the probability that the PM is done before failure is

$$\sum_{k=0}^{\infty} \sum_{j=0}^{N-1} \alpha^j \{p_j[H(kT)] \mathrm{e}^{-\beta[H((k+1)T) - H(kT)]} - p_j[H((k+1)T)]\}$$

$$= 1 - \sum_{k=0}^{\infty} [\overline{F}_\beta(kT) - \overline{F}_\beta((k+1)T)] \sum_{j=0}^{N-1} p_j[\alpha H(kT)]. \tag{6.54}$$

Similarly, the probability that a unit undergoes repair before PM is

$$\sum_{k=0}^{\infty} \sum_{j=0}^{N-1} \alpha^j p_j[H(kT)] \sum_{i=0}^{\infty} (1-\alpha^i) p_i[H((k+1)T) - H(kT)]$$

$$= \sum_{k=0}^{\infty} [\overline{F}_\beta(kT) - \overline{F}_\beta((k+1)T)] \sum_{j=0}^{N-1} p_j[\alpha H(kT)]. \tag{6.55}$$

It is evident that $(6.54) + (6.55) = 1$. Similarly, the mean time to PM or repair is

162   6 Preventive Maintenance

$$\sum_{k=0}^{\infty}[(k+1)T]\sum_{j=0}^{N-1}\alpha^{j}\{p_{j}[H(kT)]e^{-\beta[H((k+1)T)-H(kT)]}-p_{j}[H((k+1)T)]\}$$

$$+\sum_{k=0}^{\infty}\sum_{j=0}^{N-1}\alpha^{j}p_{j}[H(kT)]\sum_{i=0}^{\infty}\alpha^{i}\beta\int_{kT}^{(k+1)T}tp_{i}[H(t)-H(kT)]h(t)\,\mathrm{d}t$$

$$=\sum_{k=0}^{\infty}\int_{kT}^{(k+1)T}\overline{F}_{\beta}(t)\,\mathrm{d}t\sum_{j=0}^{N-1}p_{j}[\alpha H(kT)]. \qquad (6.56)$$

Therefore, if we neglect all costs resulting from restarts then the expected cost rate is, from (6.54), (6.55), and (6.56),

$$C_{2}(N)=\frac{(c_{1}-c_{2})\sum_{k=0}^{\infty}[\overline{F}_{\beta}(kT)-\overline{F}_{\beta}((k+1)T)]\sum_{j=0}^{N-1}p_{j}[\alpha H(kT)]+c_{2}}{\sum_{k=0}^{\infty}\int_{kT}^{(k+1)T}\overline{F}_{\beta}(t)\,\mathrm{d}t\sum_{j=0}^{N-1}p_{j}[\alpha H(kT)]}$$

$$(N=1,2,\dots), \qquad (6.57)$$

where c_1 = cost of repair and c_2 = cost of PM.

It is assumed that $c_1 > c_2$ because the repair cost would be higher than the PM cost in the actual field, and $\mu_\beta \equiv \int_0^\infty \overline{F}_\beta(t)\,\mathrm{d}t < \infty$ is the finite mean time to need repair. Let

$$q_{2}(N)\equiv\frac{\sum_{k=0}^{\infty}[\overline{F}_{\beta}(kT)-\overline{F}_{\beta}((k+1)T)]p_{N}[\alpha H(kT)]}{\sum_{k=0}^{\infty}\int_{kT}^{(k+1)T}\overline{F}_{\beta}(t)\,\mathrm{d}t\,p_{N}[\alpha H(kT)]} \qquad (N=1,2,\dots).$$

Theorem 6.9. When $h(t)$ is strictly increasing, $q_2(N)$ is also strictly increasing and $\lim_{N\to\infty} q_2(N) \equiv \beta h(\infty)$.

Proof. We use the following notations.

$$B_{k}\equiv e^{-\alpha H(kT)}\int_{kT}^{(k+1)T}\overline{F}_{\beta}(t)\,\mathrm{d}t \qquad (k=0,1,2,\dots)$$

$$C_{k}\equiv e^{-\alpha H(kT)}[\overline{F}_{\beta}(kT)-\overline{F}_{\beta}((k+1)T)]. \qquad (k=0,1,2,\dots).$$

Then, $q_2(N)$ is written as

$$q_{2}(N)=\frac{\sum_{k=0}^{\infty}[H(kT)]^{N}C_{k}}{\sum_{k=0}^{\infty}[H(kT)]^{N}B_{k}}.$$

In a similar way to that of Theorem 6.7, it is easy to prove that when $h(t)$ is strictly increasing,

$$\beta h(kT) < \frac{C_k}{B_k} < \beta h((k+1)T), \qquad q_2(N+1)-q_2(N) > 0.$$

Thus, we have

6.3 Modified Discrete Preventive Maintenance Policies

$$\lim_{k \to \infty} \frac{C_k}{B_k} = \beta h(\infty)$$

and for any positive number n,

$$\beta h(nT) \leq \lim_{N \to \infty} q_2(N) \leq \beta h(\infty)$$

which completes the proof. ∎

From this theorem, it is easy to see that

$$C_2(\infty) \equiv \lim_{N \to \infty} C_2(N) = \frac{c_1}{\mu_\beta}. \tag{6.58}$$

We derive an optimum number N^* that minimizes $C_2(N)$ in (6.57).

Theorem 6.10. Suppose that $h(t)$ is continuous and strictly increasing.

(i) If $\beta \mu_\beta h(\infty) > c_1/(c_1 - c_2)$ then there exists a finite and unique minimum that satisfies

$$L_2(N) \geq \frac{c_2}{c_1 - c_2} \qquad (N = 1, 2, \dots) \tag{6.59}$$

and the resulting cost rate is

$$(c_1 - c_2)q_2(N^* - 1) < C_2(N^*) \leq (c_1 - c_2)q_2(N^*), \tag{6.60}$$

where

$$L_2(N) \equiv q_2(N) \sum_{k=0}^{\infty} \int_{kT}^{(k+1)T} \overline{F}_\beta(t)\, dt \sum_{j=0}^{N-1} p_j[\alpha H(kT)]$$

$$- \sum_{k=0}^{\infty} [\overline{F}_\beta(kT) - \overline{F}_\beta((k+1)T)] \sum_{j=0}^{N-1} p_j[\alpha H(kT)] \quad (N = 1, 2, \dots).$$

(ii) If $\beta \mu_\beta h(\infty) \leq c_1/(c_1 - c_2)$ then $N^* = \infty$; i.e., we should do no PM, and $C_2(\infty)$ is given in (6.58).

Proof. From the inequality $C_2(N+1) \geq C_2(N)$, we have

$$(c_1 - c_2) \left[\sum_{k=0}^{\infty} \{\overline{F}_\beta(kT) - \overline{F}_\beta((k+1)T)\} \sum_{j=0}^{N} p_j[\alpha H(kT)] \right.$$

$$\left. \times \sum_{k=0}^{\infty} \int_{kT}^{(k+1)T} \overline{F}_\beta(t)\, dt \sum_{j=0}^{N-1} p_j[\alpha H(kT)] \right.$$

164   6 Preventive Maintenance

$$-\sum_{k=0}^{\infty}[\overline{F}_\beta(kT) - \overline{F}_\beta((k+1)T)] \sum_{j=0}^{N-1} p_j[\alpha H(kT)]$$

$$\times \sum_{k=0}^{\infty} \int_{kT}^{(k+1)T} \overline{F}_\beta(t)\,\mathrm{d}t \sum_{j=0}^{N} p_j[\alpha H(kT)] \Bigg]$$

$$\geq c_2 \sum_{k=0}^{\infty} \int_{kT}^{(k+1)T} \overline{F}_\beta(t)\,\mathrm{d}t\, p_N[\alpha H(kT)].$$

Dividing both sides by $(c_1 - c_2) \sum_{k=0}^{\infty} \int_{kT}^{(k+1)T} \overline{F}_\beta(t)\,\mathrm{d}t\, p_N[\alpha H(kT)]$ implies

$$L_2(N) \geq \frac{c_2}{c_1 - c_2}.$$

Using Theorem 6.9,

$$L_2(N+1) - L_2(N)$$

$$= [q_2(N+1) - q_2(N)] \sum_{k=0}^{\infty} \int_{kT}^{(k+1)T} \overline{F}_\beta(t)\,\mathrm{d}t \sum_{j=0}^{N} p_j[\alpha H(kT)] > 0$$

$$\lim_{N \to \infty} L_2(N) = \beta \mu_\beta h(\infty) - 1.$$

Therefore, similar to Theorem 6.8, we have the results of Theorem 6.10. ∎

Example 6.3. A computer system stops at certain faults according to a Weibull distribution with shape parameter 2; i.e., $H(t) = \lambda t^2$. The restart succeeds with probability α and fails with probability β [37]. If the total number of successes of restarts exceeds a specified number N, the PM can be done at the next planned time. Then, from Theorem 6.10, there always exists an optimum number N^* $(1 \leq N^* < \infty)$ that satisfies (6.59).

Table 6.4 gives N^* for $T = 24$ hours, 48 hours, $\alpha = 0.8, 0.85, 0.90, 0.95$, and $c_1/c_2 = 1.5, 2.0, 3.0$ when $1/\lambda = 720$ hours. Also, the mean time to N faults is given by

$$\mu_N \equiv \int_0^\infty t p_{N-1}[H(t)] h(t)\,\mathrm{d}t = \frac{\Gamma(N + \frac{1}{2})}{\sqrt{\lambda}}.$$

For example, when $T = 24$ and $\alpha = 0.9$, $\mu_{N^*} = 133, 84, 52$ hours for $N^* = 25, 10, 4$ hours, respectively, and on the average, we may employ PM at about 1 time per 6 days, 1 time per 4 days, and 1 time per 3 days for cost rates $c_1/c_2 = 1.5, 2.0$, and 3.0. ∎

Up to now, we have assumed that the times for repairs and PMs are negligible. If the PM and the repair require the mean times θ_2 and θ_1, respectively, then the expected cost rate in (6.57) can be rewritten as

Table 6.4. Optimum number N^* that minimizes $C_2(N)$ when $H(t) = \lambda t^2$ and $1/\lambda = 720$ hours

α	T = 24 hours c_1/c_2			T = 48 hours c_1/c_2		
	1.5	2.0	3.0	1.5	2.0	3.0
0.80	12	5	1	9	5	1
0.85	18	7	3	18	7	1
0.90	25	10	4	21	8	1
0.95	52	20	9	42	19	7

$$C_2(N) = \frac{(c_1\theta_1 - c_2\theta_2)\sum_{k=0}^{\infty}[\overline{F}(kT) - \overline{F}((k+1)T)] \times \sum_{j=0}^{N-1} p_j[\alpha H(kT)] + c_2\theta_2}{\sum_{k=0}^{\infty}\int_{kT}^{(k+1)T}\overline{F}_\beta(t)\,dt \sum_{j=0}^{N-1} p_j[\alpha H(kT)] + (\theta_1 - \theta_2)\sum_{k=0}^{\infty}[\overline{F}(kT) - \overline{F}((k+1)T)]\sum_{j=0}^{N-1}p_j[\alpha H(kT)] + \theta_2}$$
$$(N = 1, 2, \ldots), \qquad (6.61)$$

where c_1 = cost per unit of time for repair and c_2 = cost per unit of time for PM.

6.3.3 Other PM Models

(1) Number of Uses

Uses of a unit occur at a nonhomogeneous Poisson process with an intensity function $h(t)$. A unit deteriorates with use and fails at a certain number of uses. The probability that a unit does not fail at use j is α_j $(0 < \alpha_j < 1)$ $(j = 1, 2, \ldots)$. Then, if the total number of uses exceeds N, the PM of a unit is done at the next planned time $(k+1)T$.

By the method similar to that of Section 6.3.2, the probability that the PM is done before failure is

$$\sum_{k=0}^{\infty}\sum_{j=0}^{N-1} p_j[H(kT)] \sum_{i=N-j}^{\infty} \Phi_{i+j}(\alpha) p_i[H((k+1)T) - H(kT)]$$

and the probability that a unit fails is

$$\sum_{k=0}^{\infty}\sum_{j=0}^{N-1} p_j[H(kT)] \sum_{i=0}^{\infty}[\Phi_j(\alpha) - \Phi_{i+j}(\alpha)] p_i[H((k+1)T) - H(kT)],$$

where $\Phi_i(\alpha) \equiv \alpha_1\alpha_2\ldots\alpha_i$ $(i = 1, 2, \ldots)$ and $\Phi_0(\alpha) \equiv 1$. Furthermore, the mean time to PM or failure is

$$\sum_{k=0}^{\infty}[(k+1)T]\sum_{j=0}^{N-1}p_j[H(kT)]\sum_{i=N-j}^{\infty}\Phi_{i+j}(\alpha)p_i[H((k+1)T)-H(kT)]$$

$$+\sum_{k=0}^{\infty}\sum_{j=0}^{N-1}p_j[H(kT)]\sum_{i=0}^{\infty}[\Phi_{i+j}(\alpha)-\Phi_{i+j+1}(\alpha)]\int_{kT}^{(k+1)T}tp_i[H(t)-H(kT)]h(t)\,dt$$

$$=\sum_{k=0}^{\infty}\sum_{j=0}^{N-1}p_j[H(kT)]\sum_{i=0}^{\infty}\Phi_{i+j}(\alpha)\int_{kT}^{(k+1)T}p_i[H(t)-H(kT)]\,dt.$$

Therefore, the expected cost rate is

$$C_3(N)=\frac{(c_1-c_2)\sum_{k=0}^{\infty}\sum_{j=0}^{N-1}p_j[H(kT)]\sum_{i=0}^{\infty}[\Phi_j(\alpha)-\Phi_{i+j}(\alpha)]\times p_i[H((k+1)T)-H(kT)]+c_2}{\sum_{k=0}^{\infty}\sum_{j=0}^{N-1}p_j[H(kT)]\sum_{i=0}^{\infty}\Phi_{i+j}(\alpha)\int_{kT}^{(k+1)T}p_i[H(t)-H(kT)]\,dt}$$

$$(N=1,2,\dots),\qquad(6.62)$$

where c_1 = cost of failure and c_2 = cost of PM. In particular, when $\Phi_i(\alpha)=\alpha^i$, $C_3(N)$ agrees with $C_2(N)$ in (6.57).

(2) Number of Shocks

Shocks occur at a nonhomogeneous Poisson process with an intensity function $h(t)$ [38]. A unit fails at a certain number of shocks due to damage done by shocks. When the total amount of damage has exceeded a failure level Z ($0<Z<\infty$), a unit fails. If the total number of shocks exceeds N before failure, the PM is done at the next planned time $(k+1)T$ [20].

Let $G(x)$ be the distribution of an amount of damage produced by each shock. Then, the probability that a unit fails at shock j is $G^{(j-1)}(Z)-G^{(j)}(Z)$, where $G^{(j)}(Z)$ denotes the j-fold Stieltjes convolution of G with itself. Thus, replacing $\Phi_i(\alpha)$ in (6.62) with $G^{(i)}(Z)$ formally, the expected cost rate can be derived as

$$C_4(N)=\frac{(c_1-c_2)\sum_{k=0}^{\infty}\sum_{j=0}^{N-1}p_j[H(kT)]\sum_{i=0}^{\infty}[G^{(j)}(Z)-G^{(i+j)}(Z)]\times p_i[H((k+1)T)-H(kT)]+c_2}{\sum_{k=0}^{\infty}\sum_{j=0}^{N-1}p_j[H(kT)]\sum_{i=0}^{\infty}G^{(i+j)}(Z)\int_{kT}^{(k+1)T}p_i[H(t)-H(kT)]\,dt}$$

$$(N=1,2,\dots).\qquad(6.63)$$

(3) Number of Unit Failures

Consider a parallel redundant system with n ($n\geq 2$) units in which each unit has an identical failure distribution $F(t)$. The system fails when all of n units have failed. If the total number of unit failures exceeds N ($1\leq N\leq n-1$) before system failure, the PM is done at time $(k+1)T$. The probability that the system fails before PM is

$$\sum_{k=0}^{\infty} \sum_{j=0}^{N-1} \binom{n}{j} [F(kT)]^j [F((k+1)T) - F(kT)]^{n-j}$$

and the probability that the PM is done before system failure is

$$\sum_{k=0}^{\infty} \sum_{j=0}^{N-1} \binom{n}{j} [F(kT)]^j \sum_{i=N-j}^{n-j-1} \binom{n-j}{i} [F((k+1)T) - F(kT)]^i [\overline{F}((k+1)T)]^{n-j-i}.$$

Furthermore, the mean time to system failure or PM is

$$\sum_{k=0}^{\infty} [(k+1)T] \sum_{j=0}^{N-1} \binom{n}{j} [F(kT)]^j \sum_{i=N-j}^{n-j-1} \binom{n-j}{i} [F((k+1)T) - F(kT)]^i$$
$$\times [\overline{F}((k+1)T)]^{n-j-i}$$
$$+ \sum_{k=0}^{\infty} \sum_{j=0}^{N-1} \binom{n}{j} [F(kT)]^j \int_{kT}^{(k+1)T} t \, d[F(t) - F(kT)]^{n-j}$$
$$= \sum_{k=0}^{\infty} \sum_{j=0}^{N-1} \binom{n}{j} [F(kT)]^j \int_{kT}^{(k+1)T} \{[\overline{F}(kT)]^{n-j} - [\overline{F}(kT) - \overline{F}(t)]^{n-j}\} \, dt.$$

Therefore, the expected cost rate is

$$C_5(N) = \frac{(c_1-c_2)\sum_{k=0}^{\infty}\sum_{j=0}^{N-1}\binom{n}{j}[F(kT)]^j[F((k+1)T)-F(kT)]^{n-j} + nc_0 + c_2}{\sum_{k=0}^{\infty}\sum_{j=0}^{N-1}\binom{n}{j}[F(kT)]^j \times \int_{kT}^{(k+1)T}\{[\overline{F}(kT)]^{n-j} - [\overline{F}(kT)-\overline{F}(t)]^{n-j}\}\,dt}$$
$$(N = 1, 2, \ldots, n-1), \quad (6.64)$$

where c_0 = cost of one unit, c_1 = cost of system failure, and c_2 = cost of PM.

It is very difficult to discuss analytically optimum policies for the above models, however, it would be easy to calculate $C_i(N)$ ($i = 3, 4, 5$) with a computer and obtain optimum numbers N^* numerically. By making some modifications in these models, they could be applied to actual models and become interesting theoretical studies as well.

References

1. Morse PM (1958) Queues, Inventories, and Maintenance. J Wiley & Sons, New York.
2. Nakagawa T (1977) Optimum preventive maintenance policies for repairable systems. IEEE Trans Reliab R-26:168–173.
3. Okumoto K, Osaki S (1977) Optimum policies for a standby system with preventive maintenance. Oper Res Q 28:415–423.

4. Sherif YS, Smith ML (1981) Optimal maintenance models for systems subject to failure – A review. Nav Res Logist Q 28:47–74.
5. Jardine AKS, Buzacott JA (1985) Equipment reliability and maintenance. Eur J Oper Res 19:285–296.
6. Reineke DM, Murdock Jr WP, Pohl EA, Rehmert I (1999) Improving availability and cost performance for complex systems with preventive maintenance. In: Proceedings Annual Reliability and Maintainability Symposium:383–388.
7. Liang TY (1985) Optimum piggyback preventive maintenance policies. IEEE Trans Reliab R-34:529–538.
8. Aven T (1990) Availability formulae for standby systems of similar units that are preventively maintained. IEEE Trans Reliab R-39:603–606.
9. Smith MAJ, Dekker R (1997) Preventive maintenance in a 1 out of n systems: the uptime, downtime and costs. Eur J Oper Res 99:565–583.
10. Silver EA, Fiechter CN (1995) Preventive maintenance with limited historical data. Eur J Oper Res 82:125-144.
11. Scarf PA (1997) On the application of mathematical models in maintenance. Eur J Oper Res 99:493–506.
12. Chockie A, Bjorkelo K (1992) Effective maintenance practices to manage system aging. In: Proceedings Annual Reliability and Maintainability Symposium:166–170.
13. Smith AM (1992) Preventive-maintenance impact on plant availability. In: Proceedings Annual Reliability and Maintainability Symposium:177–180.
14. Susova GM, Petrov AN (1992) Markov model-based reliability and safety evaluation for aircraft maintenance-system optimization. In: Proceedings Annual Reliability and Maintainability Symposium:29–36.
15. Kumar UD, Crocker J (2003) Maintainability and maintenance – A case study on mission critical aircraft and engine components. In: Blischke WR, Murthy DNP (eds) Case Studies in Reliability and Maintenance. J Wiley & Sons, Hoboken, NJ:377–398
16. Mine H, Nakagawa T (1977) Interval reliability and optimum preventive maintenance policy. IEEE Trans Reliab R-26:131–133.
17. Nakagawa T, Osaki S (1974) Optimum preventive maintenance policies for a 2-unit redundant system. IEEE Trans Reliab R-23:86–91.
18. Nakagawa T, Osaki S (1974) Optimum preventive maintenance policies maximizing the mean time to the first system failure for a two-unit standby redundant system. Optim Theor Appl 14:115–129.
19. Nakagawa T, Osaki S (1976) A summary of optimum preventive maintenance policies for a two-unit standby redundant system. Z Oper Res 20:171–187.
20. Nakagawa T (1986) Modified discrete preventive maintenance policies. Nav Res Logist Q 33:703–715.
21. Barlow RE, Hunter LC (1961) Reliability analysis of a one-unit system. Oper Res 9:200–208.
22. Jardine AKS (1970) Equipment replacement strategies. In: Jardine AKS (ed) Operational Research in Maintenance. Manchester University Press, New York.
23. Nakagawa T (1978) Reliability analysis of standby repairable systems when an emergency occurs. Microelectron Reliab 17:461–464.
24. Rozhdestvenskiy DV, Fanarzhi GN (1970) Reliability of a duplicated system with renewal and preventive maintenance. Eng Cybernet 8:475–479.
25. Osaki S, Asakura T (1970) A two-unit standby redundant system with preventive maintenance. J Appl Prob 7:641–648.

26. Berg M (1976) Optimal replacement policies for two-unit machines with increasing running costs I. Stoch Process Appl 4:89–106.
27. Berg M (1977) Optimal replacement policies for two-unit machines with running costs II. Stoch Process Appl 5:315–322.
28. Berg M (1978) General trigger-off replacement procedures for two-unit systems. Nav Res Logist Q 25:15–29.
29. Teixeira de Almedia A, Campello de Souza FM (1993) Decision theory in maintenance strategy for a 2-unit redundant standby system. IEEE Trans Reliab R-42:401–407.
30. Gupta PP, Kumar A (1981) Operational availability of a complex system with two types of failure under different repair preemptions. IEEE Trans Reliab R-30:484–485.
31. Pullen KW, Thomas MU (1986) Evaluation of an opportunistic replacement policy for a 2-unit system. IEEE Trans Reliab R-35:320–324.
32. Barros A, Bérenguer C, Grall A (2003) Optimization of replacement times using imperfect monitoring information. IEEE Trans Reliability 52:523–533.
33. Nakagawa T (2002) Two-unit redundant models. In: Osaki S (ed) Stochastic Models in Reliability and Maintenance. Springer, New York:165–191.
34. Castillo X, McConner SR, Siewiorek DP (1982) Derivation and calibration of a transfer error reliability model. IEEE Trans Comput C31:658–671.
35. Sandoh H, Hirakoshi H, Nakagawa T (1998) A new modified discrete preventive maintenance policy and its application to hard disk management. J Qual Maint Eng 4:284–290.
36. Ito K, Nakagawa T (2004) Comparison of cyclic and delayed maintenance for a phased array radar. J Oper Res Soc Jpn 47:51–61.
37. Nakagawa T, Nishi K, Yasui K (1984) Optimum preventive maintenance policies for a computer system with restart. IEEE Trans Reliab R-33:272–276.
38. Esary JD, Marshall AW, Proschan F (1973) Shock models and wear processes. Ann Prob 1:627–649.

7
Imperfect Preventive Maintenance

The maintenance of an operating unit after failure is costly, and sometimes, it requires a long time to repair failed units. It would be an important problem to determine when to maintain preventively the unit before it fails. However, it would be not wise to maintain the unit too often. From this viewpoint, commonly considered maintenance policies are preventive replacement for units with no repair as described in Chapters 3 through 5 and preventive maintenance for units with repair discussed in Chapter 6. It may be wise to maintain units to prevent failures when their failure rates increase with age.

The usual preventive maintenance (PM) of the unit is done before failure at a specified time T after its installation. The mean time to failure (MTTF), the availability, and the expected cost are derived as the reliability measures for maintained units. Optimum PM policies that maximize or minimize these measures have been summarized in Chapter 6. All models have assumed that *the unit after PM becomes as good as new*. Actually, this assumption might not be true. The unit after PM usually might be younger at PM, and occasionally, it might be worse than before PM because of faulty procedures, *e.g.*, wrong adjustments, bad parts, and damage done during PM. Generally, the improvement of the unit by PM would depend on the resources spent for PM.

It was first assumed in [1] that the inspection to detect failures may not be perfect. Similar models such that inspection, test, and detection of failures are uncertain were treated in [2,3]. The imperfect PM where the unit after PM is not like new with a certain probability was considered, and the optimum PM policies that maximize the availability or minimize the expected cost were discussed in [4–7]. In addition, the PM policies with several reliability levels were presented in [8].

It is imperative to check a computer system and remove as many unit faults, failures, and degradations as possible, by providing fault-tolerant techniques. Imperfect maintenance for a computer system was first treated in [9]. The MTTF and availability were obtained in [10–12] in the case where although the system is usually renewed after PM, it sometimes remains unchanged. The imperfect test of intermittent faults incurred in digital systems was studied in [13].

Two imperfect PM models of the unit were considered [14, 15]: the age becomes x units of time younger at each PM and the failure rate is reduced in proportion to that before PM or to the PM cost. The improvement factor in failure rate after maintenance [16, 17] and the system degradation with time where the PM restores the hazard function to the same shape [18] were introduced. Furthermore, the PM policy that slows the degradation rate was considered in [19].

On the other hand, it was assumed in [20–22] that a failed unit becomes as good as new after repair with a certain probability, and some properties of its failure distribution were investigated. Similar imperfect repair models were generalized by [23–31]. Also, the stochastic properties of imperfect repair models with PM were derived in [32, 33]. Multivariate distributions and their probabilistic quantities of these models were derived in [34–36]. The improvement factors of imperfect PM and repair were statistically estimated in [37–40]. The PM was classified into four terms of its effect [41]: perfect maintenance, minimal maintenance, imperfect maintenance, and worse maintenance. Some chapters [42–44] of recently published books summarized many results of imperfect maintenance.

This chapter summarizes our results of imperfect maintenance models that could be applied to actual systems and would be helpful for further studies in research fields. It is assumed in Section 7.1 that the operating unit is replaced at failure or is maintained preventively at time T. Then, the unit after PM has the same failure rate as before PM with a certain probability. The expected cost rate is obtained and an optimum PM policy that minimizes it is discussed analytically [5]. Section 7.2 considers several imperfect PM models with minimal repair at failures: (1) the unit after PM becomes as good as new with a certain probability; (2) the age becomes younger at each PM; and (3) the age or failure rate after PM reduces in proportion to that before PM. The expected cost rates of four models are obtained and optimum policies for each model are derived [15].

Section 7.3 considers a modified inspection model where the unit after inspection becomes like new with a certain probability. The MTTF, the expected number of inspections, and the total expected cost are obtained [45, 46]. Furthermore, an imperfect inspection model with two human errors is proposed. Section 7.4 considers the imperfect PM of a computer system that is maintained at periodic times [12]. The MTTF and the availability are obtained, and optimum policies that maximize them are discussed. Finally, Section 7.5 suggests a sequential imperfect PM model where the PM is done at successive times and the age or failure rate reduces in proportion to that before PM. The expected cost rates are obtained and optimum policies that minimize them are discussed [47]. It is shown in numerical examples that optimum intervals are uniquely determined when the failure time has a Weibull distribution.

The following notation is used throughout this chapter. A unit begins to operate at time 0, and has the failure distribution $F(t)$ $(t \geq 0)$ with finite mean μ and its density function $f(t) \equiv \mathrm{d}F(t)/\mathrm{d}t$. Furthermore, the failure rate $h(t) \equiv f(t)/\overline{F}(t)$ and the cumulative hazard function $H(t) \equiv \int_0^t h(u)\mathrm{d}u$, where $\overline{\Phi} \equiv 1 - \Phi$.

7.1 Imperfect Maintenance Policy

All models have assumed until now that a unit after any PM becomes as good as new. Actually, this assumption might not be true. It sometimes occurs that a unit after PM is worse than before PM because of faulty procedures, e.g., wrong adjustments, bad parts, and damage done during PM. To include this, it is assumed that the failure rate after PM is the same as before PM with a certain probability, and a unit is not like new. This section derives the expected cost rate of the model with imperfect PM, and discusses an optimum policy that minimizes it.

Consider the imperfect PM policy for a one-unit system that should operate for an infinite time span.

1. The operating unit is repaired at failure or is maintained preventively at time T $(0 < T \leq \infty)$, whichever occurs first, after its installation or previous PM.
2. The unit after repair becomes as good as new.
3. The unit after PM has the same failure rate as it had before PM with probability p $(0 \leq p < 1)$ and becomes as good as new with probability $q \equiv 1 - p$.
4. Cost of each repair is c_1 and cost of each PM is c_2.
5. The repair and PM times are negligible.

Consider one cycle from time $t = 0$ to the time that the unit becomes as good as new by either repair or perfect PM. Then, the expected cost of one cycle is given by the sum of the repair cost and PM cost;

$$\widehat{C}(T;p) = c_1 \Pr\{\text{unit is repaired at failure}\} + c_2 \Pr\{\text{expected number of PMs per one cycle}\}. \quad (7.1)$$

The probability that the unit is repaired at failure is

$$\sum_{j=1}^{\infty} p^{j-1} \int_{(j-1)T}^{jT} \mathrm{d}F(t) = 1 - q \sum_{j=1}^{\infty} p^{j-1} \overline{F}(jT) \quad (7.2)$$

and the expected number of PMs including perfect PM per one cycle is

$$\sum_{j=1}^{\infty} (j-1) p^{j-1} \int_{(j-1)T}^{jT} \mathrm{d}F(t) + q \sum_{j=1}^{\infty} j p^{j-1} \overline{F}(jT) = \sum_{j=1}^{\infty} p^{j-1} \overline{F}(jT). \quad (7.3)$$

7 Imperfect Preventive Maintenance

Furthermore, the mean time of one cycle is

$$\sum_{j=1}^{\infty} p^{j-1} \int_{(j-1)T}^{jT} t \, dF(t) + q \sum_{j=1}^{\infty} p^{j-1} (jT) \overline{F}(jT) = \sum_{j=1}^{\infty} p^{j-1} \int_{(j-1)T}^{jT} \overline{F}(t) \, dt. \tag{7.4}$$

Thus, substituting (7.2) and (7.3) into (7.1), and dividing it by (7.4), the expected cost rate is, from (3.3),

$$C(T;p) = \frac{c_1 \left[1 - q \sum_{j=1}^{\infty} p^{j-1} \overline{F}(jT)\right] + c_2 \sum_{j=1}^{\infty} p^{j-1} F(jT)}{\sum_{j=1}^{\infty} p^{j-1} \int_{(j-1)T}^{jT} \overline{F}(t) \, dt}. \tag{7.5}$$

We clearly have

$$C(0;p) \equiv \lim_{T \to 0} C(T;p) = \infty, \qquad C(\infty;p) \equiv \lim_{T \to \infty} C(T;p) = \frac{c_1}{\mu} \tag{7.6}$$

which is the expected cost for the case where no PM is done and the unit is repaired only at failure.

We seek an optimum PM time T^* that minimizes $C(T;p)$. Let

$$H(t;p) \equiv \frac{\sum_{j=1}^{\infty} p^{j-1} j f(jt)}{\sum_{j=1}^{\infty} p^{j-1} j \overline{F}(jt)}. \tag{7.7}$$

Then, differentiating $C(T;p)$ with respect to T and setting it equal to zero,

$$H(T;p) \sum_{j=1}^{\infty} p^{j-1} \int_{(j-1)T}^{jT} \overline{F}(t) \, dt - q \sum_{j=1}^{\infty} p^{j-1} F(jT) = \frac{c_2}{c_1 q - c_2}, \tag{7.8}$$

where $c_1 q - c_2 \neq 0$. Denoting the left-hand side of (7.8) by $Q(T;p)$, we easily have that if $H(t;p)$ is strictly increasing then $Q(T;p)$ is also strictly increasing from 0 and

$$Q(\infty;p) \equiv \lim_{T \to \infty} Q(T;p) = \mu H(\infty;p) - 1. \tag{7.9}$$

It is assumed that $H(t;p)$ is strictly increasing in t for any p. Then, we have the following optimum policy.

(i) If $c_1 q > c_2$ and $H(\infty;p) > c_1 q / [\mu (c_1 q - c_2)]$ then there exists a finite and unique T^* that satisfies (7.8), and the resulting cost rate is

$$C(T^*;p) = \left(c_1 - \frac{c_2}{q}\right) H(T^*;p). \tag{7.10}$$

(ii) If $c_1 q > c_2$ and $H(\infty;p) \leq c_1 q / [\mu (c_1 q - c_2)]$, or $c_1 q \leq c_2$ then $T^* = \infty$; i.e., no PM should be done, and the expected cost is given in (7.6).

Table 7.1. Optimum PM time T^* and expected cost rate $C(T^*;p)$ for p when $c_1 = 5$ and $c_2 = 1$

p	T^*	$C(T^*;p)$
0.00	1.31	2.27
0.01	1.32	2.27
0.05	1.36	2.30
0.10	1.43	2.34
0.15	1.52	2.37
0.20	1.64	2.40
0.25	1.80	2.43
0.30	2.02	2.45
0.35	2.33	2.47
0.40	2.79	2.49

When $p = 0$, *i.e.*, the PM is perfect, the model corresponds to a standard age replacement policy, and the above results agree with those of Chapter 3.

Example 7.1. Suppose that $F(t)$ is a gamma distribution with order 2; *i.e.*, $F(t) = 1 - (1+t)\mathrm{e}^{-t}$. Then, $H(t;p)$ in (7.7) is

$$H(t;p) = \frac{t(1+p\mathrm{e}^{-t})}{1 - p\mathrm{e}^{-t} + t(1+p\mathrm{e}^{-t})}$$

which is strictly increasing from 0 to 1. Thus, if $c_1 q > 2c_2$ then there exists a finite and unique T^* that satisfies (7.8), and otherwise, $T^* = \infty$.

Table 7.1 gives the optimum PM time T^* and the expected cost rate $C(T^*;p)$ for $p = 0.0 \sim 0.4$ when $c_1 = 5$ and $c_2 = 1$. Both T^* and $C(T^*;p)$ are increasing when the probability p of imperfect PM is large. The reason is that it is better to repair a failed unit than to perform PM when p is large. ∎

7.2 Preventive Maintenance with Minimal Repair

Earlier results of optimum PM policies have been summarized in Chapter 6. However, almost all models have assumed that a unit becomes as good as new after any PM. In practice, this assumption often might not be true. A unit after PM usually might be younger at PM, and occasionally, it might become worse than before PM because of faulty procedures.

This section considers the following four imperfect PM models with minimal repair at failures.

(1) The unit after PM has the same failure rate as before PM or becomes as good as new with a certain probability q.
(2) The age becomes x units of time younger at each PM.
(3) The age or failure rate after PM reduces to at or $bh(t)$ when it was t or $h(t)$ before PM, respectively.

(4) The age or failure rate is reduced to the original one at the beginning of all PMs in proportion to the PM cost.

For each model, we obtain the expected cost rates and discuss optimum PM policies that minimize them. A numerical example is finally given when the failure time has a Weibull distribution.

(1) Model A – Probability

Consider the periodic PM policy for a one-unit system that should operate for an infinite time span.

1. The operating unit is maintained preventively at times kT ($k = 1, 2, \dots$), and undergoes only minimal repair at failures between PMs (see Chapter 4).
2. The failure rate $h(t)$ remains undisturbed by minimal repair.
3. The unit after PM has the same failure rate as it had before PM with probability p ($0 \le p < 1$) and becomes as good as new with probability $q \equiv 1 - p$.
4. Cost of each minimal repair is c_1 and cost of each PM is c_2.
5. The minimal repair and PM times are negligible.
6. The failure rate $h(t)$ is strictly increasing.

Consider one cycle from time $t = 0$ to the time that the unit becomes as good as new by perfect PM. Then, the total expected cost of one cycle is

$$\sum_{j=1}^{\infty} p^{j-1} q \left[c_1 \int_0^{jT} h(t)\,dt + jc_2 \right] = c_1 q \sum_{j=1}^{\infty} p^{j-1} \int_0^{jT} h(t)\,dt + \frac{c_2}{q} \tag{7.11}$$

and its mean time is

$$\sum_{j=1}^{\infty} jTp^{j-1}q = \frac{T}{q}. \tag{7.12}$$

Thus, dividing (7.11) by (7.12) and arranging them, the expected cost rate is

$$C_A(T; p) = \frac{1}{T}\left[c_1 q^2 \sum_{j=1}^{\infty} p^{j-1} \int_0^{jT} h(t)\,dt + c_2 \right]. \tag{7.13}$$

We seek an optimum PM time T^* that minimizes $C_A(T; p)$. Differentiating $C_A(T; p)$ with respect to T and setting it equal to zero,

$$\sum_{j=1}^{\infty} p^{j-1} \int_0^{jT} t\,dh(t) = \frac{c_2}{c_1 q^2} \tag{7.14}$$

whose left-hand side is strictly increasing from 0 to $\int_0^{\infty} t\,dh(t)$, which may be possibly infinity. It is clearly seen that $\int_0^{\infty} t\,dh(t) \to \infty$ as $h(t) \to \infty$.

Therefore, we have the following optimum policy.

7.2 Preventive Maintenance with Minimal Repair

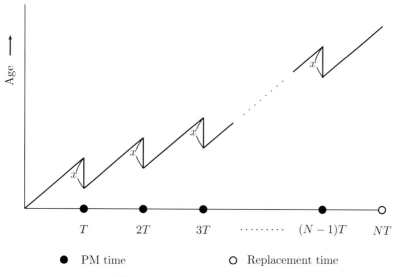

Fig. 7.1. Process of Model B

(i) If $\int_0^\infty t\, dh(t) > c_2/(c_1 q^2)$ then there exists a finite and unique T^* that satisfies (7.14), and the resulting cost rate is

$$C_A(T^*; p) = c_1 q^2 \sum_{j=1}^{\infty} p^{j-1} j h(jT^*). \tag{7.15}$$

(ii) If $\int_0^\infty t\, dh(t) \le c_2/(c_1 q^2)$ then $T^* = \infty$, and the expected cost rate is

$$C_A(\infty; p) \equiv \lim_{T \to \infty} C_A(T; p) = c_1 q^2 h(\infty).$$

(2) Model B – Age

The process in Model B is shown in Figure 7.1.

3. The age becomes x units younger at each PM, where x $(0 \le x \le T)$ is constant and previously specified. Furthermore, the unit is replaced if it operates for the time interval NT $(N = 1, 2, \ldots, \infty)$.
4. Cost of each minimal repair is c_1, cost of each PM is c_2, and cost of replacement at time NT is c_3 with $c_3 > c_2$.

1, 2, 5, 6. Same as the assumptions of Model A.

The expected cost rate is easily given by

$$C_B(N; T, x) = \frac{1}{NT} \left[c_1 \sum_{j=0}^{N-1} \int_{j(T-x)}^{T+j(T-x)} h(t)\, dt + (N-1)c_2 + c_3 \right]$$

$$(N = 1, 2, \ldots). \tag{7.16}$$

It is trivial that the expected cost rate is decreasing in x because the failure rate $h(t)$ is increasing.

We seek an optimum replacement number N^* ($1 \leq N^* \leq \infty$) that minimizes $C_B(N; T, x)$ for specified $T > 0$ and x. From the inequality $C_B(N+1; T, x) \geq C_B(N; T, x)$, we have

$$L(N; T, x) \geq \frac{(c_3 - c_2)}{c_1} \quad (N = 1, 2, \ldots), \tag{7.17}$$

where

$$L(N; T, x) \equiv N \int_{N(T-x)}^{T+N(T-x)} h(t)\, dt - \sum_{j=0}^{N-1} \int_{j(T-x)}^{T+j(T-x)} h(t)\, dt$$

$$= \sum_{j=0}^{N-1} \int_0^T \{h[t + N(T-x)] - h[t + j(T-x)]\}\, dt \quad (N = 1, 2, \ldots).$$

In addition, we have

$$L(N+1; T, x) - L(N; T, x)$$
$$= (N+1) \int_0^T \{h[t + (N+1)(T-x)] - h[t + N(T-x)]\}\, dt > 0.$$

Therefore, we have the following optimum policy.

(i) If $L(\infty; T, x) \equiv \lim_{N \to \infty} L(N; T, x) > (c_3 - c_2)/c_1$ then there exists a finite and unique minimum N^* that satisfies (7.17).
(ii) If $L(\infty; T, x) \leq (c_3 - c_2)/c_1$ then $N^* = \infty$, and the expected cost rate is

$$C_B(\infty; T, x) \equiv \lim_{N \to \infty} C_B(N; T, x) = c_1 h(\infty) + \frac{c_2}{T}.$$

We clearly have $N^* < \infty$ if $h(t) \to \infty$ as $t \to \infty$.

(3) Model C – Rate

It is assumed that:

3. The age after PM reduces to at ($0 < a \leq 1$) when it was t before PM; i.e., the age becomes $t(1-a)$ units of time younger at each PM. Furthermore, the unit is replaced if it operates for NT.

1, 2, 4, 5, 6. Same as the assumptions of Model B.

The expected cost rate is

$$C_C(N; T, a) = \frac{1}{NT} \left[c_1 \sum_{j=0}^{N-1} \int_{A_j T}^{(A_j+1)T} h(t)\, dt + (N-1)c_2 + c_3 \right]$$

$$(N = 1, 2, \ldots), \tag{7.18}$$

7.2 Preventive Maintenance with Minimal Repair

where $A_j \equiv a + a^2 + \cdots + a^j$ $(j = 1, 2, \ldots)$ and $A_0 \equiv 0$.

We can have similar results to Model B. From the inequality $C_C(N+1; T, a) \geq C_C(N; T, a)$,

$$L(N; T, a) \geq \frac{c_3 - c_2}{c_1} \quad (N = 1, 2, \ldots), \quad (7.19)$$

where

$$L(N; T, a) \equiv N \int_{A_N T}^{(A_N+1)T} h(t)\,dt - \sum_{j=0}^{N-1} \int_{A_j T}^{(A_j+1)T} h(t)\,dt \quad (N = 1, 2, \ldots)$$

which is strictly increasing in N.

Therefore, we have the following optimum policy.

(i) If $L(\infty; T, a) > (c_3 - c_2)/c_1$ then there exists a finite and unique minimum N^* that satisfies (7.19).
(ii) If $L(\infty; T, a) \leq (c_3 - c_2)/c_1$ then $N^* = \infty$.

If the age after the jth PM reduces to $a_j t$ when it was t before the jth PM, we have the expected cost $C_c(N; T, a_j)$ by denoting that $A_j \equiv a_1 + a_1 a_2 + \cdots + a_1 a_2 \ldots a_j$ $(j = 1, 2, \ldots)$ and $A_0 \equiv 0$.

Next, it is assumed that:

3. The failure rate after PM reduces to $bh(t)$ $(0 < b \leq 1)$ when it was $h(t)$ before PM.

The expected cost rate is

$$C_C(N; T, b) = \frac{1}{NT}\left[c_1 \sum_{j=0}^{N-1} b^j \int_{jT}^{(j+1)T} h(t)\,dt + (N-1)c_2 + c_3\right]$$

$$(N = 1, 2, \ldots) \quad (7.20)$$

and (7.19) is rewritten as

$$L(N; T, b) \geq \frac{c_3 - c_2}{c_1} \quad (N = 1, 2, \ldots), \quad (7.21)$$

where

$$L(N; T, b) \equiv N b^N \int_{NT}^{(N+1)T} h(t)\,dt - \sum_{j=0}^{N-1} b^j \int_{jT}^{(j+1)T} h(t)\,dt \quad (N = 1, 2, \ldots)$$

which is strictly increasing in N.

If the failure rate becomes $h_j(t)$ for $jT \leq t < (j+1)T$ between the jth and $(j+1)$th PMs, the expected cost rate in (7.20) is written in the general form

$$C_c(N; T) = \frac{1}{NT}\left[c_1 \sum_{j=0}^{N-1} \int_{jT}^{(j+1)T} h_j(t)\,dt + (N-1)c_2 + c_3\right].$$

(4) Model D – Cost

Models B and C have assumed that the age reduced by PM is independent of PM cost. In this model, it is assumed that:

3. The age or failure rate after PM is reduced in proportion to PM cost c_2.
4. Cost of each minimal repair is c_1 and cost of each PM is c_2. Furthermore, the cost c_0 with $c_0 \geq c_2$ is the initial cost of the unit.

1, 2, 5, 6. Same as the assumptions of Model A.

First, suppose that the age after PM reduces to $[1 - (c_2/c_0)](x + T)$ at each PM when it was $x + T$ immediately before PM. If the operation of the unit enters into the steady-state then we have the equation

$$\left(1 - \frac{c_2}{c_0}\right)(x+T) = x, \quad i.e., \quad x = \left(\frac{c_0}{c_2} - 1\right)T. \tag{7.22}$$

Thus, the expected cost rate is

$$C_D(T; c_0) = \frac{1}{T}\left[c_1 \int_0^T h(t+x)\,dt + c_2\right]$$

$$= \frac{1}{T}\left[c_1 \int_{[(c_0/c_2)-1]T}^{(c_0/c_2)T} h(t)\,dt + c_2\right]. \tag{7.23}$$

Differentiating $C_D(T; c_0)$ with respect to T and setting it equal to zero,

$$\int_{[(c_0/c_2)-1]T}^{(c_0/c_2)T} t\,dh(t) = \frac{c_2}{c_1}. \tag{7.24}$$

Next, suppose that the failure rate after PM reduces to $[1-(c_2/c_0)]h(x+T)$ at each PM where it was $h(x + T)$ before PM. In the steady-state, we have

$$\left(1 - \frac{c_2}{c_0}\right)h(x+T) = h(x) \tag{7.25}$$

and the expected cost rate is

$$\tilde{C}_D(T; c_0) = \frac{1}{T}\left[c_1 \int_0^T h(t+x)\,dt + c_2\right]. \tag{7.26}$$

Thus, the age after PM is computed from (7.25), and hence, an optimum PM time T^* is computed by substituting x into (7.26) and changing T to minimize it.

We have considered four imperfect PM models and have obtained the expected cost rates. It is noted that all models are identical and agree with the standard model in Section 4.2 when $p = 0$ in Model A, $N = 1$ in Models B and C, and $c_0 = c_2$ in Model D.

7.2 Preventive Maintenance with Minimal Repair

Example 7.2. We finally consider an example when the failure time has a Weibull distribution and show how to determine optimum PM times. When $F(t) = 1 - \exp(-\lambda t^m)$ $(\lambda > 0, m > 1)$, we have the following results for each model.

(1) Model A

The expected cost rate is, from (7.13),

$$C_A(T;p) = \frac{1}{T}[c_1 q \lambda T^m g(m) + c_2],$$

where $g(m) \equiv q \sum_{j=1}^{\infty} p^{j-1} j^m$ which represents the mth moment of the geometric distribution with parameter p. The optimum PM time is, from (7.14),

$$T^* = \left[\frac{c_2}{c_1 q \lambda (m-1) g(m)}\right]^{1/m}.$$

(2) Model B

The expected cost rate is, from (7.16),

$$C_B(N;T,x)$$
$$= \frac{1}{NT}\left[c_1 \lambda \sum_{j=0}^{N-1}\{[T+j(T-x)]^m - [j(T-x)]^m\} + (N-1)c_2 + c_3\right]$$

and from (7.17),

$$\sum_{j=0}^{N-1}\{[T+N(T-x)]^m - [T+j(T-x)]^m - [N(T-x)]^m + [j(T-x)]^m\}$$

$$\geq \frac{c_3 - c_2}{\lambda c_1}$$

whose left-hand side is strictly increasing in N to ∞ for $0 \leq x < T$. Thus, there exists a finite and unique minimum N^* $(1 \leq N^* < \infty)$.

(3) Model C

The expected cost rate is, from (7.18),

$$C_C(N;T,a) = \frac{1}{NT}\left[c_1 \lambda T^m \sum_{j=0}^{N-1}[(A_j+1)^m - (A_j)^m] + (N-1)c_2 + c_3\right]$$

and from (7.19),

$$T^m \sum_{j=0}^{N-1}[(A_N+1)^m - (A_j+1)^m - (A_N)^m + (A_j)^m] \geq \frac{c_3 - c_2}{\lambda c_1}$$

whose left-hand side is strictly increasing in N to ∞ because $m > 1$. Thus, there exists a finite and unique minimum N^* ($1 \leq N^* < \infty$). Furthermore, the left-hand side of the above equation is increasing in T for a fixed N and m, and hence, the optimum N^* is a decreasing function of T.

(4) Model D

The expected cost rate is, from (7.23),

$$C_D(T;c_0) = \frac{1}{T}\left\{c_1 \lambda T^m \left[\left(\frac{c_0}{c_2}\right)^m - \left(\frac{c_0}{c_2}-1\right)^m\right] + c_2\right\}$$

and the optimum PM time is, from (7.24),

$$T^* = \left[\frac{c_2}{c_1\lambda(m-1)\{(c_0/c_2)^m - [(c_0/c_2)-1]^m\}}\right]^{1/m}.$$

Similarly, the expected cost rate in (7.26) is

$$\tilde{C}_D(T;c_0) = \frac{1}{T}\{c_1\lambda T^m[D^m - (D-1)^m] + c_2\}$$

and hence, the optimum PM time is

$$T^* = \left\{\frac{c_2}{c_1\lambda(m-1)[D^m-(D-1)^m]}\right\}^{1/m},$$

where

$$D \equiv \frac{1}{1-[1-(c_2/c_0)]^{1/(m-1)}}. \blacksquare$$

7.3 Inspection with Preventive Maintenance

In this section, we check a unit periodically to see whether it is good, and at the same time, provide preventive maintenance. For example, we test a unit, and if needed, we make the overhaul and the repair or replacement of bad parts. This policy could actually be applied to the models of production machines, standby units, and preventive medical checks for diseases [23]. The standard inspection policy is explained in detail in Chapter 8.

7.3 Inspection with Preventive Maintenance

We consider a modified inspection model in which the unit after inspection has the same age as before with probability p and becomes as good as new with probability q. Then, we obtain the following reliability quantities: (1) the mean time to failure and (2) the expected number of inspections until failure detection. When the failure rate is increasing, we investigate some properties of these quantities. Furthermore, we derive the total expected cost and the expected cost rate until failure detection. Optimum inspection times that minimize the expected costs are given numerically where the failure time has a Weibull distribution. Moreover, we propose two extended cases where the age becomes younger at each inspection; *i.e.*, the age becomes x units of time younger at each inspection and the age after inspection reduces to at when it was t before inspection. Finally, we consider two types of human error at inspection and obtain the total expected cost.

7.3.1 Imperfect Inspection

Consider the periodic inspection policy with PM for a one-unit system that should operate for an infinite time span.

1. The operating unit is inspected and maintained preventively at times kT $(k = 1, 2, \dots)$ $(0 < T < \infty)$.
2. The failed unit is detected only through inspection.
3. The unit after inspection has the same failure rate as it had before inspection with probability p $(0 \le p \le 1)$ and becomes as good as new with probability $q \equiv 1 - p$.
4. Cost of each inspection is c_1 and cost of time elapsed between a failure and its detection per unit of time is c_2.
5. Inspection and PM times are negligible.

Let $l(T; p)$ be the mean time to failure of a unit. Then, we can form the renewal-type equation:

$$l(T;p) = \sum_{j=1}^{\infty} \left\{ p^{j-1} \int_{(j-1)T}^{jT} t \, dF(t) + p^{j-1} q\overline{F}(jT)[jT + l(T;p)] \right\}. \quad (7.27)$$

The first term in the bracket on the right-hand side is the mean time until it fails between $(j-1)$th and jth inspections, and the second term is the mean time until it becomes new at the jth inspection, and after that, it fails. By solving (7.27) and arranging it,

$$l(T;p) = \frac{\sum_{j=0}^{\infty} p^j \int_{jT}^{(j+1)T} \overline{F}(t) \, dt}{\sum_{j=0}^{\infty} p^j \{\overline{F}(jT) - \overline{F}[(j+1)T]\}}. \quad (7.28)$$

In particular, when $p = 0$, *i.e.*, the unit always becomes as good as new at each inspection,

184   7 Imperfect Preventive Maintenance

$$l(T;0) = \frac{1}{F(T)} \int_0^T \overline{F}(t)\,dt \qquad (7.29)$$

which agrees with (1.6) in Chapter 1. When $p=1$, i.e., the unit after inspection has the same failure rate as before inspection, $l(T;1) = \mu$ which is the mean failure time of the unit.

Next, let $M(T;p)$ be the expected number of inspections until failure detection. Then, by a similar method to that of obtaining (7.27),

$$M(T;p) = \sum_{j=1}^{\infty} \left[p^{j-1} j\{\overline{F}[(j-1)T] - \overline{F}(jT)\} + p^{j-1} q\overline{F}(jT)[j + M(T;p)] \right];$$

i.e.,

$$M(T;p) = \frac{\sum_{j=0}^{\infty} p^j \overline{F}(jT)}{\sum_{j=0}^{\infty} p^j \{\overline{F}(jT) - \overline{F}[(j+1)T]\}}. \qquad (7.30)$$

In particular,

$$M(T;0) = \frac{1}{F(T)}, \qquad M(T;1) = \sum_{j=0}^{\infty} \overline{F}(jT). \qquad (7.31)$$

It is easy to see that

$$T\overline{F}[(j+1)T] \le \int_{jT}^{(j+1)T} \overline{F}(t)\,dt \le T\overline{F}(jT)$$

because $\overline{F}(t)$ is a nonincreasing function of t. Thus, from (7.28) and (7.30),

$$T[M(T;p) - 1] \le l(T;p) \le TM(T;p). \qquad (7.32)$$

Furthermore, it has been proved in [16] that if the failure rate is increasing then both $l(T;p)$ and $M(T;p)$ are decreasing functions of p for a fixed T. From this result, we have the inequalities

$$\mu \le l(T;p) \le \frac{1}{F(T)} \int_0^T \overline{F}(t)\,dt \qquad (7.33)$$

$$\sum_{j=0}^{\infty} \overline{F}(jT) \le M(T;p) \le \frac{1}{F(T)}, \qquad (7.34)$$

where all equalities hold when F is exponential.

The total expected cost until failure detection is (see Equation (8.1) in Chapter 8),

7.3 Inspection with Preventive Maintenance   185

Table 7.2. Optimum inspection time T^* for p and m when $c_1 = 10$ and $c_2 = 1$

p	\multicolumn{5}{c}{m}
	1.0	1.5	2.0	2.5	3.0
0.00	97	171	236	289	330
0.01	97	170	234	286	328
0.05	97	168	228	275	314
0.10	97	164	219	262	295
0.20	97	158	204	237	260
0.30	97	151	189	214	231
0.40	97	144	175	195	207

$$C(T;p) = \sum_{j=1}^{\infty}\left\{ p^{j-1}\int_{(j-1)T}^{jT}[c_1 j + c_2(jT-t)]\,\mathrm{d}F(t) \right.$$

$$\left. + p^{j-1} q \overline{F}(jT)[c_1 j + C(T;p)] \right\}.$$

Solving the above renewal equation with respect to $C(T;p)$, we have

$$C(T;p) = \frac{(c_1 + c_2 T)\sum_{j=0}^{\infty} p^j \overline{F}(jT) - c_2 \sum_{j=0}^{\infty} p^j \int_{jT}^{(j+1)T} \overline{F}(t)\,\mathrm{d}t}{\sum_{j=0}^{\infty} p^j \{\overline{F}(jT) - \overline{F}[(j+1)T]\}}$$

$$= (c_1 + c_2 T)M(T;p) - c_2 l(T;p). \quad (7.35)$$

It is easy to see that $\lim_{T\to 0} C(T;p) = \lim_{T\to\infty} C(T;p) = \infty$. Thus, there exists a finite and positive T^* that minimizes the expected cost $C(T;p)$. Also, from the relation of (7.32), we have

$$\frac{c_1 l(T;p)}{T} \leq C(T;p) \leq c_1 M(T;p) + c_2 T. \quad (7.36)$$

Example 7.3. We give a numerical example when the failure time has a Weibull distribution with shape parameter m ($m \geq 1$). Suppose that $\overline{F}(t) = \exp[-(\lambda t)^m]$, $1/\lambda = 500$, $c_1 = 10$, and $c_2 = 1$. Table 7.2 presents the optimum inspection time T^* that minimizes the expected cost $C(T;p)$ for several values of p and m. It is noted that optimum times T^* are independent of p for the particular case of $m = 1$. Except for $m = 1$, they are small when p is large. The reason is that when the failure rate increases with age, it is better to inspect early for large p. ∎

7.3.2 Other Inspection Models

Consider two inspection models with PM where the age becomes younger at each inspection. It is assumed that the age becomes x ($0 \leq x \leq T$) units of time younger at each inspection. Then, the probability that the unit does not fail until time t is

$$\overline{S}(t;T,x) = \overline{\lambda}[k(T-x);t-kT]\prod_{j=0}^{k-1}\overline{\lambda}[j(T-x);T] \quad \text{for } kT \le t < (k+1)T,$$
(7.37)

where $\lambda(t;x) \equiv [F(t+x)-F(t)]/\overline{F}(t)$ is the probability that the unit with age t fails during $(t,t+x]$. Thus, the mean time to failure is

$$\begin{aligned} l(T;x) &= \sum_{k=0}^{\infty}\int_{kT}^{(k+1)T}\overline{S}(t;T,x)\,\mathrm{d}t \\ &= \sum_{k=0}^{\infty}\left\{\prod_{j=0}^{k-1}\overline{\lambda}[j(T-x);T]\right\}\frac{\int_{k(T-x)}^{k(T-x)+T}\overline{F}(t)\,\mathrm{d}t}{\overline{F}[k(T-x)]}, \end{aligned}$$
(7.38)

where $\prod_{0}^{-1} \equiv 1$, and the expected number of inspections until failure detection is

$$\begin{aligned} M(T;x) &= \sum_{k=0}^{\infty}k\lambda[k(T-x);T]\prod_{j=0}^{k-1}\overline{\lambda}[j(T-x);T] \\ &= \sum_{k=0}^{\infty}\prod_{j=0}^{k}\overline{\lambda}[j(T-x);T]. \end{aligned}$$
(7.39)

Next, it is assumed that the age after inspection reduces to at ($0 \le a \le 1$) where it was t before inspection. Then, in similar ways to those of obtaining (7.38) and (7.39),

$$l(T;a) = \sum_{k=0}^{\infty}\left\{\prod_{j=0}^{k-1}\overline{\lambda}[A_jT;T]\right\}\frac{\int_{A_kT}^{(A_k+1)T}\overline{F}(t)\,\mathrm{d}t}{\overline{F}(A_kT)}$$
(7.40)

$$M(T;a) = \sum_{k=0}^{\infty}\prod_{j=0}^{k}\overline{\lambda}[A_jT;T],$$
(7.41)

where $A_j \equiv a + a^2 + \cdots + a^j$ ($j = 1, 2, \ldots$) and $A_0 \equiv 0$.

Note that the mean times $l(T;\cdot)$ and the expected numbers $M(T;\cdot)$ of three models are equal in both cases of $p = a = 0$ and $x = T$ (i.e., the unit becomes as good as new by perfect inspection), and $p = a = 1$ and $x = 0$ (i.e., the unit has the same age by imperfect inspection). Furthermore, substituting (7.38), (7.39) and (7.40), (7.41) into (7.35), respectively, we obtain two expected costs until failure detection.

7.3.3 Imperfect Inspection with Human Error

It is well known that a high percentage of failures in most systems is directly due to human error [48]. There are the following types of human error when we inspect a standby unit at periodic times kT $(k = 1, 2, \ldots)$ [2, 49–51]:

1. Type A human error: The unit in a good state, *i.e.*, in a normal condition, is judged to be bad and is repaired.
2. Type B human error: The unit in a bad state, *i.e.*, in a failed state, is judged to be good.

It is assumed that the probabilities of type A error and type B error are α and β, respectively, where $0 \le \alpha + \beta < 1$. Then, the expected number of inspections until a failed unit is detected is

$$\sum_{j=0}^{\infty} j \beta^{j-1}(1-\beta) = \frac{1}{1-\beta}.$$

Consider one cycle from time $t = 0$ to the time when a failed unit is detected by perfect inspection or a good unit is repaired by type A error, whichever occurs first. Then, the total expected cost of one cycle is given by

$$C(T; \alpha, \beta) = \sum_{j=0}^{\infty} (1-\alpha)^j \left[\int_{jT}^{(j+1)T} c_1 \left(j + \frac{1}{1-\beta}\right) \mathrm{d}F(t) \right.$$

$$+ \alpha c_1 (j+1) \overline{F}((j+1)T) + \int_{jT}^{(j+1)T} c_2 \left(jT + \frac{T}{1-\beta} - t\right) \mathrm{d}F(t) \Bigg]$$

$$= (c_1 + c_2 T) \left\{ \frac{1}{1-\beta} \sum_{j=0}^{\infty} (1-\alpha)^j [\overline{F}(jT) - \overline{F}((j+1)T)] \right.$$

$$\left. + \sum_{j=0}^{\infty} (1-\alpha)^j \overline{F}((j+1)T) \right\} - c_2 \sum_{j=0}^{\infty} (1-\alpha)^j \int_{jT}^{(j+1)T} \overline{F}(t) \, \mathrm{d}t. \quad (7.42)$$

When $\alpha = \beta = 0$, *i.e.*, the inspection is perfect, Equation (7.42) is equal to that of a standard periodic inspection policy (see Section 8.1).

In particular, when $F(t) = 1 - \mathrm{e}^{-\lambda t}$, the expected cost is rewritten as

$$C(T; \alpha, \beta) = (c_1 + c_2 T) \frac{(1 - \mathrm{e}^{-\lambda T})/(1-\beta) + \mathrm{e}^{-\lambda T}}{1 - (1-\alpha)\mathrm{e}^{-\lambda T}} - \frac{c_2}{\lambda} \frac{1 - \mathrm{e}^{-\lambda T}}{1 - (1-\alpha)\mathrm{e}^{-\lambda T}}. \quad (7.43)$$

Differentiating $C(T; \alpha, \beta)$ with respect to T and setting it equal to zero,

$$\frac{\mathrm{e}^{\lambda T} - 1}{\lambda} [1 - \beta(1-\alpha)\mathrm{e}^{-\lambda T}] - (1 - \alpha - \beta)T = \frac{c_1}{c_2}(1 - \alpha - \beta). \quad (7.44)$$

Note that the left-hand side of (7.44) is strictly increasing from 0 to ∞. Therefore, there exists a finite and unique T^* that satisfies (7.44).

7.4 Computer System with Imperfect Maintenance

Periodic maintenance of a computer system is imperative in order to inspect and remove as many component faults, failures, and degradations as possible. In most cases, it has been assumed that the system becomes like new and operates normally after maintenance. However, the system occasionally becomes worse for one or more of the following reasons:

(1) Hidden faults and failures that are not detected during maintenance;
(2) Human errors such as wrong adjustments and further damage done during maintenance; or
(3) Replacement with faulty parts.

It is useful to develop an imperfect maintenance strategy for a computer system.

This section considers a system that is maintained at periodic times kT ($k = 1, 2, \ldots$). Due to imperfect PM, one of the following results occurs: the system is not changed, is renewed, or is put in a failed state and needs repair. The MTTF and availability of the system are derived by the usual probability calculations. Furthermore, we calculate an optimum PM time T^* that maximizes the availability, and show that T^* is determined by a unique solution of an equation under certain conditions. A numerical example is given for a triple redundant system that fails when two or more units have failed.

A computer system begins to operate at time 0 and should operate for an infinite time span.

1. The system is maintained preventively at periodic times kT ($k = 1, 2, \ldots$) ($0 < T \leq \infty$).
2. The failed system is repaired immediately when it fails, and becomes as good as new after repair.
3. One of the following cases after PM results.
   (a) The system is not changed with probability p_1; viz, PM is imperfect.
   (b) The system becomes as good as new with probability p_2; viz, PM is perfect.
   (c) The system fails with probability p_3; viz, PM fails, where $p_1 + p_2 + p_3 = 1$ and $p_2 > 0$.
4. The mean times to repair actual failure in case 2 and maintenance failure in (c) are β_1 and β_2 with $\beta_1 \geq \beta_2$, respectively.
5. The PM time is negligible.

The probability that the system is renewed by repair upon actual failure is

$$\sum_{j=1}^{\infty} p_1^{j-1} \int_{(j-1)T}^{jT} \mathrm{d}F(t) = (1 - p_1) \sum_{j=1}^{\infty} p_1^{j-1} F(jT), \qquad (7.45)$$

the probability that the system is renewed by perfect maintenance is

7.4 Computer System with Imperfect Maintenance

$$p_2 \sum_{j=1}^{\infty} p_1^{j-1} \overline{F}(jT), \qquad (7.46)$$

and the probability that the system is renewed by repair after maintenance failure is

$$p_3 \sum_{j=1}^{\infty} p_1^{j-1} \overline{F}(jT), \qquad (7.47)$$

where $(7.45) + (7.46) + (7.47) = 1$.

Furthermore, the mean time of one cycle from time $t = 0$ to the time when the system is renewed by either repair or perfect maintenance is

$$\sum_{j=1}^{\infty} p_1^{j-1} \int_{(j-1)T}^{jT} t \, dF(t) + (p_2 + p_3) \sum_{j=1}^{\infty} jT p_1^{j-1} \overline{F}(jT)$$

$$= (1 - p_1) \sum_{j=1}^{\infty} p_1^{j-1} \int_0^{jT} \overline{F}(t) \, dt. \qquad (7.48)$$

Therefore, the mean time to failure is

$$l(T; p_1, p_2, p_3) = \sum_{j=1}^{\infty} \left\{ p_1^{j-1} \int_{(j-1)T}^{jT} t \, dF(t) \right.$$

$$\left. + p_1^{j-1} \overline{F}(jT) \left[p_2(jT + l(T; p_1, p_2, p_3)) + p_3 jT \right] \right\};$$

i.e.,

$$l(T; p_1, p_2, p_3) = \frac{(1 - p_1) \sum_{j=1}^{\infty} p_1^{j-1} \int_0^{jT} \overline{F}(t) \, dt}{1 - p_2 \sum_{j=1}^{\infty} p_1^{j-1} \overline{F}(jT)} \qquad (7.49)$$

which agrees with (5) of [11] when $p_3 = 0$, and (9) of [13].

The availability is, from (6.10) in Chapter 6,

$$A(T; p_1, p_2, p_3) = \frac{(1 - p_1) \sum_{j=1}^{\infty} p_1^{j-1} \int_0^{jT} \overline{F}(t) \, dt}{\left[\begin{array}{c} (1 - p_1) \sum_{j=1}^{\infty} p_1^{j-1} \int_0^{jT} \overline{F}(t) \, dt + \beta_2 p_3 \sum_{j=1}^{\infty} p_1^{j-1} \overline{F}(jT) \\ + \beta_1 (1 - p_1) \sum_{j=1}^{\infty} p_1^{j-1} F(jT) \end{array} \right]} \qquad (7.50)$$

which agrees with (10) of [11] when $p_3 = 0$.

First, we seek an optimum PM time T_1^* that maximizes MTTF $l(T; p_1, p_2, p_3)$ in (7.49). It is evident that

7 Imperfect Preventive Maintenance

$$l(0; p_1, p_2, p_3) \equiv \lim_{T \to 0} l(T; p_1, p_2, p_3) = 0$$

$$l(\infty; p_1, p_2, p_3) \equiv \lim_{T \to \infty} l(T; p_1, p_2, p_3) = \mu. \tag{7.51}$$

Thus, there exists some positive T_1^* $(0 < T_1^* \le \infty)$ that maximizes $l(T; p_1, p_2, p_3)$. Differentiating $l(T; p_1, p_2, p_3)$ with respect to T and setting it equal to zero, we have

$$H(T; p_1) \sum_{j=1}^{\infty} p_1^{j-1} \int_0^{jT} \overline{F}(t)\,dt + \sum_{j=1}^{\infty} p_1^{j-1} \overline{F}(jT) = \frac{1}{p_2}, \tag{7.52}$$

where

$$H(T; p_1) \equiv \frac{\sum_{j=1}^{\infty} p_1^{j-1} j f(jT)}{\sum_{j=1}^{\infty} p_1^{j-1} j \overline{F}(jT)}.$$

It can be shown that the left-hand side of (7.52) is strictly increasing from $1/(1-p_1)$ to $\mu H(\infty; p_1)/(1-p_1)$ when $H(t; p_1)$ is strictly increasing. Thus, the optimum policy is:

(i) If $H(T; p_1)$ is strictly increasing and $H(\infty; p_1) > (1-p_1)/(\mu p_2)$ then there exists a finite and unique T_1^* that satisfies (7.52), and the resulting MTTF is

$$l(T_1^*; p_1, p_2, p_3) = \frac{1-p_1}{p_2 H(T_1^*; p_1)}. \tag{7.53}$$

(ii) If $H(T; p_1)$ is nonincreasing, or $H(T; p_1)$ is strictly increasing and $H(\infty; p_1) \le (1-p_1)/(\mu p_2)$, then $T_1^* = \infty$; viz, no PM should be done, and the MTTF is given in (7.51).

Next, we seek an optimum PM time T_2^* that maximizes the availability $A(T; p_1, p_2, p_3)$ in (7.50). Differentiating $A(T; p_1, p_2, p_3)$ with respect to T and setting it equal to zero imply

$$H(T; p_1) \sum_{j=1}^{\infty} p_1^{j-1} \int_0^{jT} \overline{F}(t)\,dt + \sum_{j=1}^{\infty} p_1^{j-1} \overline{F}(jT) = \frac{\beta_1}{\beta_1(1-p_1) - \beta_2 p_3}. \tag{7.54}$$

Note that $\beta_1(1-p_1) > \beta_2 p_3$ because $\beta_1 \ge \beta_2$.

Thus, we have a similar optimum policy to the previous case. Also, it is of interest that $T_1^* \ge T_2^*$ because $\beta_1/[\beta_1(1-p_1) - \beta_2 p_3] \le 1/p_2$.

Example 7.4. Consider a triple redundant system that consists of three units, and fails when two or more units have failed. This system is a 2-out-of-3 system and is applied to the design of a fail-safe system. The failure distribution of the system is $\overline{F}(t) = 3e^{-2t} - 2e^{-3t}$, and the mean time to failure is $\mu = 5/6$. In addition, we have

$$H(t;p_1) = \frac{6\sum_{j=1}^{\infty} p_1^{j-1} j(e^{-2jt} - e^{-3jt})}{\sum_{j=1}^{\infty} p_1^{j-1} j(3e^{-2jt} - 2e^{-3jt})}$$

$$H(0;p_1) = 0, \qquad H(\infty;p_1) = 2$$

$$\frac{\mathrm{d}H(t;p_1)}{6\,\mathrm{d}t} = \frac{1}{D}\left[6\sum_{j=1}^{\infty} p_1^{j-1} j^2(e^{-2jt} - e^{-3jt}) \sum_{j=1}^{\infty} p_1^{j-1} j(e^{-2jt} - e^{-3jt}) \right.$$

$$\left. - \sum_{j=1}^{\infty} p_1^{j-1} j^2(2e^{-2jt} - 3e^{-3jt}) \sum_{j=1}^{\infty} p_1^{j-1} j(3e^{-2jt} - 2e^{-3jt}) \right]$$

$$= \frac{1}{D}\left[\sum_{j=1}^{\infty}\sum_{i=1}^{\infty} p_1^{i+j-2} (i^2 j)(3e^{-it} - 2e^{-jt}) e^{-2(i+j)t} \right] > 0,$$

where

$$D \equiv \left[\sum_{j=1}^{\infty} p_1^{j-1} j(3e^{-2jt} - 2e^{-3jt}) \right]^2.$$

Thus, $H(t;p_1)$ is strictly increasing from 0 to 2.

Therefore, if $1 - p_1 > (5/2)(\beta_2/\beta_1)p_3$ then there exists a finite and unique T_2^* that satisfies

$$\frac{H(T;p_1)}{6}\left\{ \sum_{j=1}^{\infty} p_1^{j-1}[9(1 - e^{-2jT}) - 4(1 - e^{-3jT})] \right\}$$

$$+ \sum_{j=1}^{\infty} p_1^{j-1}(3e^{-2jT} - 2e^{-3jT}) = \frac{\beta_1}{\beta_1(1 - p_1) - \beta_2 p_3}$$

and otherwise, $T_2^* = \infty$.

Table 7.3 shows the optimum PM time T_2^* ($\times 10^2$) for $p_1 = 10^{-3}$, 10^{-2}, 10^{-1}, $p_3 = 10^{-4}$, 10^{-3}, 10^{-2}, 10^{-1}, and $\beta_2/\beta_1 = 0.1, 1.0$. For example, when $p_1 = 0.1$, $p_3 = 0.01$, and $\beta_2/\beta_1 = 0.1$, $T_2^* = 1.72 \times 10^{-2}$. If the MTTF of each unit is 10^4 hours then $T_2^* = 172$ hours. These results indicate that the system should be maintained about once a week. Furthermore, it is of great interest that the optimum T_2^* depends considerably on the product of β_2/β_1 and p_3, but depends little on p_1. When $(\beta_2/\beta_1)p_3 = 10^{-4}, 10^{-3}, 10^{-2}, 10^{-1}$, the approximate optimum times T_2^* are $0.005, 0.018, 0.06, 0.28$, respectively. ∎

7.5 Sequential Imperfect Preventive Maintenance

We consider the following two PM policies, by introducing improvement factors [15,52] in failure rate and age for a sequential PM policy [53,54]: the PM

Table 7.3. Optimum PM time T_2^* ($\times 10^2$) to maximize availability $A(T; p_1, p_2, p_3)$ for p_1, p_2 and β_2/β_1

	$\beta_2/\beta_1 = 0.1$			$\beta_2/\beta_1 = 1.0$		
p_3	p_1					
	10^{-3}	10^{-2}	10^{-1}	10^{-3}	10^{-2}	10^{-1}
10^{-4}	0.183	0.181	0.166	0.582	0.578	0.529
10^{-3}	0.582	0.578	0.529	1.88	1.87	1.72
10^{-2}	1.88	1.87	1.72	6.42	6.37	5.98
10^{-1}	6.42	6.37	5.98	28.1	28.1	27.6

is done at fixed intervals T_k ($k = 1, 2, \ldots, N-1$) and is replaced at the Nth PM; if the system fails between PMs, it undergoes only minimal repair. The PM is imperfect as follows.

(1) The age after the kth PM reduces to $a_k t$ when it was t before PM.
(2) The failure rate after the kth PM becomes $b_k h(t)$ when it was $h(t)$ in the period of the kth PM.

The imperfect PM model that combines two policies was considered in [55].

The expected cost rates of two models are obtained and optimum sequences $\{T_k^*\}$ are derived. When the failure time has a Weibull distribution, optimum policies are computed explicitly.

(1) Model A – Age

Consider the sequential PM policy for a one-unit system for an infinite time span. It is assumed that (see Figure 7.2):

1. The PM is done at fixed intervals T_k ($k = 1, 2, \ldots, N-1$) and is replaced at the Nth PM; i.e., the unit is maintained preventively at successive times $T_1 < T_1 + T_2 < \cdots < T_1 + T_2 + \cdots + T_{N-1}$ and is replaced at time $T_1 + T_2 + \cdots + T_N$, where $T_0 \equiv 0$.
2. The unit undergoes only minimal repair at failures between replacements and becomes as good as new at replacement.
3. The age after the kth PM reduces to $a_k t$ when it was t before PM; i.e., the unit with age t becomes $t(1 - a_k)$ units of time younger at the kth PM, where $0 = a_0 < a_1 \leq a_2 \leq \cdots \leq a_N < 1$.
4. Cost of each minimal repair is c_1, cost of each PM is c_2, and cost of replacement at the Nth PM is c_3.
5. The times for PM, repair, and replacement are negligible.

The unit is aged from $a_{k-1}(T_{k-1} + a_{k-2}T_{k-2} + \cdots + a_{k-2}a_{k-3}\ldots a_2 a_1 T_1)$ after the $(k-1)$th PM to $T_k + a_{k-1}(T_{k-1} + a_{k-2}T_{k-2} + \cdots + a_{k-2}a_{k-3}\ldots a_2 a_1 T_1)$ before the kth PM, i.e., from $a_{k-1} Y_{k-1}$ to Y_k, where $Y_k \equiv T_k + a_{k-1} T_{k-1} + \cdots + a_{k-1} a_{k-2} + \cdots + a_2 a_1 T_1$ ($k = 1, 2, \ldots$), which is the age immediately before the kth PM. Thus, the expected cost rate is

7.5 Sequential Imperfect Preventive Maintenance

$$C_A(Y_1, Y_2, \ldots, Y_N) = \frac{c_1 \sum_{k=1}^{N} \int_{a_{k-1}Y_{k-1}}^{Y_k} h(t)\,dt + (N-1)c_2 + c_3}{\sum_{k=1}^{N-1}(1-a_k)Y_k + Y_N} \qquad (N = 1, 2, \ldots) \qquad (7.55)$$

because $T_k = Y_k - a_{k-1}Y_{k-1}$ and $\sum_{k=1}^{N} T_k = \sum_{k=1}^{N-1}(1-a_k)Y_k + Y_N$.

To find an optimum sequence $\{Y_k\}$ that minimizes $C_A(Y_1, Y_2, \ldots, Y_N)$, differentiating $C_A(Y_1, Y_2, \ldots, Y_N)$ with respect to Y_k and setting it equal to zero,

$$\frac{h(Y_k) - a_k h(a_k Y_k)}{1 - a_k} = h(Y_N) \qquad (k = 1, 2, \ldots, N-1) \qquad (7.56)$$

$$c_1 h(Y_N) = C_A(Y_1, Y_2, \ldots, Y_N). \qquad (7.57)$$

Suppose that Y_N $(0 < Y_N < \infty)$ is fixed. If $h(t)$ is strictly increasing then there exists some Y_k $(0 < Y_k < Y_N)$ that satisfies (7.56), because

$$\frac{h(0) - a_k h(0)}{1 - a_k} < h(Y_N), \qquad \frac{h(Y_N) - a_k h(a_k Y_N)}{1 - a_k} > h(Y_N).$$

Furthermore, if $dh(t)/dt$ is also strictly increasing then a solution to (7.56) is unique.

Thus, substituting each Y_k into (7.57), its equation becomes a function of Y_N only which is

$$h(Y_N)\left[\sum_{k=1}^{N-1}(1-a_k)Y_k + Y_N\right] - \sum_{k=1}^{N}\int_{a_{k-1}Y_{k-1}}^{Y_k} h(t)\,dt = \frac{(N-1)c_2 + c_3}{c_1}, \qquad (7.58)$$

where each Y_k $(k = 1, 2, \ldots, N-1)$ is given by some function of Y_N. If there exists a solution Y_N to (7.58) then a sequence $\{Y_k\}$ minimizes the expected cost $C_A(Y_1, Y_2, \ldots, Y_N)$.

Finally, suppose that $Y_1, Y_2, \ldots, Y_N$ are determined from (7.56) and (7.58). Then, from (7.57), the resulting cost rate is $c_1 h(Y_N)$, which is a function of N. To complete an optimum PM schedule, we may seek an optimum number N^* that minimizes $h(Y_N)$.

From the above discussion, we can specify the computing procedure for obtaining the optimum PM schedule.

1. Solve (7.56) and express Y_k $(k = 1, 2, \ldots, N-1)$ by a function of Y_N.
2. Substitute Y_k into (7.58) and solve it with respect to Y_N.
3. Determine N^* that minimizes $h(Y_N)$.
4. Compute T_k^* $(k = 1, 2, \ldots, N^*)$ from $T_k = Y_k - a_{k-1}Y_{k-1}$.

(2) Model B – Failure rate

3. The failure rate after the kth PM becomes $b_k h(t)$ when it was $h(t)$ before PM; i.e., the unit has the failure rate $B_k h(t)$ in the kth PM period, where $1 = b_0 < b_1 \leq b_2 \leq \cdots \leq b_{N-1}$, $B_k \equiv \prod_{j=0}^{k-1} b_j$ ($k = 1, 2, \ldots, N$) and $1 = B_1 < B_2 < \cdots < B_N$.

1, 2, 4, 5. Same as the assumptions of Model A.
The expected cost rate is

$$C_B(T_1, T_2, \ldots, T_N) = \frac{c_1 \sum_{k=1}^{N} B_k \int_0^{T_k} h(t)\,dt + (N-1)c_2 + c_3}{T_1 + T_2 + \cdots + T_N}$$
$$(N = 1, 2, \ldots). \quad (7.59)$$

Differentiating $C_B(T_1, T_2, \ldots, T_N)$ with respect to T_k and setting it equal to zero, we have

$$B_1 h(T_1) = B_2 h(T_2) = \cdots = B_N h(T_N) \quad (7.60)$$
$$c_1 B_k h(T_k) = C_B(T_1, T_2, \ldots, T_N) \quad (k = 1, 2, \ldots, N). \quad (7.61)$$

When the failure rate is strictly increasing to infinity, we can specify the computing procedure for obtaining an optimum schedule.

1. Solve $B_k h(T_k) = D$ and express T_k ($k = 1, 2, \ldots, N$) by a function of D.
2. Substitute T_k into (7.60) and solve it with respect to D.
3. Determine N^* that minimizes D.

Example 7.5. Suppose that the failure time has a Weibull distribution; i.e., $h(t) = m t^{m-1}$ for $m > 1$. From the computing procedure of Model A, by solving (7.56), we have

$$Y_k = \left(\frac{1 - a_k}{1 - a_k^m}\right)^{1/(m-1)} Y_N \quad (k = 1, 2, \ldots, N-1). \quad (7.62)$$

Substituting Y_k into (7.58) and arranging it,

$$Y_N = \left[\frac{(N-1)c_2 + c_3}{(m-1)c_1 \sum_{k=0}^{N-1} d_k}\right]^{1/m}, \quad (7.63)$$

where

$$d_k \equiv (1 - a_k)\left(\frac{1 - a_k}{1 - a_k^m}\right)^{1/(m-1)} \quad (k = 0, 1, 2, \ldots, N-1).$$

Next, we consider the problem that minimizes

7.5 Sequential Imperfect Preventive Maintenance

$$C_A(N) \equiv \frac{(N-1)c_2 + c_3}{\sum_{k=0}^{N-1} d_k} \quad (N = 1, 2, \dots) \tag{7.64}$$

which is the same problem as minimizing $h(Y_N)$, i.e., $C_A(Y_1, Y_2, \dots, Y_N)$. From the inequality $C_A(N+1) \geq C_A(N)$, we have

$$L_A(N) \geq \frac{c_3}{c_2} \quad (N = 1, 2, \dots), \tag{7.65}$$

where

$$L_A(N) \equiv \sum_{k=0}^{N-1} \frac{d_k}{d_N} - (N-1) \quad (N = 1, 2, \dots). \tag{7.66}$$

If d_k is decreasing in k then $L_A(N)$ is increasing in N. Thus, there exists a finite and unique minimum N^* that satisfies (7.65) if $L_A(\infty) > c_3/c_2$.

We show that d_k is decreasing in k from the assumption that $a_k < a_{k+1}$. Let $g(x) \equiv (1-x)^m/(1-x^m)$ $(0 < x < 1)$ for $m > 1$. Then, $g(x)$ is decreasing from 1 to 0, and hence,

$$\frac{(1-a_k)^m}{1-a_k^m} > \frac{(1-a_{k+1})^m}{1-a_{k+1}^m}$$

which follows that $d_k > d_{k+1}$. Furthermore, if $a_k \to 1$ as $k \to \infty$ then

$$\lim_{k \to \infty} d_k = \lim_{x \to 1} [g(x)]^{1/(m-1)} = 0;$$

i.e., $L_A(N) \to \infty$ as $N \to \infty$, and a finite N^* exists uniquely.

Therefore, if $a_k \to 1$ as $k \to \infty$ then an N^* is a finite and unique minimum that satisfies (7.65), and the optimum intervals are $T_k^* = Y_k - a_{k-1}Y_{k-1}$ ($k = 1, 2, \dots, N^*$), where Y_k and Y_N are given in (7.62) and (7.63).

For Model B, by solving $B_k h(T_k) = D$, we have

$$T_k = \left(\frac{D}{mB_k}\right)^{1/(m-1)} \quad (k = 1, 2, \dots, N). \tag{7.67}$$

Substituting T_k into (7.61) and arranging it,

$$D = \left[\frac{(N-1)c_2 + c_3}{c_1\left(1 - \frac{1}{m}\right)\sum_{k=1}^{N}[(1/mB_k)]^{1/(m-1)}}\right]^{(m-1)/m} \tag{7.68}$$

which is a function of N. Let us denote D by $D(N)$. Then, from the inequality $D(N+1) \geq D(N)$, an N^* to minimize D is given by a unique minimum that satisfies

$$L_B(N) \geq \frac{c_3}{c_2} \quad (N = 1, 2, \dots), \tag{7.69}$$

196   7 Imperfect Preventive Maintenance

Table 7.4. Optimum N^* and PM intervals of Model A when $c_1/c_2 = 3$

c_3/c_2	2	5	10	20	40
N^*	1	2	4	7	11
T_1	0.54	0.82	1.07	1.40	1.84
T_2		0.82	0.43	0.56	0.74
T_3			0.28	0.36	0.48
T_4			0.92	0.27	0.35
T_5				0.21	0.28
T_6				0.18	0.23
T_7				1.13	0.20
T_8					0.17
T_9					0.15
T_{10}					0.14
T_{11}					1.45

Table 7.5. Optimum N^* and PM intervals of Model B when $c_1/c_2 = 3$

c_3/c_2	2	5	10	20	40
N^*	2	3	4	5	6
T_1	0.77	1.06	1.37	1.82	2.45
T_2	0.52	0.71	0.92	1.21	1.64
T_3		0.43	0.55	0.73	0.98
T_4			0.31	0.42	0.56
T_5				0.23	0.31
T_6					0.17

where

$$L_B(N) \equiv \sum_{k=1}^{N} \left(\frac{B_{N+1}}{B_k}\right)^{1/(m-1)} - (N-1) \quad (N = 1, 2, \dots)$$

which is increasing in N because B_k is increasing in k. Also, if $B_k \to \infty$ as $k \to \infty$ then $L_B(N) \to \infty$ as $N \to \infty$, and hence, a finite N^* exists uniquely in (7.69), and the optimum intervals T_k^* ($k = 1, 2, \dots, N^*$) are given in (7.67) and (7.68).

Tables 7.4 and 7.5 present the optimum number N^* and the PM intervals $T_1^*, T_2^*, \dots, T_N^*$ for $c_3/c_2 = 2, 5, 10, 20, 40$, where $c_1/c_2 = 3$, $m = 2$, and $a_k = k/(k+1)$, $b_k = 1 + k/(k+1)$ ($k = 0, 1, 2, \dots$). These examples show that $T_1^* > T_2^* > \dots > T_N^*$ for Model B, but $T_1^* > T_N^* > T_2^*$ for $c_3/c_2 = 10, 20, 40$ of Model A. This indicates that it would be reasonable to do frequent PM with age, but it would be better to do the last PM as late as possible because the system should be replaced at the next PM. Figure 7.2 shows the graph of Model A for time and age when $c_3/c_2 = 10$. ∎

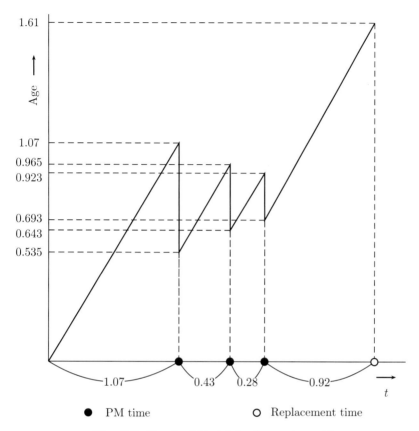

Fig. 7.2. Graph of Model A when $c_3/c_2 = 10$

References

1. Weiss GH (1962) A problem in equipment maintenance. Manage Sci 8:266–277.
2. Coleman JJ, Abrams IJ (1962) Mathematical model for operational readiness. Oper Res 10:126–133.
3. Noonan GC, Fain CG (1962) Optimum preventive maintenance policies when immediate detection of failure is uncertain. Oper Res 10:407–410.
4. Chan PKW, Downs T (1978) Two criteria for preventive maintenance. IEEE Trans Reliab R-27:272–273.
5. Nakagawa T (1979) Optimal policies when preventive maintenance is imperfect. IEEE Trans Reliab R-28:331–332.
6. Nakagawa T (1979), Imperfect preventive-maintenance, IEEE Trans Reliab R-28:402.
7. Murthy DNP, Nguyen DG (1981) Optimal age-policy with imperfect preventive maintenance. IEEE Trans Reliab R-30:80–81.
8. Zhao YX (2003) On preventive maintenance policy of a critical reliability level for system subject to degradation. Reliab Eng Syst Saf 79:301–308.

9. Ingle AD, Siewiorek DP (1977) Reliability models for multiprocess systems with and without periodic maintenance. 7th International Symposium Fault-Tolerant Computing:3–9.
10. Helvic BE (1980) Periodic maintenance on the effect of imperfectness. 10th International Symposium Fault-Tolerant Computing:204–206.
11. Yak YW, Dillon TS, Forward KE (1984) The effect of imperfect periodic maintenance of fault-tolerant computer system. 14th International Symposium Fault-Tolerant Computing:66–70.
12. Nakagawa T, Yasui K (1987) Optimum policies for a system with imperfect maintenance. IEEE Trans Reliab R-36:631–633.
13. Chung KJ (1995) Optimal test-times for intermittent faults. IEEE Trans Reliab 44:645–647.
14. Nakagawa T (1980) Mean time to failure with preventive maintenance. IEEE Trans Reliab R-29:341.
15. Nakagawa T (1980) A summary of imperfect preventive maintenance policies with minimal repair. RAIRO Oper Res 14:249–255.
16. Lie CH, Chun YH (1986) An algorithm for preventive maintenance policy. IEEE Trans Reliab R-35:71–75.
17. Zhang F, Jardine AKS (1998) Optimal maintenance models with minimal repair, periodic overhaul and complete renewal. IIE Trans 30:1109-1119.
18. Canfield RV (1986) Cost optimization of periodic preventive maintenance. IEEE Trans Reliab R-35:78–81.
19. Park DH, Jung GM, Yum JK (2000) Cost minimization for periodic maintenance policy of a system subject to slow degradation. Reliab Eng Syst Saf 68:105-112.
20. Brown M, Proschan F (1983) Imperfect repair. J Appl Prob 20:851–859.
21. Fontenot RA, Proschan F (1984) Some imperfect maintenance models. In: Abdel-Hameed MS, Çinlar E, Quinn J (eds) Reliability Theory and Models. Academic, Orlando, FL:83–101.
22. Bhattacharjee MC (1987) New results for the Brown–Proschan model of imperfect repair. J Statist Plan Infer 16:305–316.
23. Ebrahimi N (1985) Mean time to achieve a failure-free requirement with imperfect repair. IEEE Trans Reliab R-34:34–37.
24. Natvig B (1990) On information based minimal repair and the reduction in remaining system lifetime due to the failure of a specific module. J Appl Prob 27:365–375.
25. Zhao M (1994) Availability for repairable components and series systems. IEEE Trans Reliab 43:329–334.
26. Block HW, Borges WS, Savits TH (1985) Age-dependent minimal repair. J Appl Prob 22:370–385.
27. Abdel-Hameed MS (1987) An imperfect maintenance model with block replacements. Appl Stoch Models Data Analysis 3:63–72.
28. Kijima M (1989) Some results for repairable systems with general repair. J Appl Prob 26:89–102.
29. Stadje W, Zuckerman D (1991) Optimal maintenance strategies for repairable systems with general degree of repair. J Appl Prob 28:384–396.
30. Makis V, Jardine AKS (1993) A note on optimal replacement policy under general repair. Eur J Oper Res 69:75–82.
31. Doyen L, Gaudoin O (2004) Classes of imperfect repair models based on reduction of failure intensity or virtual age. Reliab Eng Syst Saf 84:45–46.

32. Wang H, Pham H (1996) Optimal age-dependent preventive maintenance policies with imperfect maintenance. Int J Reliab Qual Saf Eng 3:119–135.
33. Li HJ, Shaked M (2003) Imperfect repair models with preventive maintenance. J Appl Prob 40:1043–1059.
34. Shaked M, Shanthikumar JG (1986) Multivariate imperfect repair. Oper Res 34:437–448.
35. Sheu SH, Griffith WS (1991) Multivariate age-dependent imperfect repair. Nav Res Logist 38:839–850.
36. Sheu SH, Griffith WS (1992) Multivariate imperfect repair. J Appl Prob 29:947–956.
37. Malik MAK (1979) Reliable preventive maintenance scheduling. AIIE Trans 11:221–228.
38. Whitaker LR, Samaniego FJ (1989) Estimating the reliability of systems subject to imperfect repair. J Amer Statist Assoc 84:301–309.
39. Guo R, Love CE (1992) Statistical analysis of an age model for imperfectly repaired systems. Qual Reliab Eng Inter 8:133–146.
40. Shin I, Lin TJ, Lie CH (1996) Estimating parameters of intensity function and maintenance effect for reliable unit. Reliab Eng Syst Saf 54:1-10.
41. Pulcini G (2003) Mechanical reliability and maintenance models. In: Pham H (ed) Handbook of Reliability Engineering. Springer, London:317–348.
42. Nakagawa T (2000) Imperfect preventive maintenance models. In: Ben-Daya M, Duffuaa SO, Raouf A (eds) Maintenance, Modeling and Optimization. Kluwer Academic, Boston:201–214.
43. Nakagawa T (2002) Imperfect preventive maintenance models. In: Osaki S (ed) Stochastic Models in Reliability and Maintenance. Springer, New York:125–143.
44. Wang H, Pham H (2003) Optimal imperfect maintenance models. In: Pham H (ed) Handbook of Reliability Engineering. Springer, London:397–414.
45. Nakagawa T (1980) Replacement models with inspection and preventive maintenance. Microelectron Reliab 20:427–433.
46. Nakagawa T (1984) Periodic inspection policy with preventive maintenance. Nov Res Logist Q 31:33–40.
47. Nakagawa T (1988) Sequential imperfect preventive maintenance policies. IEEE Trans Reliab 37:295–298.
48. Dhillon BS (1986) Human Reliability with Human Factors. Pergamon, New York.
49. Gertsbakh I (1977) Models of Preventive Maintenance. North-Holland, Amsterdam.
50. Badia FG, Berrade MD, Campos CA (2001) Optimization on inspection intervals based on cost. J Appl Prob 38:872–881.
51. Badia FG, Berrade MD, Campos CA (2002) Optimal inspection and preventive maintenance of units with revealed and unrevealed failures. Reliab Eng Syst Saf 78:157–163.
52. Ng YW, Avizienis A (1980) A unified reliability model for fault-tolerant computers. IEEE Trans Comp C-29:1002–1011.
53. Nakagawa T (1986) Periodic and sequential preventive maintenance policies. J Appl Prob 23:536–542.
54. Nguyen DC, Murthy DN (1981) Optimal preventive maintenance policies for repairable systems. Oper Res 29:1181–1194.
55. Lin D, Zuo MJ, Yam RCM (2001) Sequential imperfect preventive maintenance models with two categories of failure modes. Nav Res Logist 48:172–183.

8
Inspection Policies

System reliability can be improved by providing some standby units. Especially, even a single standby unit plays an important role in the case where failures of an operating unit are costly and/or dangerous. A typical example is the case of standby electric generators in nuclear power plants, hospitals, and other public facilities. It is, however, extremely serious if a standby generator fails at the very moment of electric power supply stoppage. Hence, frequent inspections are necessary to avoid such unfavorable situations.

Similar examples can be found in army defense systems, in which all weapons are on standby, and hence, must be checked at suitable times. For example, missiles are stored for a great part of their lifetimes after delivery. However, their reliabilities are known to decrease with time because some parts deteriorate with time. Thus, it would be important to test the functions of missiles as to whether they can operate normally. We need to check them periodically to monitor their reliabilities and to repair them if necessary.

Earlier work has been done on the problem of checking a single unit. The optimum schedules of inspections that minimize two expected costs until failure detection and per unit of time were summarized in [1]. The modified models where checking times are nonnegligible, a unit is inoperative during checking times, and checking hastens failures and failure symptoms, were considered in [2–5]. Furthermore, the availability of a periodic inspection model [6] and the mean duration of hidden faults [7,8] were derived. The downtime cost of checking intervals for a continuous production process [9,10] and two types of inspection [11,12] were proposed. The optimum inspection policies for more complicated systems were discussed in [13–20]. A good survey of optimization problems for inspection models was made in [21].

It was difficult to compute an optimum solution of the algorithm presented by [1] before high-power computers were popular. Nearly optimum inspection policies were considered in [22–28]. A continuous inspection intensity was introduced and the approximate checking interval was derived in [29,30]. Using these approximate methods, some modified inspection models were discussed and compared with other methods [31–37].

All failures cannot be detected upon inspection. The imperfect inspection models were treated in [38–41], and the parameter of an exponential failure distribution was estimated in [42]. Furthermore, optimum inspection models for a unit with hidden failure [43] were discussed in [44]. In such models, even if a unit fails, it continues to operate in hidden failure, and then, it fails. Such a type of failure is called unrevealed fault [45], pending failure [25], or fault latency [47].

Most faults occur intermittently in digital systems. The optimum periodic tests for intermittent faults were discussed in [48–50]. A simple algorithm to compute an optimum time was developed in [51], and random test for fault detection in combinational circuits was introduced in [52].

It is especially important to check and maintain standby and protective units. The optimum inspection models for standby units [53–57] and protective devices [59–61] were presented. Also, the following inspection maintenance to actual systems was done: building, industrial plant, and underwater structure [62–64]; combustion turbine units and standby equipment in dormant systems and nuclear generating stations [65–67]; productive equipment [68]; fail-safe structure [69]; manufacturing station [70]; automatic trips and warning instruments [71]; bearing [72]; and safety-critical systems [73]. Moreover, the delay time models were reviewed in [74, 75], where a defect arises and becomes a failure after its delay time, and were applied to plant maintenance [76].

This chapter reviews the results of [1] and mainly summarizes our own results of inspection models. In Section 8.1, we briefly mention the results of [1], and consider the inspection model with finite number of checks [77]. In Section 8.2, we summarize four approximate inspection policies [31–35, 78]. In Section 8.3, we derive two optimum inspection policies for a standby unit as an example of an electric generator [53]. In Section 8.4, we consider the inspection policy for a storage system required to achieve a high reliability, and derive an optimum checking number until overhaul that minimizes the expected cost rate [80–83]. In Section 8.5, we discuss optimum testing times for intermittent faults [49, 50]. Finally, in Section 8.6, we rewrite the results of a standard model for inspection policies for units that have to be operating for a finite interval [84, 85]. It is shown that the proposed partition method is a useful technique for analyzing maintenance policies for a finite interval. The inspection with preventive maintenance and random inspection is covered in Sections 7.3 and 9.3, respectively.

8.1 Standard Inspection Policy

A unit should operate for an infinite time span and is checked at successive times T_k ($k = 1, 2, \dots$), where $T_0 \equiv 0$ (see Figure 8.1). Any failure is detected at the next checking time and is replaced immediately. A unit has a failure distribution $F(t)$ with finite mean μ whose failure rate $h(t)$ is not unchanged by any check. It is assumed that all times needed for checks and replacement

8.1 Standard Inspection Policy

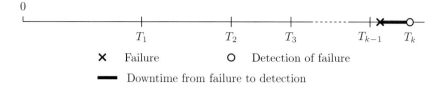

Fig. 8.1. Process of sequential inspection with checking time T_k

are negligible. Let c_1 be the cost of one check and c_2 be the loss cost per unit of time for the time elapsed between a failure and its detection at the next checking time, and c_3 be the replacement cost of a failed unit. Then, the total expected cost until replacement is

$$C_1(T_1, T_2, \ldots) \equiv \sum_{k=0}^{\infty} \int_{T_k}^{T_{k+1}} [c_1(k+1) + c_2(T_{k+1} - t)] \, dF(t) + c_3$$

$$= \sum_{k=0}^{\infty} [c_1 + c_2(T_{k+1} - T_k)] \overline{F}(T_k) - c_2\mu + c_3, \qquad (8.1)$$

where throughout this chapter, we use the notation $\overline{\Phi} \equiv 1 - \Phi$.

Differentiating the expected cost $C_1(T_1, T_2, \ldots)$ with T_k and putting it equal to zero,

$$T_{k+1} - T_k = \frac{F(T_k) - F(T_{k-1})}{f(T_k)} - \frac{c_1}{c_2} \qquad (k = 1, 2, \ldots), \qquad (8.2)$$

where f is a density function of F. The optimum checking intervals are decreasing when f is PF$_2$ (Pólya frequency function of order 2), and Algorithm 1 for computing the optimum inspection schedule is given in [1].

Algorithm 1

1. Choose T_1 to satisfy $c_1 = c_2 \int_0^{T_1} F(t) dt$.
2. Compute $T_2, T_3, \ldots$ recursively from (8.2).
3. If any $\delta_k > \delta_{k-1}$, reduce T_1 and repeat, where $\delta_k \equiv T_{k+1} - T_k$. If any $\delta_k < 0$, increase T_1 and repeat.
4. Continue until $T_1 < T_2 < \ldots$ are determined to the degree of accuracy required.

Clearly, because the mean time to replacement time is $\sum_{k=0}^{\infty}(T_{k+1} - T_k)\overline{F}(T_k)$, the expected cost rate is, from (3.3) in Chapter 3,

$$C_2(T_1, T_2, \ldots) \equiv \frac{c_1 \sum_{k=0}^{\infty} \overline{F}(T_k) - c_2\mu + c_3}{\sum_{k=0}^{\infty} (T_{k+1} - T_k)\overline{F}(T_k)} + c_2. \qquad (8.3)$$

In particular, when a unit is checked at periodic times and the failure time is exponential, i.e., $T_k = kT$ ($k = 0, 1, 2, \ldots$) and $F(t) = 1 - e^{-\lambda t}$, the total expected cost is

$$C_1(T) = \frac{c_1 + c_2 T}{1 - e^{-\lambda T}} - \frac{c_2}{\lambda} + c_3. \tag{8.4}$$

The optimum checking time T^* to minimize (8.4) is given by a unique solution that satisfies

$$e^{\lambda T} - (1 + \lambda T) = \frac{\lambda c_1}{c_2}. \tag{8.5}$$

Similarly, the expected cost rate is

$$C_2(T) = \frac{c_1 - (c_2/\lambda - c_3)(1 - e^{-\lambda T})}{T} + c_2. \tag{8.6}$$

When $c_2/\lambda > c_3$, the optimum T^* is given by solving

$$1 - (1 + \lambda T)e^{-\lambda T} = \frac{c_1}{c_2/\lambda - c_3}. \tag{8.7}$$

The following total expected cost for a continuous production system was proposed in [9].

$$\begin{aligned}\tilde{C}_1(T_1, T_2, \dots) &\equiv \sum_{k=0}^{\infty} \int_{T_k}^{T_{k+1}} [c_1(k+1) + c_2(T_{k+1} - T_k)] \, dF(t) + c_3 \\ &= c_1 \sum_{k=0}^{\infty} \overline{F}(T_k) + c_2 \sum_{k=0}^{\infty} (T_{k+1} - T_k)[\overline{F}(T_k) - \overline{F}(T_{k+1})] + c_3.\end{aligned} \tag{8.8}$$

In this case, Equation (8.2) can rewritten as

$$T_{k+1} - 2T_k + T_{k-1} = \frac{\overline{F}(T_{k+1}) - 2\overline{F}(T_k) + \overline{F}(T_{k-1})}{f(T_k)} - \frac{c_1}{c_2}$$
$$(k = 1, 2, \dots). \tag{8.9}$$

In general, it would be important to consider the availability more than the expected cost in some production systems [86, 87]. Let β_1 be the time of one check and β_3 be the replacement time of a failed unit. Then, the availability is, from **(3)** of Section 2.1.1,

$$A(T_1, T_2, \dots) \equiv \frac{\int_0^{\infty} \overline{F}(t) \, dt}{\sum_{k=0}^{\infty} [\beta_1 + T_{k+1} - T_k] \overline{F}(T_k) + \beta_3}.$$

Thus, the policy maximizing $A(T_1, T_2, \dots)$ is the same one as minimizing $C_1(T_1, T_2, \dots)$ in (8.1) by replacing $c_i = \beta_i$ $(i = 1, 3)$ and $c_2 = 1$.

Next, we consider the inspection model with a finite number of checks, because a system such as missiles involves some parts that have to be replaced when the total operating times of checks have exceeded a prespecified time of quality warranty. A unit is checked at times T_k $(k = 1, 2, \dots, N-1)$ and

8.1 Standard Inspection Policy

is replaced at time T_N ($N = 1, 2, \ldots$). The periodic inspection policy was suggested in [86], where a system is maintained preventively at the Nth check or is replaced at failure, whichever occurs first. We may consider replacement as preventive maintenance or overhaul.

In the above finite inspection model, the expected cost when a failure is detected and a unit is replaced at time T_k ($k = 1, 2, \ldots, N$) is

$$\sum_{k=1}^{N} \int_{T_{k-1}}^{T_k} [c_1 k + c_2(T_k - t) + c_3] \, \mathrm{d}F(t)$$

and the expected cost when a unit is replaced without failure at time T_N is

$$(c_1 N + c_3)\overline{F}(T_N).$$

Thus, the total expected cost until replacement is

$$\sum_{k=0}^{N-1} [c_1 + c_2(T_{k+1} - T_k)]\overline{F}(T_k) - c_2 \int_0^{T_N} \overline{F}(t) \, \mathrm{d}t + c_3.$$

Similarly, the mean time to replacement is

$$\sum_{k=1}^{N} \int_{T_{k-1}}^{T_k} T_k \, \mathrm{d}F(t) + T_N \overline{F}(T_N) = \sum_{k=0}^{N-1} (T_{k+1} - T_k)\overline{F}(T_k).$$

Therefore, the expected cost rate is

$$C_2(T_1, T_2, \ldots, T_N) = \frac{c_1 \sum_{k=0}^{N-1} \overline{F}(T_k) - c_2 \int_0^{T_N} \overline{F}(t) \, \mathrm{d}t + c_3}{\sum_{k=0}^{N-1} (T_{k+1} - T_k)\overline{F}(T_k)} + c_2. \quad (8.10)$$

In particular, when $T_k = kT$ ($k = 1, 2, \ldots, N$) and $F(t) = 1 - e^{-\lambda t}$, the expected cost rate is

$$C_2(T) = \frac{c_1}{T} - \frac{1}{\lambda T}(1 - e^{-\lambda T})\left(c_2 - \frac{c_3 \lambda}{1 - e^{-\lambda NT}}\right) + c_2. \quad (8.11)$$

Differentiating $C_2(T)$ with respect to T and putting it to 0, we have

$$\left(\frac{c_2}{\lambda} - \frac{c_3}{1 - e^{-\lambda NT}}\right)[1 - (1 + \lambda T)e^{-\lambda T}] - \frac{c_3 \lambda NT e^{-\lambda NT}(1 - e^{-\lambda T})}{(1 - e^{-\lambda NT})^2} = c_1. \quad (8.12)$$

Denoting the left-hand side of (8.12) by $Q_N(T)$, $\lim_{T \to 0} Q_N(T) = -c_3/N$ and $\lim_{T \to \infty} Q_N(T) = c_2/\lambda - c_3$. First, we prove that $Q_N(T)$ is an increasing function of T for $c_2/\lambda > c_1 + c_3$. It is noted that the first term in $Q_N(T)$ is strictly increasing in T. Differentiating $-Te^{-\lambda NT}(1 - e^{-\lambda T})/(1 - e^{-\lambda NT})^2$ with respect to T,

$$A[\lambda NT(1 - e^{-\lambda T})(1 + e^{-\lambda NT}) - (1 - e^{-\lambda NT})(1 - e^{-\lambda T} + \lambda Te^{-\lambda T})],$$

where $A \equiv e^{-\lambda NT}/(1 - e^{-\lambda NT})^3 > 0$ for $T > 0$. Denoting the quantity in the bracket of the above equation by $L_N(T)$,

$$L_1(T) = (1 - e^{-\lambda T})(\lambda T - 1 + e^{-\lambda T}) > 0$$
$$L_{N+1}(T) - L_N(T) = (1 - e^{-\lambda T})[\lambda T(1 - Ne^{-\lambda NT} + Ne^{-\lambda(N+1)T})$$
$$- (1 - e^{-\lambda T})e^{-\lambda NT}]$$
$$> (1 - e^{-\lambda T})^2[1 - (N+1)e^{-\lambda NT} + Ne^{-\lambda(N+1)T}] > 0.$$

Hence, $L_N(T)$ is strictly increasing in N. Thus, $L_N(T)$ is always positive for any N, and the second term of $Q_N(T)$ is an increasing function of T, which completes the proof. Therefore, there exists a finite and unique T_N^* ($0 < T_N^* < \infty$) that satisfies (8.12) for $c_2/\lambda > c_1 + c_3$, and it minimizes $C_2(T)$ in (8.11).

Next, we investigate properties of T_N^*. We prove that $Q_N(T)$ is also an increasing function of N as follows. From (8.12),

$$Q_{N+1}(T) - Q_N(T) = c_3(1 - e^{-\lambda T})[1 - E_N(T)]$$
$$\times \left[\frac{1 - (1 + \lambda T)e^{-\lambda T}}{E_N(T)E_{N+1}(T)} + \lambda T\left(\frac{N}{E_N(T)^2} - \frac{(N+1)e^{-\lambda T}}{E_{N+1}(T)^2}\right)\right],$$

where $E_N(T) \equiv 1 - e^{-\lambda NT}$. The first term in the bracket of the above equation is positive. The second term can be rewritten as

$$\frac{N}{E_N(T)^2} - \frac{(N+1)e^{-\lambda T}}{E_{N+1}(T)^2} = \frac{NE_{N+1}(T)^2 - (N+1)e^{-\lambda T}E_N(T)^2}{E_N(T)^2 E_{N+1}(T)^2}$$

and the numerator of the right-hand side is

$$NE_{N+1}(T)^2 - (N+1)e^{-\lambda T}E_N(T)^2$$
$$= e^{-\lambda T}[N(e^{\lambda T} - 1)(1 - e^{-\lambda(2N+1)T}) - (1 - e^{-\lambda NT})^2] > 0.$$

Hence, $Q_N(T)$ is a strictly increasing function of N because $Q_{N+1}(T) - Q_N(T) > 0$. Thus, T_N^* decreases when N increases. When $N = 1$, we have from (8.12),

$$1 - (1 + \lambda T)e^{-\lambda T} = \frac{(c_1 + c_3)\lambda}{c_2} \tag{8.13}$$

and when $N = \infty$,

$$1 - (1 + \lambda T)e^{-\lambda T} = \frac{c_1\lambda}{c_2 - c_3\lambda}. \tag{8.14}$$

Because $[(c_1 + c_3)\lambda]/c_2 > c_1\lambda/(c_2 - c_3\lambda)$, we easily find that $T_\infty^* < T_N^* \le T_1^*$, where T_1^* and T_∞^* are the respective solutions of (8.13) and (8.14).

8.2 Asymptotic Inspection Schedules

Table 8.1. Optimum checking time T_N^* when $c_1 = 10$, $c_2 = 1$, and $c_3 = 100$

N	$\lambda = 1.0 \times 10^{-3}$				$\lambda = 1.1 \times 10^{-3}$				$\lambda = 1.2 \times 10^{-3}$			
						m						
	1.0	1.1	1.2	1.3	1.0	1.1	1.2	1.3	1.0	1.1	1.2	1.3
1	564	436	355	309	543	423	347	307	526	412	341	307
2	396	315	259	223	380	304	251	219	367	294	245	217
3	328	268	224	193	314	258	216	188	303	249	210	185
4	289	243	206	178	277	233	198	173	267	225	192	169
5	264	228	195	170	253	218	188	165	243	210	181	161
6	246	217	189	165	236	208	181	160	226	200	174	156
7	233	210	184	162	223	200	176	157	214	192	170	154
8	222	204	181	161	212	194	173	156	204	186	167	153
9	214	200	179	160	204	190	171	155	196	182	165	152
10	207	196	178	160	197	187	170	155	189	179	163	152

The condition of $c_2/\lambda > c_1 + c_3$ means that the total loss cost until the whole life of a unit is higher than the sum of costs of checks and replacements. This would be realistic in the actual field.

Example 8.1. We compute the optimum checking time T_N^* that minimizes $C_2(T)$ in (8.11) when $F(t) = 1 - \exp(-\lambda t^m)$ ($m \geq 1$). When $m = 1$, it corresponds to an exponential case. Table 8.1 shows the optimum time T_N^* for $\lambda = 1.0 \times 10^{-3}$, 1.1×10^{-3}, 1.2×10^{-3}/hour, $m = 1.0, 1.1, 1.2, 1.3$ and $N = 1, 2, \ldots, 10$ when $c_1 = 10$, $c_2 = 1$, and $c_3 = 100$. This indicates that T_N^* decreases when λ, m, and N increase, and that a unit should be checked once every several weeks. ∎

8.2 Asymptotic Inspection Schedules

The computing procedure for obtaining the optimum inspection schedule was specified in [1]. Unfortunately, it is difficult to compute Algorithm 1 numerically, because the computations are repeated until the procedures are determined to the required degree by changing the first checking time. To avoid this, a nearly optimum inspection policy that depends on a single parameter p was suggested in [22]. This policy was used for Weibull and gamma distribution cases [23, 24]. Furthermore, the procedure of introducing a continuous intensity $n(t)$ of checks per unit of time was proposed in [29, 30]. This section summarizes four approximate calculations of optimum checking procedures.

(1) Periodic Inspection

When a unit is checked at periodic times kT ($k = 1, 2, \ldots$), the total expected cost is, from (8.1),

$$C_1(T) = \frac{c_1}{T}[E\{D\} + \mu] + c_2 E\{D\} + c_3, \tag{8.15}$$

where $E\{D\} \equiv \sum_{k=0}^{\infty} \int_0^T [F(t+kT) - F(kT)]\,dt$, which is the mean duration of time elapsed between a failure and its detection.

Suppose that $F(t)$ has the piecewise linear approximation:

$$F(t+kT) - F(kT) = \frac{t}{T}[F((k+1)T) - F(kT)] \quad (0 \le t \le T). \tag{8.16}$$

Then, $E\{D\} = T/2$; i.e., the mean duration of undetected failure is half the time between the checking times. The result is also given when the failure times between successive checking times are independent and distributed uniformly. In this case, the optimum checking time is $\tilde{T}_1 = \sqrt{(2c_1\mu)/c_2}$. This time is also derived from (8.5) by putting $e^{\lambda T} \approx 1 + \lambda T + (\lambda T)^2/2$ approximately and $\lambda = 1/\mu$.

(2) Munford and Shahani's Method

The asymptotic method for computing the optimum schedule was proposed in [22]. When a unit is operating at time T_{k-1}, the probability that it fails in an interval $(T_{k-1}, T_k]$ is constant for all k; i.e.,

$$\frac{F(T_k) - F(T_{k-1})}{\overline{F}(T_{k-1})} \equiv p \quad (k = 1, 2, \ldots). \tag{8.17}$$

This represents that the probability that a unit with age T_{k-1} fails in interval $(T_{k-1}, T_k]$ is given by a constant p. Noting that $F(T_1) = p$, Equation (8.17) can be solved for T_k, and we have

$$\overline{F}(T_k) = q^k \quad \text{or} \quad T_k = \overline{F}^{-1}(q^k) \quad (k = 1, 2, \ldots), \tag{8.18}$$

where $q \equiv 1 - p$ $(0 < p < 1)$. Thus, from (8.1), the total expected cost is

$$C_1(p) = \frac{c_1}{p} + c_2 \sum_{k=1}^{\infty} T_k q^{k-1} p - c_2 \mu + c_3. \tag{8.19}$$

We seek p that minimizes $C_1(p)$ in (8.19). It was assumed in [28] that p is not constant and is an increasing function of the checking number.

(3) Keller's Method

An inspection intensity $n(t)$ is defined as follows [29]: $n(t)dt$ denotes the probability that a unit is checked at interval $(t, t+dt)$ (see Figure 8.2). From this definition, when a unit is checked at times T_k, we have the relation

$$\int_0^{T_k} n(t)\,dt = k \quad (k = 1, 2, \ldots). \tag{8.20}$$

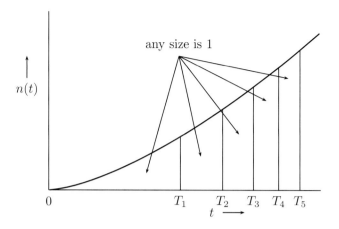

Fig. 8.2. Inspection intensity $n(t)$

Furthermore, suppose that the mean time from the failure at time t to its detection at time $t + a$ is half of a checking interval, the same as obtained in case **(1)**. Then, we have

$$\int_t^{t+a} n(u)\,du = \frac{1}{2}$$

which can be approximately written as

$$\int_t^{t+a} n(u)\,du \approx an(t) = \frac{1}{2}$$

and hence, $a = 1/[2n(t)]$. By the same arguments, we can easily see that the next checking interval, when a unit was checked at time T_k, is $1/n(T_k)$ approximately.

Therefore, the total expected cost in (8.1) is given by

$$C(n(t)) = \int_0^\infty \left[c_1 \int_0^t n(u)\,du + \frac{c_2}{2n(t)} \right] dF(t) + c_3$$

$$= \int_0^\infty \overline{F}(t) \left[c_1 n(t) + \frac{c_2 h(t)}{2n(t)} \right] dt + c_3, \qquad (8.21)$$

where $h(t) \equiv f(t)/\overline{F}(t)$ which is the failure rate. Differentiating $C(n(t))$ with $n(t)$ and putting it to zero,

$$n(t) = \sqrt{\frac{c_2 h(t)}{2c_1}}. \qquad (8.22)$$

Thus, from (8.20), the optimum checking time is given by the equation:

210   8 Inspection Policies

$$k = \int_0^{T_k} \sqrt{\frac{c_2}{2c_1} h(t)} \, dt \quad (k = 1, 2, \dots). \tag{8.23}$$

The inspection intensity $n(t)$ was also obtained in [36] by solving the Euler equation in (8.21), and using $n(t)$, the optimum policies for models with imperfect inspection were derived in [88].

In particular, when $F(t) = 1 - e^{-\lambda t}$, the interval between checks is constant, and is $\sqrt{2c_1/(\lambda c_2)}$ which agrees with the result of case **(1)**. It is of great interest that a function $\sqrt{2c_1/(\lambda c_2)}$ evolves into the same form as an optimum order time of a classical inventory control model [89], by denoting c_1 and c_2 as the ordering cost per order and holding cost per unit of time, respectively, and λ as the constant demand rate for an inventory unit.

(4) Nakagawa and Yasui's Method

When T_n is sufficiently large, we may assume approximately [79]

$$T_{n+1} - T_n + \varepsilon = T_n - T_{n-1}. \tag{8.24}$$

It is easy to see that if f is PF$_2$ then $\varepsilon \geq 0$ because the optimum checking intervals are decreasing [1]. Further substituting the relation (8.24) into (8.2),

$$\frac{c_1}{c_2} - \varepsilon = \frac{\int_{T_{n-1}}^{T_n} [f(t) - f(T_n)] \, dt}{f(T_n)} \geq 0 \tag{8.25}$$

because $f(t) \geq f(T_n)$ for $t \leq T_n$ and large T_n. Thus, we have $0 \leq \varepsilon \leq c_1/c_2$.

From the above discussion, we can specify the computation for obtaining the asymptotic inspection schedule.

Algorithm 2

1. Choose an appropriate ε from $0 < \varepsilon < c_1/c_2$.
2. Determine a checking time T_n after sufficient time for required accuracy.
3. Compute T_{n-1} to satisfy

$$T_n - T_{n-1} - \varepsilon = \frac{F(T_n) - F(T_{n-1})}{f(T_n)} - \frac{c_1}{c_2}.$$

4. Compute $T_{n-1} > T_{n-2} > \dots$ recursively from (8.2).
5. Continue until $T_k < 0$ or $T_{k+1} - T_k > T_k$.

Example 8.2. Suppose that the failure time has a Weibull distribution with a shape parameter m; i.e., $F(t) = 1 - \exp[-(\lambda t)^m]$.

(1) *Periodic inspection.* The optimum checking time is

$$\lambda \widetilde{T}_1 = \left[\frac{2\lambda c_1}{c_2} \Gamma\left(1 + \frac{1}{m}\right) \right]^{1/2}.$$

Table 8.2. Comparisons of Nakagawa, Barlow, Munford, and Keller policies when $F(t) = 1 - \exp[-(\lambda t)^2]$, $1/\lambda = 500$, and $c_1/c_2 = 10$.

k	Nakagawa $T_n = 1500$ $\varepsilon = 5$	$\varepsilon = 4.5$	Barlow		Munford $p = 0.215$	Keller
1	219.6	207.1	205.6	205.6	246.0	177.8
2	318.7	308.9	307.6	307.6	347.9	282.3
3	402.0	393.5	392.3	392.3	426.1	369.9
4	476.4	468.7	467.5	467.5	492.0	448.1
5	544.8	537.6	536.4	536.5	550.1	520.0
6	608.7	601.9	600.7	600.8	602.6	587.2
7	669.1	662.6	661.5	661.6	650.9	650.8
8	726.6	720.4	719.2	719.4	695.8	711.4
9	781.7	775.8	774.6	774.8	738.0	769.5
10	834.8	829.1	827.8	828.2	777.9	825.5
11	886.1	880.6	879.3	879.7	815.9	879.6
12	935.8	930.5	929.1	929.7	852.2	932.2
13	984.1	979.0	977.4	978.3	887.0	983.3
14	1031.1	1026.2	1024.5	1025.6	920.5	1033.1

(2) *Munford and Shahani's method*. From (8.19), we obtain p that minimizes

$$g(p) = \frac{\lambda c_1}{p c_2} + \left(\log \frac{1}{q}\right)^{1/m} \sum_{k=1}^{\infty} k^{1/m} q^{k-1} p$$

and the optimum checking intervals are

$$\lambda T_k = \left(k \log \frac{1}{q}\right)^{1/m} \quad (k = 1, 2, \dots).$$

(3) *Keller's method*. From (8.23),

$$T_k = \left[(m+1)k\sqrt{\frac{c_1}{2m\lambda^m c_2}}\right]^{2/(m+1)} \quad (k = 1, 2, \dots).$$

In particular, when $m = 1$, $T_k = k\sqrt{2c_1/(\lambda c_2)}$.

Table 8.2 shows the comparisons of the methods of Barlow et al., Munford et al., Keller, and Nakagawa et al., when $m = 2$, $1/\lambda = 500$, $c_1/c_2 = 10$. Nakagawa and Yasui's method gives a fairly good approximation of Barlow's one. In particular, when we choose $\varepsilon = 4.5$, the results are almost the same as the sequence of optimum checking times. The computation of Keller's method is very easy, and this method would be very useful for obtaining checking times in the actual field. ∎

8.3 Inspection for a Standby Unit

In this section, we consider an inspection policy for a single standby electric generator. We check a standby generator frequently to guarantee the upper bound of the probability that it has failed at the time of the electric power supply stoppage, but to reduce unnecessary costs do not check it too frequently.

The details of the model are described as follows.

(1) The failure time of a standby generator has a general distribution $F(t)$ and its failure is detected only at the next checking time.
(2) A failed standby generator, which was detected at some check, undergoes repair immediately and its repair time has a general distribution $G(t)$.
(3) The time required for the check is negligible and a standby generator becomes as good as new upon inspection or repair.
(4) The next checking time is scheduled at constant time T ($0 < T \le \infty$) after either the prior check or the repair completion.
(5) Costs c_0 and c_1 are incurred for each repair and check, respectively, and cost c_2 is incurred for the failure of a generator when the electric power supply stops, where $c_2 > c_0 \ge c_1$.
(6) The policy terminates with the time of electric power supply stoppage, which occurs according to an exponential distribution $(1 - e^{-\alpha t})$.

Under the assumptions above, we consider two optimization problems: (a) an optimum checking time T^* that minimizes the expected cost until the time of electric power supply stoppage, and (b) the largest $\overline{T}$ such that the probability that a generator has failed at the time of electric power supply stoppage is not greater than a prespecified value ε.

To obtain the expected cost of the inspection model as described above, we derive the expected numbers of checks and repairs of a standby electric generator, and the probability that it has failed at the time of electric power supply stoppage.

As an initial condition, it is assumed for convenience that a generator goes into standby and is good at time 0. Furthermore, for simplicity of equations, we define $D(t) \equiv 0$ for $t < T$ and $\equiv 1$ for $t \ge T$; i.e., $D(t)$ is a degenerate distribution at time T.

Let $H(t)$ be the distribution of the recurrence time to the state that a standby generator is good upon inspection or repair completion. Then, we have

$$H(t) = \int_0^t \overline{F}(u)\,\mathrm{d}D(u) + \left[\int_0^t F(u)\,\mathrm{d}D(u)\right] * G(t), \qquad (8.26)$$

where the asterisk represents the Stieltjes convolution. Equation (8.26) can be explained by the first term on the right-hand side being the probability that a standby generator is good upon inspection until time t, and the second term the probability that a failed generator is detected at a check and its repair is completed until time t.

8.3 Inspection for a Standby Unit

In addition, let $M_0(t)$ and $M_1(t)$ be the expected numbers of repairs of a failed generator and of checks of a standby generator during $(0, t]$, respectively. Then, the following renewal-type equations are given by

$$M_0(t) = \int_0^t F(u)\,\mathrm{d}D(u) + H(t) * M_0(t) \tag{8.27}$$

$$M_1(t) = D(t) + H(t) * M_1(t). \tag{8.28}$$

Thus, forming the Laplace–Stieltjes (LS) transforms of (8.26), (8.27), and (8.28), respectively, we have

$$H^*(s) = e^{-sT}[\overline{F}(T) + F(T)G^*(s)] \tag{8.29}$$

$$M_0^*(s) = \frac{e^{-sT}F(T)}{1 - H^*(s)}, \qquad M_1^*(s) = \frac{e^{-sT}}{1 - H^*(s)}, \tag{8.30}$$

where throughout this section, we denote the LS transform of the function by the corresponding asterisk; e.g., $G^*(s) \equiv \int_0^\infty e^{-st}\,\mathrm{d}G(t)$ for $s > 0$.

Next, let $P(t)$ denote the probability that a standby generator has failed at time t; i.e., a standby generator, which is not good, will be detected at the next check or a failed generator, which was detected at the prior check, is now under repair. Then, the probability that a standby generator is good at time t is given by

$$\overline{P}(t) = \overline{F}(t)\overline{D}(t) + H(t) * \overline{P}(t).$$

Forming the LS transform of $P(t)$, we have

$$1 - P^*(s) = \frac{\int_0^T s e^{-st}\overline{F}(t)\,\mathrm{d}t}{1 - H^*(s)}. \tag{8.31}$$

We consider the total expected cost until the time of electric power supply stoppage. Note that the inspection model of a standby generator may involve at least the following three costs: the costs c_0 and c_1 incurred by each repair and each check, respectively, and the cost c_2 incurred by failure of a standby generator when the electric power supply stops.

Suppose that the electric power supply stops at time t. Then, the total expected cost during $(0, t]$ is given by

$$\widetilde{C}(t) = c_0 M_0(t) + c_1 M_1(t) + c_2 P(t).$$

Thus, dropping the condition that the electric power supply stops at time t from assumption (6), we have the expected cost:

$$C_1(T) \equiv \int_0^\infty \widetilde{C}(t)\alpha e^{-\alpha t}\,\mathrm{d}t = c_0 M_0^*(\alpha) + c_1 M_1^*(\alpha) + c_2 P^*(\alpha)$$

which is a function of T. Using (8.30) and (8.31), $C_1(T)$ can be written as

$$C_1(T) = \frac{e^{-\alpha T}[c_0 F(T) + c_1] - c_2 \int_0^T \alpha e^{-\alpha T}\overline{F}(t)\,dt}{1 - e^{-\alpha T}[\overline{F}(T) + F(T)G^*(\alpha)]} + c_2. \tag{8.32}$$

It is evident that

$$C_1(0) \equiv \lim_{T \to 0} C_1(T) = \infty, \qquad C_1(\infty) \equiv \lim_{T \to \infty} C_1(T) = c_2 F^*(\alpha)$$

which represents the expected cost for the case where no inspection is made.

We seek an optimum checking time T_1^* that minimizes the expected cost $C_1(T)$ given in (8.32). Differentiating $\log C_1(T)$ with respect to T, we have, for large T,

$$\frac{d[\log C_1(T)]}{dT} \approx \alpha e^{-\alpha T}\left[\frac{c_2 G^*(\alpha) - c_0 - c_1}{c_2 F^*(\alpha)} - G^*(\alpha)\right].$$

Thus, if the quantity in the bracket on the right-hand side is positive; i.e.,

$$c_2 G^*(\alpha)[1 - F^*(\alpha)] > c_0 + c_1, \tag{8.33}$$

then there exists at least some finite T such that $C_1(\infty) > C_1(T)$, and hence, it is better to check a standby generator at finite time T.

In general, it is difficult to discuss analytically an optimum checking time T^* that minimizes $C_1(T)$. In particular, consider the case where $F(t) = 1 - e^{-\lambda t}$ and $G(t) \equiv 1$ for $t \geq 0$; i.e., the failure time is exponential and the repair time is negligible. Then, the resulting cost is

$$C_1(T) = \frac{e^{-\alpha T}[c_0(1 - e^{-\lambda T}) + c_1] + c_2[1 - e^{-\alpha T} - \frac{\alpha}{\alpha + \lambda}(1 - e^{-(\alpha + \lambda)T})]}{1 - e^{-\alpha T}}. \tag{8.34}$$

Differentiating $C_1(T)$ with respect to T and setting it equal to zero,

$$c_0 e^{-\lambda T}\left[1 + \frac{\lambda}{\alpha}(1 - e^{-\alpha T})\right] + c_2\left[1 - e^{-\lambda T} - \frac{\lambda}{\alpha + \lambda}(1 - e^{-(\alpha + \lambda)T})\right] = c_0 + c_1, \tag{8.35}$$

where the left-hand side is strictly increasing in the case of $c_2 > [(\alpha + \lambda)/\alpha]c_0$, and conversely, nonincreasing in the case of $c_2 \leq [(\alpha + \lambda)/\alpha]c_0$. Further note that the left-hand side is c_0 as $T \to 0$ and $[\alpha/(\alpha + \lambda)]c_2$ as $T \to \infty$.

Therefore, we have the following results from the above discussion.

(i) If $c_2 > [(\alpha + \lambda)/\alpha](c_1 + c_0)$ then there exists a finite checking time T_1^* that satisfies (8.35), and the resulting cost is

$$C_1(T^*) = c_2 - c_1 - c_0 - \left(c_2 - c_0\frac{\alpha + \lambda}{\alpha}\right)e^{-\lambda T^*}. \tag{8.36}$$

(ii) If $c_2 \leq [(\alpha + \lambda)/\alpha](c_1 + c_0)$ then $T_1^* = \infty$; i.e., no inspection is made, and $C_1(\infty) = c_2[\lambda/(\alpha + \lambda)]$.

Note that the inequality of $c_2 > [(\alpha + \lambda)/\alpha](c_1 + c_0)$ has been already derived from (8.33).

It is also of interest to make the probability as small as possible by checks, that a standby generator has failed at the time of electric power supply stoppage. If the probability is prespecified, we can compute a checking time $\overline{T}_1$ such that $P^*(\alpha) \leq \varepsilon$; i.e.,

$$\frac{\int_0^T e^{-\alpha t}\, dF(t) - e^{-\alpha T} F(T) G^*(\alpha)}{1 - e^{-\alpha T}[\overline{F}(T) + F(T) G^*(\alpha)]} \leq \varepsilon. \tag{8.37}$$

For instance, if the repair time is negligible, i.e., $G^*(\alpha) = 1$, then the left-hand side of (8.37) is strictly increasing in T. Hence, there exists a unique checking time $\overline{T}$ that satisfies

$$\frac{\int_0^T F(t) \alpha e^{-\alpha t}\, dt}{1 - e^{-\alpha T}} = \varepsilon \tag{8.38}$$

for sufficiently small $\varepsilon > 0$.

Until now, we have assumed that a standby generator becomes as good as new upon inspection. Next, we make the same assumption as the previous ones except that the failure rate of a standby generator remains undisturbed by any inspection. This assumption would be more plausible than the previous model in practice, however, the analysis becomes more difficult. Then, the expected cost until the time of electric power supply stoppage is [53]

$$C_2(T) = \frac{c_0 \sum_{k=1}^{\infty} e^{-\alpha kT}[\overline{F}((k-1)T) - \overline{F}(kT)] + c_1 \sum_{k=1}^{\infty} e^{-\alpha kT} \overline{F}((k-1)T) - c_2[1 - F^*(\alpha)]}{1 - G^*(\alpha) \sum_{k=1}^{\infty} e^{-\alpha kT}[\overline{F}((k-1)T) - \overline{F}(kT)]} + c_2. \tag{8.39}$$

It is evident that $C_2(0) = \infty$ and $C_2(\infty) = c_2 F^*(\alpha)$. Furthermore, for large T,

$$\frac{d[\log C_2(T)]}{dT} \approx \alpha e^{-\alpha T}\left[\frac{c_2 G^*(\alpha) - c_0 - c_1}{c_2 F^*(\alpha)} - G^*(\alpha)\right].$$

Thus, if $c_2 G^*(\alpha)[1 - F^*(\alpha)] > c_0 + c_1$, then there exists at least some finite T such that $C_2(\infty) > C_2(T)$, which agrees with the results of the previous model.

It is very difficult to obtain analytically an optimum time T_2^* that minimizes $C_2(T)$ in (8.39). It is noted, however, that the expected cost $C_2(T)$ agrees with (8.34) in the special case of $F(t) = 1 - e^{-\lambda t}$ and $G(t) \equiv 1$ for $t \geq 0$.

Example 8.3. We give a numerical example where $\overline{F}(t) = (1 + \lambda t)e^{-\lambda t}$ and $\overline{G}(t) = (1 + \theta t)e^{-\theta t}$, both of which are the gamma distribution with shape parameter 2. Table 8.3 shows the optimum checking times T_1^* and T_2^* for the mean failure time $2/\lambda$ and cost c_2, when $c_0 = 30$ dollars, $c_1 = 3$ dollars, $1/\theta = 12$ hours, and $1/\alpha = 1460$ hours; i.e., the electric power supply stops 6 times a year on the average. It has been shown that both of the checking

Table 8.3. Dependent of mean failure time $2/\lambda$ and cost c_2 in optimum checking times T_1^* and T_2^* when $c_0 = 30$, $c_1 = 3$, $1/\theta = 12$, and $1/\alpha = 1460$

	$c_2 = 150$		$c_2 = 250$		$c_2 = 350$	
$2/\lambda$	T_1^*	T_2^*	T_1^*	T_2^*	T_1^*	T_2^*
1200	292	480	249	308	224	241
1600	368	535	311	354	279	280
2000	439	594	369	399	330	318
2400	507	656	424	445	379	356
2800	572	720	477	491	425	393
3200	635	783	528	537	469	430
3600	697	848	578	582	512	467
4000	757	914	626	628	554	503

times are increasing if $2/\lambda$ is increasing and are decreasing if c_2 is increasing. In addition, T_1^* becomes greater than T_2^* when c_2 and $2/\lambda$ are large enough. ∎

8.4 Inspection for a Storage System

A system such as missiles is in storage for a long time from delivery to the actual usage and has to hold a high mission reliability when it is used [90]. After a system is transported to each firing operation unit via the depot, it is installed on a launcher and is stored in a warehouse for a great part of its lifetime, and waits for its operation. Therefore, missiles are often called dormant systems.

However, the reliability of a storage system goes down with time because some kinds of electronic and electric parts of a system degrade with time [91–95]. The periodic inspection of stored electronic equipment was studied and how to compute its reliability after ten years of storage was shown in [96]. We should test and maintain a storage system at periodic times to hold a high reliability, because it is impossible to check whether a storage system can operate normally.

In most inspection models, it has been assumed that the function test can clarify all system failures. However, a missile is exposed to a very severe flight environment and some kinds of failures are revealed only in such severe conditions. That is, some failures of a missile cannot be detected by the function test on the ground. To solve this problem, we assume that a system is divided into two independent units: Unit 1 becomes new after every test because all failures of unit 1 are detected by the function test and are removed completely by maintenance, but unit 2 degrades steadily with time from delivery to overhaul because all failures of unit 2 cannot be detected by any test. The reliability of a system deteriorates gradually with time as the reliability of unit 2 deteriorates steadily.

8.4 Inspection for a Storage System

This section considers a system in storage that is required to achieve a higher reliability than a prespecified level q $(0 < q \le 1)$. To hold the reliability, a system is tested and is maintained at periodic times NT $(N = 1, 2, \ldots)$, and is overhauled if the reliability becomes equal to or lower than q. A test number N^* and the time $N^*T + t_0$ until overhaul, are derived when a system reliability is just equal to q. Using them, the expected cost rate $C(T)$ until overhaul is obtained, and an optimum test time T^* that minimizes it is computed. Finally, numerical examples are given when failure times of units have exponential and Weibull distributions. Two extended models were considered in [82, 97], where a system is also replaced at time $(N+1)T$, and may be degraded at each inspection, respectively.

A system consists of unit 1 and unit 2, where the failure time of unit i has a cumulative hazard function $H_i(t)$ $(i = 1, 2)$. When a system is tested at periodic times NT $(N = 1, 2, \ldots)$, unit 1 is maintained and is like new after every test, and unit 2 is not done; i.e., its hazard rate remains unchanged by any tests.

From the above assumptions, the reliability function $R(t)$ of a system with no inspection is

$$R(t) = e^{-H_1(t) - H_2(t)}. \tag{8.40}$$

If a system is tested and maintained at time t, the reliability just after test is

$$R(t_{+0}) = e^{-H_2(t)}.$$

Thus, the reliabilities just before and after the Nth test are, respectively,

$$R(NT_{-0}) = e^{-H_1(T) - H_2(NT)}, \qquad R(NT_{+0}) = e^{-H_2(NT)}. \tag{8.41}$$

Next, suppose that the overhaul is performed if a system reliability is equal to or lower than q. Then, if

$$e^{-H_1(T) - H_2(NT)} > q \ge e^{-H_1(T) - H_2[(N+1)T]} \tag{8.42}$$

then the time to overhaul is $NT + t_0$, where t_0 $(0 < t_0 \le T)$ satisfies

$$e^{-H_1(t_0) - H_2(NT + t_0)} = q. \tag{8.43}$$

This shows that the reliability is greater than q just before the Nth test and is equal to q at time $NT + t_0$.

Let c_1 and c_2 be the test and the overhaul costs, respectively. Then, denoting the time interval $[0, NT + t_0]$ as one cycle, the expected cost rate until overhaul is given by

$$C(T, N) = \frac{Nc_1 + c_2}{NT + t_0}. \tag{8.44}$$

We consider two particular cases where the cumulative hazard functions $H_i(t)$ are exponential and Weibull ones. A test number N^* that satisfies (8.42), and t_0 that satisfies (8.43), are computed. Using these quantities, we compute the expected cost $C(T, N)$ until overhaul and seek an optimum test time T^* that minimizes it.

(1) Exponential Case

When the failure time of units has an exponential distribution, *i.e.*, $H_i(t) = \lambda_i t$ $(i = 1, 2)$, Equation (8.42) is rewritten as

$$\frac{1}{Na+1}\log\frac{1}{q} \leq \lambda T < \frac{1}{(N-1)a+1}\log\frac{1}{q}, \qquad (8.45)$$

where $\lambda \equiv \lambda_1 + \lambda_2$ and $a \equiv H_2(T)/[H_1(T)+H_2(T)] = \lambda_2/\lambda$ $(0 < a < 1)$ which represents an efficiency of inspection [90], and is widely adopted in practical reliability calculation of a storage system.

When a test time T is given, a test number N^* that satisfies (8.45) is determined. Particularly, if $\log(1/q) \leq \lambda T$ then $N^* = 0$, and N^* diverges as λT tends to 0. In this case, Equation (8.43) is

$$N^*\lambda_2 T + \lambda t_0 = \log\frac{1}{q}. \qquad (8.46)$$

From (8.46), we can compute t_0 easily. Thus, the total time to overhaul is

$$N^*T + t_0 = N^*(1-a)T + \frac{1}{\lambda}\log\frac{1}{q} \qquad (8.47)$$

and the expected cost rate is

$$C(T, N^*) = \frac{N^* c_1 + c_2}{N^*(1-a)T + \frac{1}{\lambda}\log\frac{1}{q}}. \qquad (8.48)$$

When a test time T is given, we compute N^* from (8.45) and $N^*T + t_0$ from (8.47). Substituting these values into (8.48), we have $C(T, N^*)$. Changing T from 0 to $\log(1/q)/[\lambda(1-a)]$, because λT is less than $\log(1/q)/(1-a)$ from (8.45), we can compute an optimum T^* that minimizes $C(T, N^*)$. In the particular case of $\lambda T \geq \log(1/q)/(1-a)$, $N^* = 0$ and the expected cost rate is $C(T, 0) = c_2/t_0 = \lambda c_2 / \log(1/q)$.

Example 8.4. Table 8.4 presents the optimum number N^* and the total time $\lambda(N^*T + t_0)$ to overhaul for λT when $a = 0.1$ and $q = 0.8$. For example, when λT increases from 0.203 to 0.223, $N^* = 1$ and $\lambda(N^*T + t_0)$ increases from 0.406 to 0.424. In accordance with the decrease in λT, both N^* and $\lambda(N^*T + t_0)$ increase as shown in (8.45) and (8.47), respectively.

Table 8.5 gives the optimum number N^* and time λT^* that minimize the expected cost $C(T, N)$ for c_2/c_1, a and q, and the resulting total time $\lambda(N^*T^* + t_0)$ and the expected cost rate $C(T^*, N^*)/\lambda$ for $c_1 = 1$. These indicate that λT^* increases and $\lambda(N^*T^* + t_0)$ decreases when c_1/c_2 and a increase, and both λT^* and $\lambda(N^*T^* + t_0)$ decrease when q increases. ∎

Table 8.4. Optimum inspection number N^* and total time to overhaul $\lambda(N^*T+t_0)$ for λT when $a = 0.1$ and $q = 0.8$

λT	N^*	$\lambda(N^*T+t_0)$
$[0.223, \infty)$	0	$[0.223, \infty)$
$[0.203, 0.223)$	1	$[0.406, 0.424)$
$[0.186, 0.203)$	2	$[0.558, 0.588)$
$[0.172, 0.186)$	3	$[0.687, 0.725)$
$[0.159, 0.172)$	4	$[0.797, 0.841)$
$[0.149, 0.159)$	5	$[0.893, 0.940)$
$[0.139, 0.149)$	6	$[0.976, 1.026)$
$[0.131, 0.139)$	7	$[1.050, 1.102)$
$[0.124, 0.131)$	8	$[1.116, 1.168)$
$[0.117, 0.124)$	9	$[1.174, 1.227)$
$[0.112, 0.117)$	10	$[1.227, 1.280)$

Table 8.5. Optimum inspection time λT^*, total time to overhaul $\lambda(N^*T+t_0)$, and expected cost rate $C(T^*)/\lambda$

c_2/c_1	a	q	N^*	λT^*	$\lambda(N^*T^*+t_0)$	$C(T^*, N^*)/\lambda$
10	0.1	0.8	8	0.131	1.168	15.41
50	0.1	0.8	19	0.080	1.586	43.51
10	0.5	0.8	2	0.149	0.372	32.27
10	0.1	0.9	7	0.062	0.552	32.63

(2) Weibull Case

When the failure time of units has a Weibull distribution; i.e., $H_i(t) = (\lambda_i t)^m$ ($i = 1, 2$), Equations (8.42) and (8.43) are rewritten as

$$\left\{\frac{1}{a[(N+1)^m - 1] + 1} \log \frac{1}{q}\right\}^{1/m} \leq \lambda T < \left\{\frac{1}{a[N^m - 1] + 1} \log \frac{1}{q}\right\}^{1/m} \tag{8.49}$$

$$(1-a)t_0^m + a(NT+t_0)^m = \frac{1}{\lambda^m} \log \frac{1}{q}, \tag{8.50}$$

respectively, where $\lambda^m \equiv \lambda_1^m + \lambda_2^m$ and

$$a \equiv \frac{H_2(T)}{H_1(T) + H_2(T)} = \frac{\lambda_2^m}{\lambda_1^m + \lambda_2^m}.$$

When an inspection time T is given, N^* and t_0 are computed from (8.49) and (8.50). Substituting these values into (8.44), we have $C(T, N^*)$, and changing T from 0 to $[\log(1/q)/(1-a)]^{1/m}/\lambda$, we can compute an optimum T^* that minimizes $C(T, N^*)$.

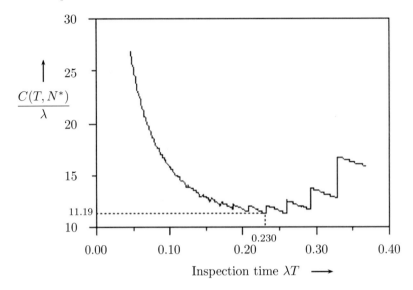

Fig. 8.3. Relation between λT and $C(T)/\lambda$ in the exponential case

Example 8.5. When the failure time of unit i has Weibull distribution $\{1 - \exp[-(\lambda_i t)^{1.5}]\}$ and $c_1 = 1$, $c_2 = 10$, $a = 0.1$, and $q = 0.8$, Figure 8.3 shows the relationship between λT and $C(T, N^*)/\lambda$, and that the optimum time is $\lambda T^* = 0.230$ and the resulting cost rate is $C(T^*, N^*)/\lambda = 11.19$. In this case, the optimum number is $N^* = 5$ and the total time is $\lambda(N^*T^* + t_0) = 1.34$. ■

8.5 Intermittent Faults

Digital systems have two types of faults from the viewpoint of operational failures: permanent faults due to hardware failures or software errors, and intermittent faults due to transient failures [98, 99]. Intermittent faults are automatically detected by the error-correcting code and corrected by the error control [100, 101] or the restart [102, 103]. However, some faults occur repeatedly, and consequently, will be permanent faults. Some tests are applied to detect and isolate faults, but it would waste time and money to do more frequent tests.

Continuous and repetitive tests for a continuous Markov model with intermittent faults were considered in [48]. Redundant systems with independent modules were treated in [46]. Furthermore, they were extended for non-Markov models [98] and redundant systems with dependent modules [104].

This section applies the inspection policy to intermittent faults where the test is planned at periodic times kT $(k = 1, 2, \ldots)$ to detect these faults (see Figure 8.4). We obtain the mean time to detect a fault and the expected

8.5 Intermittent Faults

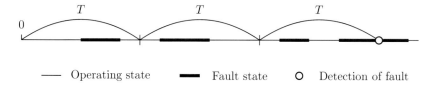

— Operating state   — Fault state   ○ Detection of fault

Fig. 8.4. Process of periodic inspection for intermittent faults

number of tests. In addition, we discuss optimum times T^* that minimize the expected cost until fault detection, and maximize the probability of detecting the first fault. An imperfect test model where faults are detected with probability p was treated in [50].

Suppose that faults occur intermittently; i.e., a unit repeats the operating state (State 0) and fault state (State 1) alternately. The times of respective operating and fault states are independent and have identical exponential distributions $(1 - e^{-\lambda t})$ and $(1 - e^{-\theta t})$ with $\theta > \lambda$. The periodic test to detect faults is planned at times kT $(k = 1, 2, \dots)$. It is assumed that the faults of a unit are investigated only through test which is perfect; i.e., faults are always detected by test when they occur and are isolated. The time required for test is negligible.

The transition probabilities $P_{0j}(t)$ from state 0 to state j $(j = 0, 1)$ are, from Section 2.1,

$$P_{00}(t) = \frac{\theta}{\lambda + \theta} + \frac{\lambda}{\lambda + \theta} e^{-(\lambda + \theta)t}, \qquad P_{01}(t) = \frac{\lambda}{\lambda + \theta}(1 - e^{-(\lambda + \theta)t}).$$

Using the above equations, we have the following reliability quantities. The expected number $M(T)$ of tests to detect a fault is

$$M(T) = \sum_{j=0}^{\infty} (j+1)[P_{00}(T)]^j P_{01}(T) = \frac{1}{P_{01}(T)}, \qquad (8.51)$$

the mean time $l(T)$ to detect a fault is

$$l(T) = \sum_{j=0}^{\infty} (j+1)T[P_{00}(T)]^j P_{01}(T) = \frac{T}{P_{01}(T)}, \qquad (8.52)$$

the probability $P_0(T)$ that the first occurrence of faults is detected at the first test is

$$P_0(T) = \int_0^T e^{-\theta(T-t)} \lambda e^{-\lambda t}\, dt = \frac{\lambda}{\theta - \lambda}(e^{-\lambda T} - e^{-\theta T}), \qquad (8.53)$$

the probability $P_1(T)$ that the first occurrence of faults is detected at some test is

$$P_1(T) = P_0(T) + e^{-\lambda T} P_1(T),$$

i.e.,
$$P_1(T) = \frac{\lambda}{\theta - \lambda} \frac{e^{-\lambda T} - e^{-\theta T}}{1 - e^{-\lambda T}}, \tag{8.54}$$

and the probability $Q_N(T)$ that some fault is detected until the Nth test is
$$Q_N(T) = 1 - [P_{00}(T)]^N. \tag{8.55}$$

Using the above quantities, we consider the following four optimum policies. The expected cost until fault detection is, from (8.51) and (8.52),
$$C(T) \equiv c_1 M(T) + c_2 l(T) = \frac{c_1 + c_2 T}{P_{01}(T)}, \tag{8.56}$$

where $c_1 =$ cost of one test and $c_2 =$ operational cost rate of a unit. We seek an optimum time T_1^* that minimizes $C(T)$. Differentiating $C(T)$ with respect to T and setting it equal to zero imply
$$\frac{1}{\lambda + \theta}(e^{(\lambda+\theta)T} - 1) - T = \frac{c_1}{c_2}. \tag{8.57}$$

The left-hand side of (8.57) is strictly increasing from 0 to infinity. Thus, there exists a finite and unique T_1^* that satisfies (8.57).

We derive an optimum time T_2^* that maximizes the probability $P_0(T)$. From (8.53), it is evident that $\lim_{T \to 0} P_0(T) = 0$, and
$$\frac{dP_0(T)}{dT} = \frac{\lambda}{\theta - \lambda}(\theta e^{-\theta T} - \lambda e^{-\lambda T}).$$

Thus, by putting $dP_0(T)/dT = 0$ because $\theta > \lambda$, an optimum T_2^* is
$$T_2^* = \frac{\log \theta - \log \lambda}{\theta - \lambda}. \tag{8.58}$$

Furthermore, we derive a maximum time T_3^* that satisfies $P_1(T) \geq q_1$; i.e., the probability that the first occurrence of faults is detected at some test is greater than a specified q_1 ($0 < q_1 < 1$). It is evident that $\lim_{T \to 0} P_1(T) = 1$, $\lim_{T \to \infty} P_1(T) = 0$, and
$$\frac{dP_1(T)}{dT} = \frac{\lambda}{\theta - \lambda} \frac{e^{-(\lambda+\theta)T}}{(1 - e^{-\lambda T})^2}[\theta(e^{\lambda T} - 1) - \lambda(e^{\theta T} - 1)] < 0.$$

Thus, $P_1(T)$ is strictly decreasing from 1 to 0, and hence, there exists a finite and unique T_3^* that satisfies $P_1(T) = q_1$.

Next, suppose that the testing times T_i ($i = 1, 2, 3$) are determined from the above results. The probability that a fault is detected until the Nth test is greater than q_2 ($0 < q_2 < 1$) is $Q_N(T) \geq q_2$. Thus, a minimum number N^* that satisfies $[P_{00}(T_i^*)]^N \leq 1 - q_2$ is

8.5 Intermittent Faults   223

Table 8.6. Optimum time T_1^* to minimize $C(T)$ and maximum time T_3^* to satisfy $P_1(T) \geq q_1$

θ/λ	\multicolumn{5}{c}{T_1^* c_1/c_2}	\multicolumn{5}{c}{T_3^* q_1 (%)}
	1	5	10	50	100	50	60	70	80	90
1.2	0.80	1.39	1.70	2.50	2.87	1.29	0.96	0.68	0.43	0.20
1.5	0.85	1.49	1.82	2.70	3.10	1.33	0.99	0.69	0.44	0.20
2.0	0.90	1.60	1.97	2.93	3.37	1.38	1.02	0.71	0.44	0.21
5.0	1.03	1.86	2.30	3.49	4.03	1.49	1.07	0.74	0.45	0.21
10.0	1.09	1.97	2.45	3.73	4.32	1.54	1.10	0.75	0.46	0.21
50.0	1.14	2.07	2.59	3.95	4.59	1.58	1.12	0.76	0.46	0.21

Table 8.7. Optimum time T_2^* to maximize $P_0(T)$ and minimum number N^* such that $Q_N(T_2^*) \geq q_2$

θ/λ	T_2^*	\multicolumn{5}{c}{N^* q_2 (%)}
		50	60	70	80	90
1.2	1.09	2	2	3	4	5
1.5	1.22	2	2	3	4	6
2.0	1.39	3	3	4	5	7
5.0	2.01	5	6	8	10	14
10.0	2.56	8	11	14	19	26
50.0	4.00	36	48	62	83	119

$$N^* = \left[\frac{\log(1-q_2)}{\log P_{00}(T_i^*)} \right] + 1 \qquad (8.59)$$

where $[x]$ denotes the greatest integer contained in x.

Example 8.6. Suppose that $\theta/\lambda = 1.2, 1.5, 2.0, 5.0, 10.0, 50.0$; *i.e.*, all times are relative to the mean fault time $1/\theta$. Table 8.6 presents the optimum time T_1^* that minimizes the expected cost $C(T)$ in (8.56) for $c_1/c_2 = 1, 5, 10, 50, 100$, and the maximum time T_3^* that satisfies $P_1(T) \geq q_1$ for $q_1 = 50, 60, 70, 80, 90$ (%). Table 8.7 shows the optimum time T_2^* that maximizes $P_0(T)$ and minimum number N^* that satisfies $Q_N(T_2^*) \geq q_2$.

For example, when $\theta/\lambda = 10$ and $c_1/c_2 = 10$, the optimum time is $T_1^* = 2.45$. In particular, when $1/\lambda = 24$ hours and $1/\theta = 2.4$ hours, the test should be done at about every 6 ($\doteq 2.45 \times 2.4$) hours. To maximize the probability of detecting the first fault at the first test, $T_2^* = 2.01$ for $\theta/\lambda = 5.0$. If the same test in this case is repeated ten times, a fault is detected with more than 80% probability from Table 8.7. Furthermore, if the test is done at $T_3^* = 0.45$, the probability of detecting the first fault is more than 80% from Table 8.6.

We have adopted the testing time T_1^* in cost, and T_2^* and T_3^* in probabilities of detecting the first occurrence of faults. In particular, the result of $T_2^* =$

$(\log \theta - \log \lambda)/(\theta - \lambda)$ is quite simple. If λ and θ vary a little, we can compute T_2^* easily and should make the next test at time T_2^*. These testing strategies could be applied to real digital systems by suitable modifications. ∎

8.6 Inspection for a Finite Interval

Most units would be operating for a finite interval. Practically, the working time of units is finite in actual fields. Very few papers treated with replacements for a finite time span. The optimum sequential policy [1] and the asymptotic costs [105, 106] of age replacement for a finite interval were obtained.

This section summarizes inspection policies for an operating unit for a finite interval $(0, S]$ $(0 < S < \infty)$ in which its failure is detected only by inspection. Generally, it would be more difficult to compute optimum inspection policies in a finite case than those in an infinite one. We consider three inspection models of periodic and sequential inspections in Section 8.1, and asymptotic inspection in Section 8.2.

In periodic inspection, an interval S is divided equally into N parts and a unit is checked at periodic times kT $(k = 1, 2, \ldots, N)$ where $NT \equiv S$. When the failure time is exponential, we first compute a checking time in an infinite case, and using the partition method, we derive an optimum policy that shows how to compute an optimum number N^* of checks in a finite case.

In sequential inspection, we show how to compute optimum checking times. Such computations might be troublesome, because we have to solve some simultaneous equations, however, they would be easier than those of Algorithm 1 in Section 8.1 as recent personal computers have developed greatly.

In asymptotic inspection, we introduce an inspection intensity and show how to compute approximate checking times by a simpler method than that of the sequential one. Finally, we give numerical examples and show that the asymptotic inspection has a good approximation to the sequential one.

(1) Periodic Inspection

Suppose that a unit has to be operating for a finite interval $(0, S]$ and fails according to a general distribution $F(t)$ with a density function $f(t)$. To detect failures, a unit is checked at periodic times kT $(k = 1, 2, \ldots, N)$. Then, from (8.1), the total expected cost until failure detection or time S is

$$C(N) = \sum_{k=0}^{N-1} \int_{kT}^{(k+1)T} \{c_1(k+1) + c_2[(k+1)T - t]\} \, dF(t) + c_1 N \overline{F}(NT) + c_3$$

$$= \left(c_1 + \frac{c_2 S}{N}\right) \sum_{k=0}^{N-1} \overline{F}\left(\frac{kS}{N}\right) - c_2 \int_0^S \overline{F}(t) \, dt + c_3 \quad (N = 1, 2, \ldots). \quad (8.60)$$

8.6 Inspection for a Finite Interval

Table 8.8. Approximate time $\widetilde{T}$, optimum number N^*, and time $T^* = S/N^*$, and expected cost $\widetilde{C}(N^*)$ for $S = 100, 50$ and $c_1/c_2 = 2, 5, 10$ when $\lambda = 0.01$

S	c_1/c_2	$\widetilde{T}$	N^*	T^*	$\widetilde{C}(N^*)/c_2$
	2	19.355	5	20.0	76.72
100	5	30.040	3	33.3	85.48
	10	41.622	2	50.0	96.39
	2	19.355	3	16.7	47.85
50	5	30.040	2	25.0	53.36
	10	41.622	1	50.0	60.00

It is evident that $\lim_{N \to \infty} C(N) = \infty$ and

$$C(1) = c_1 + c_2 \int_0^S F(t)\, dt + c_3.$$

Thus, there exists a finite number N^* ($1 \le N^* < \infty$) that minimizes $C(N)$.

In particular, assume that the failure time is exponential; i.e., $F(t) = 1 - e^{-\lambda t}$. Then, the expected cost $C(N)$ in (8.60) can be rewritten as

$$C(N) = \left(c_1 + \frac{c_2 S}{N}\right) \frac{1 - e^{-\lambda S}}{1 - e^{-\lambda S/N}} - \frac{c_2}{\lambda}(1 - e^{-\lambda S}) + c_3 \quad (N = 1, 2, \ldots). \quad (8.61)$$

To find an optimum number N^* that minimizes $C(N)$, we put $T = S/N$. Then, Equation (8.61) becomes

$$C(T) = (c_1 + c_2 T) \frac{1 - e^{-\lambda S}}{1 - e^{-\lambda T}} - \frac{c_2}{\lambda}(1 - e^{-\lambda S}) + c_3. \quad (8.62)$$

Differentiating $C(T)$ with respect to T and setting it equal to zero, we have

$$e^{\lambda T} - (1 + \lambda T) = \frac{\lambda c_1}{c_2} \quad (8.63)$$

which agrees with (8.5). Thus, there exists a finite and unique $\widetilde{T}$ ($0 < \widetilde{T} < \infty$) that satisfies (8.63).

Therefore, we show the following partition method.

(i) If $\widetilde{T} < S$ then we put $[S/\widetilde{T}] \equiv N$ and calculate $C(N)$ and $C(N+1)$ from (8.61), where $[x]$ denotes the greatest integer contained in x. If $C(N) \le C(N+1)$ then $N^* = N$, and conversely, if $C(N) > C(N+1)$ then $N^* = N+1$.
(ii) If $\widetilde{T} \ge S$ then $N^* = 1$.

Note that $\widetilde{T}$ gives the optimum checking time for an infinite time span in an exponential case.

Example 8.7. Table 8.8 presents the approximate checking time $\widetilde{T}$, the optimum checking number N^*, and time $T^* = S/N^*$, and the expected cost

$\widetilde C(N^*) \equiv C(N^*) + (c_2/\lambda)(1 - e^{-\lambda S}) - c_3$ for $S = 100, 50$ and $c_1/c_2 = 2, 5, 10$ when $\lambda = 0.01$. If S is large then it would be sufficient to compute approximate checking times $\widetilde T$. ∎

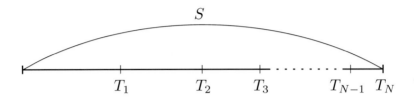

Fig. 8.5. Process of sequential inspection in a finite interval

(2) Sequential Inspection

An operating unit is checked at successive times $0 < T_1 < T_2 < \cdots < T_N$, where $T_0 \equiv 0$ and $T_N \equiv S$ (see Figure 8.5). In a similar way to that of obtaining (8.60), the total expected cost until failure detection or time S is

$$\mathbf{C}(N) = \sum_{k=0}^{N-1} \int_{T_k}^{T_{k+1}} [c_1(k+1) + c_2(T_{k+1} - t)]\, dF(t) + c_1 N \overline F(T_N) + c_3$$

$$(N = 1, 2, \dots). \quad (8.64)$$

Putting that $\partial \mathbf{C}(N)/\partial T_k = 0$, which is a necessary condition for minimizing $\mathbf{C}(N)$, we have

$$T_{k+1} - T_k = \frac{F(T_k) - F(T_{k-1})}{f(T_k)} - \frac{c_1}{c_2} \quad (k = 1, 2, \dots, N-1) \quad (8.65)$$

and the resulting minimum expected cost is

$$\widetilde{\mathbf{C}}(N) \equiv \mathbf{C}(N) + c_2 \int_0^S \overline F(t)\, dt - c_3 = \sum_{k=0}^{N-1} [c_1 + c_2(T_{k+1} - T_k)]\overline F(T_k)$$

$$(N = 1, 2, \dots). \quad (8.66)$$

For example, when $N = 3$, the checking times T_1 and T_2 are given by the solutions of equations

$$S - T_2 = \frac{F(T_2) - F(T_1)}{f(T_2)} - \frac{c_1}{c_2}$$

$$T_2 - T_1 = \frac{F(T_1)}{f(T_1)} - \frac{c_1}{c_2}$$

8.6 Inspection for a Finite Interval

Table 8.9. Checking time T_k and expected cost $\widetilde{C}(N)$ for $N = 1, 2, \ldots, 8$ when $S = 100$, $c_1/c_2 = 2$, and $F(t) = 1 - e^{-\lambda t^2}$

N	1	2	3	4	5	6	7	8
T_1	100	64.14	50.9	44.1	40.3	38.1	36.8	36.3
T_2		100	77.1	66.0	60.0	56.2	54.3	53.3
T_3			100	84.0	75.4	70.5	67.8	66.6
T_4				100	88.6	82.3	78.9	77.3
T_5					100	91.1	87.9	85.9
T_6						100	94.9	92.5
T_7							100	97.2
T_8								100
$\widetilde{C}(N)/c_2$	102.00	93.55	91.52	91.16	91.47	92.11	92.91	93.79

and the expected cost is

$$\widetilde{C}(3) = c_1 + c_2 T_1 + [c_1 + c_2(T_2 - T_1)]\overline{F}(T_1) + [c_1 + c_2(S - T_2)]\overline{F}(T_2).$$

From the above discussion, we compute T_k ($k = 1, 2, \ldots, N-1$) which satisfies (8.65), and substituting them into (8.66), we obtain the expected cost $\mathbf{C}(N)$. Next, comparing $\mathbf{C}(N)$ for all $N \geq 1$, we can get the optimum checking number N^* and times T_k^* ($k = 1, 2, \ldots, N^*$).

Example 8.8. Table 8.9 gives the checking time T_k ($k = 1, 2, \ldots, N$) and the expected cost $\widetilde{C}(N)$ for $S = 100$ and $c_1/c_2 = 2$ when $F(t) = 1 - \exp(-\lambda t^2)$. In this case, we set that the mean failure time is equal to S; i.e.,

$$\int_0^\infty e^{-\lambda t^2}\, dt = \frac{1}{2}\sqrt{\frac{\pi}{\lambda}} = S.$$

Comparing $\widetilde{C}(N)$ for $N = 1, 2, \ldots, 8$, the expected cost is minimum at $N = 4$. That is, the optimum checking number is $N^* = 4$ and optimum checking times are 44.1, 66.0, 84.0, 100. ∎

(3) Asymptotic Inspection

Suppose that $n(t)$ is an inspection intensity defined in **(3)** of Section 8.2. Then, from (8.21) and (8.64), the approximate total expected cost is

$$\mathbf{C}(n(t)) = \int_0^S \left[c_1 \int_0^t n(u)\, du + \frac{c_2}{2n(t)} \right] dF(t) + c_1 \overline{F}(S) \int_0^S n(t)\, dt + c_3. \tag{8.67}$$

Differentiating $\mathbf{C}(n(t))$ with $n(t)$ and setting it equal to zero, we have (8.22).

228   8 Inspection Policies

We compute approximate checking times $\widetilde{T}_k$ $(k = 1, 2, \ldots, N-1)$ and checking number $\widetilde{N}$, using (8.22). First, we put that

$$\int_0^S \sqrt{\frac{c_2 h(t)}{2c_1}}\, dt \equiv X$$

and $[X] \equiv N$, where $[x]$ is defined in policy (i) in **(1)**. Then, we obtain A_N $(0 < A_N \le 1)$ such that

$$A_N \int_0^S \sqrt{\frac{c_2 h(t)}{2c_1}}\, dt = N$$

and define an inspection intensity as

$$\widetilde{n}(t) = A_N \sqrt{\frac{c_2 h(t)}{2c_1}}. \tag{8.68}$$

Using (8.68), we compute checking times T_k that satisfy

$$\int_0^{T_k} \widetilde{n}(t)\, dt = k \qquad (k = 1, 2, \ldots, N), \tag{8.69}$$

where $T_0 = 0$ and $T_N = S$. Then, the total expected cost is given in (8.66).

Next, we put N by $N+1$ and do a similar computation. At last, we compare $\mathbf{C}(N)$ and $\mathbf{C}(N+1)$, and choose the small one as the total expected cost $\mathbf{C}(\widetilde{N})$ and the corresponding checking times $\widetilde{T}_k$ $(k = 1, 2, \ldots, \widetilde{N})$ as an asymptotic inspection policy.

Example 8.9. Consider a numerical example when the parameters are the same as those of Example 8.8. Then, because $\lambda = \pi/4 \times 10^4$, $n(t) = \sqrt{\lambda t/2}$, $[X] = N = 4$, and $A_N = (12/100)/\sqrt{\pi/200}$, we have that $\widetilde{n}(t) = 6\sqrt{t}/10^3$. Thus, from (8.69), checking times are

$$\int_0^{T_k} \frac{6}{1000} \sqrt{t}\, dt = \frac{1}{250} T_k^{3/2} = k \qquad (k = 1, 2, 3).$$

Also, when $N = 5$, $A_N = (15/100)/\sqrt{\pi/200}$, and $\widetilde{n}(t) = 3\sqrt{t}/4 \times 10^2$. In this case, checking times are

$$\int_0^{T_k} \frac{3}{400} \sqrt{t}\, dt = \frac{1}{200} T_k^{3/2} = k \qquad (k = 1, 2, 3, 4).$$

Table 8.10 shows the checking times and the resulting costs for $N = 4$ and 5. Because $\widetilde{\mathbf{C}}(4) < \widetilde{\mathbf{C}}(5)$, the approximate checking number is $\widetilde{N} = 4$ and its checking times $\widetilde{T}_k$ are 39.7, 63.0, 82.5, 100. These checking times are a little smaller than those in Table 8.9, however, they are closely approximate to the optimum ones. ∎

Table 8.10. Checking time $\widetilde{T}_k$ and expected cost $\widetilde{C}(N)$ for $N = 4, 5$ when $S = 100$, $c_1/c_2 = 2$, and $F(t) = 1 - e^{-\lambda t^2}$

N	4	5
1	39.7	34.2
2	63.0	54.3
3	82.5	71.1
4	100.0	86.2
5		100.0
$\widetilde{C}(N)/c_2$	91.22	91.58

References

1. Barlow RE, Proschan F (1965) Mathematical Theory of Reliability. J Wiley & Sons, New York.
2. Luss H, Kander Z (1974) Inspection policies when duration of checkings is non-negligible. Oper Res Q 25:299–309.
3. Luss H (1976) Inspection policies for a system which is inoperative during inspection periods. AIIE Trans 9:189–194.
4. Wattanapanom N, Shaw L (1979) Optimal inspection schedules for failure detection in a model where tests hasten failures. Oper Res 27:303–317.
5. B. Sengupta (1980) Inspection procedures when failure symptoms are delayed. Oper Res 28:768–776.
6. Platz O (1976) Availability of a renewable, checked system. IEEE Trans Reliab R-25:56–58.
7. Schneeweiss WG (1976) On the mean duration of hidden faults in periodically checked systems. IEEE Trans Reliab R-25:346–348.
8. Schneeweiss WG (1977) Duration of hidden faults in randomly checked systems. IEEE Trans Reliab R-26:328–330.
9. Munford AG (1981) Comparison among certain inspection policies. Manage Sci 27:260–267.
10. Luss H (1983) An inspection policy model for production facilities. Manage Sci 29:1102–1109.
11. Parmigiani G (1993) Optimal inspection and replacement policies with age-dependent failures and fallible tests. J Oper Res Soc 44:1105–1114.
12. Parmigiani G (1996) Optimal scheduling of fallible inspections. Oper Res 44:360–367.
13. Zacks S, Fenske WJ (1973) Sequential determination of inspection epochs for reliability systems with general lifetime distributions. Nav Res Logist Q 20:377–386.
14. Luss H, Kander Z (1974) A preparedness model dealing with N systems operating simultaneously. Oper Res 22:117–128.
15. Anbar D (1976) An asymptotically optimal inspection policy. Nav Res Logist Q 23:211–218.
16. Teramoto T, Nakagawa T, Motoori M (1990) Optimal inspection policy for a parallel redundant system. Microelectron Reliab 30:151–155.
17. Kander Z (1978) Inspection policies for deteriorating equipment characterized by N quality levels. Nav Res Logist Q 25:243–255.

18. Zuckerman D (1980) Inspection and replacement policies. J Appl Prob 17:168–177.
19. Zuckerman D (1989) Optimal inspection policy for a multi-unit machine. J Appl Prob 26:543–551.
20. Qiu YP (1991) A note on optimal inspection policy for stochastically deteriorating series systems. J Appl Prob 28:934–939.
21. Valdez-Flores C, Feldman RM (1989) A survey of preventive maintenance models for stochastically deteriorating single-unit systems. Nav Logist Q 36:419–446.
22. Munford AG, Shahani AK (1972) A nearly optimal inspection policy. Oper Res Q 23:373–379.
23. Munford AG, Shahani AK (1973) An inspection policy for the Weibull case. Oper Res Q 24:453–458.
24. Tadikamalla PR (1979) An inspection policy for the gamma failure distributions. J Oper Res Soc 30:77–80.
25. Sherwin DJ (1979) Inspection intervals for condition-maintained items which fail in an obvious manner. IEEE Trans Reliab R-28:85–89.
26. Schultz CR (1985) A note on computing periodic inspection policies. Manage Sci 31:1592–1596.
27. Senna V, Shahani AK (1986) A simple inspection policy for the detection of failure. Eur J Oper Res 23:222–227.
28. Chelbi A, Ait-Kadi D (1999) An optimal inspection strategy for randomly failing equipment. Reliab Eng Sys Saf 63:127–131.
29. Keller JB (1974) Optimum checking schedules for systems subject to random failure. Manage Sci 21:256–260.
30. Keller JB (1982) Optimum inspection policies. Manage Sci 28:447–450.
31. Kaio N, Osaki S (1984) Some remarks on optimum inspection policies. IEEE Trans Reliab R-33:277–279.
32. Kaio N, Osaki S (1986) Optimal inspection policies: A review and comparison. J Math Anal Appl 119:3–20.
33. Kaio N, Osaki S (1986) Optimal inspection policy with two types of imperfect inspection probabilities. Microelectron Reliab 26:935–942.
34. Kaio N, Osaki S (1988) Inspection policies: Comparisons and modifications. RAIRO Oper Res 22:387–400.
35. Kaio N, Osaki S (1989) Comparison of inspection policies. J Oper Res Soc 40:499–503.
36. Viscolani B (1991) A note on checking schedules with finite horizon. RAIRO Oper Res 25:203–208.
37. Kaio N, Dohi T, Osaki S (1994) Inspection policy with failure due to inspection. Microelectron Reliab 34:599–602.
38. Weiss GH (1962) A problem in equipment maintenance. Manage Sci 8:266–277.
39. Coleman JJ, Abrams IJ (1962) Mathematical model for operational readiness. Oper Res 10:126–138.
40. Morey RC (1967) A criterion for the economic application of imperfect inspections. Oper Res 15:695–698.
41. Apostolakis GE, Bansal PP (1977) Effect of human error on the availability of periodically inspected redundant systems. IEEE Trans Reliab R-26:220–225.
42. Srivastava MS, Wu YH (1993) Estimation & testing in an imperfect-inspection model. IEEE Trans Reliab 42:280–286.
43. Gertsbakh I (1977) Models of Preventive Maintenance. North-Holland, New York.

44. Nakagawa T (1982) Reliability analysis of a computer system with hidden failure. Policy Inf 6:43–49.
45. Phillips MJ (1979) The reliability of a system subject to revealed and unrevealed faults. Microelectron Reliab 18:495–503.
46. Koren I, Su SYH (1979) Reliability analysis of N-modular redundancy systems with intermittent and permanent faults. IEEE Trans Comput C-28:514–520.
47. Shin KG, Lee YH (1986) Measurement and application of fault latency. IEEE Trans Comput C-35:370–375.
48. Su SYH, Koren I, Malaiya YK (1978) A continuous-parameter Markov model and detection procedures for intermittent faults. IEEE Trans Comput C-27:567–570.
49. Nakagawa T, Motoori M, Yasui K (1990) Optimal testing policy for a computer system with intermittent faults. Reliab Eng Sys Saf 27:213–218.
50. Nakagawa T, Yasui K (1989) Optimal testing-policies for intermittent faults. IEEE Trans Reliab 38:577–580.
51. Chung KJ (1995) Optimal test-times for intermittent faults. IEEE Trans Reliab 44:645–647.
52. Ismaeel AA, Bhatnagar R (1997) Test for detection & location of intermittent faults in combinational circuits. IEEE Trans Reliab 46:269–274.
53. Nakagawa T (1980) Optimum inspection policies for a standby unit. J Oper Res Soc Jpn 23:13–26.
54. Thomas LC, Jacobs PA, Gaver DP (1987) Optimal inspection policies for standby systems. Commun Stat Stoch Model 3:259–273.
55. Sim SH (1987) Reliability of standby equipment with periodic testing. IEEE Trans Reliab R-36:117–123.
56. Parmigiani G (1994) Inspection times for stand-by units. J Appl Prob 31:1015–1025.
57. Vaurio JK (1995) Unavailability analysis of periodically tested standby components. IEEE Trans Reliab 44:512–517.
58. Chay SC, Mazumdar M (1975) Determination of test intervals in certain repairable standby protective systems. IEEE Trans Reliab R-24:201–205.
59. Inagaki T, Inoue K, Akashi H (1979) Improvement of supervision schedules for protective systems. IEEE Trans Reliab R-28:141–144.
60. Inagaki T, Inoue K, Akashi H (1980) Optimization of staggered inspection schedules for protective systems. IEEE Trans Reliab R-29:170–173.
61. Shima E, Nakagawa T (1984) Optimum inspection policy for a protective device. Reliab Eng 7:123–132.
62. Christer AH (1982) Modelling inspection policies for building maintenance. J Oper Res Soc 33:723–732.
63. Christer AH, Waller WM (1984) Delay time models of industrial inspection maintenance problems. J Oper Res Soc 35:401–406.
64. Christer AH, MacCallum KL, Kobbacy K, Bolland J, Hessett C (1989) A system model of underwater inspection operations. J Oper Res Soc. 40:551–565.
65. Sim SH (1984) Availability model of periodically tested standby combustion turbine units. IIE Trans 16:288–291.
66. Sim SH, Wang L (1984) Reliability of repairable redundant systems in nuclear generating stations. Eur J Oper Res 17:71–78.
67. Sim SH (1985) Unavailability analysis of periodically tested components of dormant systems. IEEE Trans Reliab R-34:88–91.

68. Turco F, Parolini P (1984) A nearly optimal inspection policy for productive equipment. Inter J Product Res 22:515–528.
69. Young PJ (1984) Inspection intervals for fail-safe structure. IEEE Trans Reliab R-33:165–170.
70. Cassandras CG, Han Y (1992) Optimal inspection policies for a manufacturing station. Eur J Oper Res 63:35–53.
71. Sherwin DJ (1995) An inspection model for automatic trips & warning instruments. In: Proceedings Annual Reliability and Maintainability Symposium:271–274.
72. Garnero MA, Beaudouin F, Delbos JP (1998) Optimization of bearing-inspection intervals. In: Proceedings Annual Reliability and Maintainability Symposium:332–338.
73. Bukowski JV (2001) Modeling and analyzing the effects of periodic inspection on the performance of safety-critical systems. IEEE Trans Reliab 50:321–329.
74. Baker R (1996) Maintenance optimisation with the delay time model. In: Özekici S (ed) Reliability and Maintenance of Complex Systems. Springer, New York:550–587.
75. Christer AH (2002) A review of delay time analysis for modelling plant maintenance. In: Osaki S (ed) Stochastic Models in Reliability and Maintenance. Springer, New York:89–123.
76. Jia X, Christer AH (2003) Case experience comparing the RCM approach to plant maintenance with a modeling approach. In: Blischke WR, Murthy DNP (eds) Case Studies in Reliability and Maintenance. J Wiley & Sons, New York:477–494.
77. Ito K, Nakagawa T (1997) An optimal inspection policy for a storage system with finite number of inspections. J Reliab Eng Assoc Jpn 19:390–396.
78. Nakagawa T, Yasui K (1979) Approximate calculation of inspection policy with Weibull failure times. IEEE Trans Reliab R-28:403–404.
79. Nakagawa T, Yasui K (1980) Approximate calculation of optimal inspection times. J Oper Res Soc 31:851–853.
80. Ito K, Nakagawa T (1992) Optimal inspection policies for a system in storage. Comput Math Appl 24:87–90.
81. Ito K, Nakagawa T (1995) An optimal inspection policy for a storage system with high reliability. Microelectron Reliab 35:875–886.
82. Ito K, Nakagawa T (1995) Extended optimal inspection policies for a system in storage. Math Comput Model 22:83–87.
83. Ito K, Nakagawa T (1995) An optimal inspection policy for a storage system with three types of hazard rate functions. J Oper Res Soc Jpn 38:423–431.
84. Nakagawa T, Mizutani S, Igaki N (2002) Optimal inspection policies for a finite interval. The Second Euro-Japan Workshop on Stochastic Risk Modelling, Insurance, Production and Reliability:334–339.
85. Mizutani S, Teramoto K, Nakagawa T (2004) A survey of finite inspection models. In: Tenth ISSAT International Conference on Reliability and Quality in Design:104–108.
86. Vaurio JK (1999) Availability and cost functions for periodically inspected preventively maintained units. Reliab Eng Sys Saf 63:133–140.
87. Biswas A, Sarkar J, Sarkar S (2003) Availability of a periodically inspected system, maintained under an imperfect-repair policy. IEEE Trans Reliab 52:311–318.

88. Leung FK (2001) Inspection schedules when the lifetime distribution of a single-unit system is completely unknown. Eur J Oper Res 132:106–115.
89. Harris FW (1915) Operations and Costs. AW Shaw Company, Chicago.
90. Bauer J et al. (1973) Dormancy and power on-off cycling effects on electronic equipment and part reliability. RADC-TR-73-248 (AD/A-768619).
91. Cottrell DF et al. (1974) Effects of dormancy on nonelectonic components and materials. RADC-TR-74-269 (AD/A-002838).
92. Malik DF, Mitchell JC (1978) Missile material reliability prediction handbook–Parts count prediction (AD/A-053403).
93. Trapp RD et al. (1981) An approach for assessing missile system dormant reliability. BDM/A-81-016-TR(AD/A-107519).
94. Smith Jr HB, Rhodes Jr C (1982) Storage reliability of missile material program-Storage reliability prediction handbook for part count prediction (AD/A-122439).
95. Menke JT (1983) Deterioration of electronics in storage. In: Proceedings National SAMPE Symposium:966–972.
96. Martinez EC (1984) Storage reliability with periodic test. In: Proceedings Annual Reliability and Maintainability Symposium:181–185.
97. Ito K, Nakagawa T (2000) Optimal inspection policies for a storage system with degradation at periodic tests. Math and Comput Model 31:191–195.
98. Malaiya YK, Su SYH (1981) Reliability measures for hardware redundancy fault-tolerant digital systems with intermittent faults. IEEE Trans Comput C-30:600–604.
99. Castillo X, McConnel SR, Siewiorek DP (1982) Derivation and calibration of a transient error reliability model. IEEE Trans Comput C-31:658–671.
100. Rao TRN (1968) Use of error correcting codes on memory words for improved reliability. IEEE Trans Reliab R-17:91–96.
101. Cox GW, Carroll BD (1978) Reliability modeling and analysis of fault-tolerant memories. IEEE Trans Reliab R-27:49–54.
102. Castillo X, Siewiorek DP (1980) A performance-reliability model for computing systems. 10 th International Symposium Fault-Tolerant Comput:187–192.
103. Nakagawa T, Nishi K, Yasui K (1984) Optimum preventive maintenance policies for a computer system with restart. IEEE Trans Reliab R-33:272–276.
104. Malaiya YK (1982) Linearly corrected intermittent failures. IEEE Trans Reliab R-31:211–215.
105. Christer AH (1978) Refined asymptotic costs for renewal reward processes. J Oper Res Soc 29:577–583.
106. Ansell J, Bendell A, Humble S (1984) Age replacement under alternative cost criteria. Manage Sci 30:358–367.

9
Modified Maintenance Models

Until now, we have dealt primarily with the basic maintenance models and their combined models. This chapter introduces modified and extended maintenance models proposed mainly by the author and our co-workers. These models further reflect the real world and present more interesting topics to theoretical researchers.

In Section 9.1, we convert the continuous models of age, periodic, and block replacements and inspection to discrete ones [1]. These would be useful for the cases where: (i) an operating unit sometimes cannot be maintained at the exact optimum time for some reason such as shortage of spare units, lack of money or workers, or inconvenience of time required to complete the maintenance, and (ii) a unit is usually maintained in idle times. We have already discussed the optimum inspection policies for a finite interval in Section 8.6. In Section 9.2, we propose the models of periodic and block replacements for a finite interval because the working times of most units would be finite in the actual field. It is shown that the optimum policies are easily given by the partition method obtained in Section 8.6, using the results of optimum policies for basic models [2, 3].

In Section 9.3, we suggest the extended models of age, periodic, and block replacements in which a unit is replaced at either a planned or random time. Furthermore, we consider the random inspection policy in which a unit is checked at both periodic and random times. These random maintenance policies would be useful for units in which maintenance should be done at the completion of their work or in their idle times [4, 5]. In Section 9.4, we consider the optimization problems of when to replace a unit with n spares, and derive an optimum replacement time that maximizes the mean time to failure [6]. In Section 9.5, we apply the modified age replacement policy in Section 9.1 to a unit with n spares; *i.e.*, we convert the continuous optimization problem in Section 9.4 to the discrete one. Finally, other maintenance policies are collected concisely in Section 9.6.

9.1 Modified Discrete Models

An operating unit sometimes cannot be replaced at the exact optimum times for some reason: shortage of spare units, lack of money or workers, or inconvenience of time required to complete the replacement. Units may be rather replaced in idle times, *e.g.*, weekend, month-end, or year-end. An intermittently used system would be preventively replaced after a certain number of uses [7,8].

This section proposes modified replacement policies that convert the standard age, periodic, block replacement, and inspection models treated in Chapters 3, 4, 5, and 8 to discrete ones. The replacement is planned only at times kT ($k = 1, 2, \ldots$), where T ($0 < T < \infty$) is previously given and refers to a day, a week, a month, a year, and so on. Then, the following replacement policies are considered.

(1) Age replacement: A unit is replaced at time NT or at failure.
(2) Periodic replacement: A unit is replaced at time NT and undergoes only minimal repair at failures.
(3) Block replacement: A unit is replaced at time NT and at failure.
(4) Inspection: A unit is replaced at time NT or at failure that is detected only through inspection.

The above four discrete replacement models are one modification of the continuous ones. These would be more economical than the usual ones if a replacement cost at time NT is less than that of the replacement time.

Suppose that the failure time of each unit is independent and has an identical distribution $F(t)$ with finite mean μ and the failure rate $h(t) \equiv f(t)/\overline{F}(t)$, where f is a density function of F and $\overline{F} \equiv 1 - F$. We obtain the expected cost rates of each model, using the usual calculus methods of replacement models, and derive optimum numbers N^* that minimize them. These are given by unique solutions of equations when the failure rate $h(t)$ is strictly increasing.

(1) Age Replacement

The time is measured only by the total operating time of a unit. It is assumed that the replacement is planned at times kT ($k = 1, 2, \ldots$) for a fixed $T > 0$; *i.e.*, the replacement is allowed only at periodic times kT. This would be more useful than the continuous-time models if replacement at the weekend is more convenient and economical than that during weekdays. A unit is replaced at time NT or at failure, whichever occurs first, where any failure is detected immediately when it fails.

From (3.4) in Chapter 3, the expected cost rate is given by

$$C_1(N) = \frac{c_1 F(NT) + c_2 \overline{F}(NT)}{\int_0^{NT} \overline{F}(t)\, dt} \qquad (N = 1, 2, \ldots), \qquad (9.1)$$

where c_1 = cost of replacement at failure, and c_2 = cost of planned replacement at time NT with $c_2 < c_1$.

Suppose that the failure rate $h(t)$ is continuous and strictly increasing with $h(\infty) \equiv \lim_{t\to\infty} h(t)$. We seek an optimum number N^* that minimizes $C_1(N)$. Forming the inequality $C_1(N+1) \geq C_1(N)$, we have

$$\frac{F((N+1)T) - F(NT)}{\int_{NT}^{(N+1)T} \overline{F}(t)\,dt} \int_0^{NT} \overline{F}(t)\,dt - F(NT) \geq \frac{c_2}{c_1 - c_2}$$

$$(N = 1, 2, \dots). \qquad (9.2)$$

From the assumption that the failure rate $h(t)$ is strictly increasing,

$$h((N+1)T) > \frac{F((N+1)T) - F(NT)}{\int_{NT}^{(N+1)T} \overline{F}(t)\,dt} > h(NT) > \frac{F(NT) - F((N-1)T)}{\int_{(N-1)T}^{NT} \overline{F}(t)\,dt}.$$

Thus, denoting the left-hand side of (9.2) by $L_1(N)$,

$$L_1(N) - L_1(N-1) =$$

$$\int_0^{NT} \overline{F}(t)\,dt \left[\frac{F((N+1)T) - F(NT)}{\int_{NT}^{(N+1)T} \overline{F}(t)\,dt} - \frac{F(NT) - F((N-1)T)}{\int_{(N-1)T}^{NT} \overline{F}(t)\,dt} \right] > 0$$

$$\lim_{N \to \infty} L_1(N) = \mu h(\infty) - 1.$$

Therefore, the optimum policy is as follows.

(i) If $h(\infty) > c_1/[(c_1 - c_2)\mu]$ then there exists a finite and unique minimum N^* that satisfies (9.2).
(ii) If $h(\infty) \leq c_1/[(c_1 - c_2)\mu]$ then $N^* = \infty$; i.e., a unit is replaced only at failure and $C_1(\infty) = c_1/\mu$.

Example 9.1. Suppose that $F(t)$ is a gamma distribution; i.e., its density function is $f(t) = [\lambda(\lambda t)^\alpha / \Gamma(\alpha)]e^{-\lambda t}$ for $\alpha > 1$ whose failure rate $h(t)$ is strictly increasing from 0 to λ. Then, Table 9.1 presents the optimum time T^* that minimizes the expected cost rate $C(T)$ in (3.4) of age replacement in Chapter 3, and the resulting cost rate $C(T^*)$, and the optimum number N^* and $C_1(N^*)$ for $\alpha = 2, 3, 4$, $T = 8, 48, 192, 2304$ when $c_1 = 10$, $c_2 = 1$, $1/\lambda = 10^3, 10^4$. It can be easily seen that N^* and $C_1(N^*)$ are approximately equal to T^*/T and $C(T^*)$, respectively, when T is small. For example, a unit works for 8 hours a day for 6 days, and is idle on Sunday. Then, when $1/\lambda = 10^3$ hours and $\alpha = 3$, a unit should be replaced at 20 weeks, i.e., 5 months, if it has not failed. ∎

Table 9.1. Comparisons with optimum time T^*, expected cost rate $C(T^*)/\lambda$, and optimum number N^*, expected cost rate $C_1(N^*)/\lambda$ when $c_1 = 10, c_2 = 1$

		$1/\lambda = 10^3$				
	$\alpha = 2$		$\alpha = 3$		$\alpha = 4$	
	T^*	$C(T^*)/\lambda$	T^*	$C(T^*)/\lambda$	T^*	$C(T^*)/\lambda$
	680.13	3.643	983.18	1.764	1400.7	1.074
T	N^*	$C_1(N^*)/\lambda$	N^*	$C_1(N^*)/\lambda$	N^*	$C_1(N^*)/\lambda$
8	85	3.643	123	1.764	175	1.074
48	14	3.643	20	1.764	29	1.074
192	4	3.657	5	1.764	7	1.075
2304	1	4.478	1	2.352	1	1.292

		$1/\lambda = 10^4$				
	$\alpha = 2$		$\alpha = 3$		$\alpha = 4$	
	T^*	$C(T^*)/\lambda$	T^*	$C(T^*)/\lambda$	T^*	$C(T^*)/\lambda$
	6801.3	3.643	9831.8	1.764	14007	1.074
T	N^*	$C_1(N^*)/\lambda$	N^*	$C_1(N^*)/\lambda$	N^*	$C_1(N^*)/\lambda$
8	850	3.643	1229	1.764	1751	1.074
48	142	3.643	205	1.764	292	1.074
192	35	3.643	51	1.764	73	1.074
2304	3	3.644	4	1.768	6	1.074

(2) Periodic Replacement

A unit is replaced at time NT and undergoes only minimal repair at failures between replacements; namely, its failure rate remains undisturbed by minimal repair. It is assumed that the repair and replacement times are negligible. The other assumptions are the same ones as age replacement.

Let $H(t)$ be a cumulative hazard function of a unit; i.e., $H(t) \equiv \int_0^t h(u)du$. Then, from (4.16) in Chapter 4, the expected cost rate is

$$C_2(N) = \frac{1}{NT}[c_1 H(NT) + c_2] \quad (N = 1, 2, \dots), \qquad (9.3)$$

where c_1 = cost of minimal repair at failure, and c_2 = cost of planned replacement at time NT.

Suppose that $h(t)$ is continuous and strictly increasing. Then, from the inequality $C_2(N+1) \geq C_2(N)$,

$$NH((N+1)T) - (N+1)H(NT) \geq \frac{c_2}{c_1} \quad (N = 1, 2, \dots). \qquad (9.4)$$

Denoting the left-hand side of (9.4) by $L_2(N)$ and $L_2(0) \equiv 0$,

$$L_2(N) - L_2(N-1) = N\int_0^T [h(t+NT) - h(t+(N-1)T)]\,dt > 0$$

$$L_2(N) > T[h(NT) - h(T)].$$

Thus, $L_2(N)$ is also strictly increasing and
$$\lim_{N \to \infty} L_2(N) \geq T[h(\infty) - h(T)].$$

If $h(t)$ is strictly increasing to infinity then there exists a finite and unique minimum $N^* (1 \leq N^* < \infty)$ that satisfies (9.4).

For example, when $F(t) = 1 - \exp[-(\lambda t)^m]$ and $H(t) = (\lambda t)^m$ for $m > 1$, an optimum N^* $(1 \leq N^* < \infty)$ is given by a unique minimum integer such that
$$N(N+1)^m - (N+1)N^m \geq \frac{c_2}{c_1(\lambda T)^m}.$$

(3) Block Replacement

A unit is replaced at time NT and at each failure. Failures of a unit are detected immediately when it fails. The other assumptions are the same ones as age replacement.

Let $M(t)$ be the renewal function of $F(t)$; i.e., $M(t) \equiv \sum_{j=1}^{\infty} F^{(j)}(t)$, where $F^{(j)}(t)$ is the j-fold Stieltjes convolution of $F(t)$. Then, from (5.1) in Chapter 5, the expected cost rate is
$$C_3(N) = \frac{1}{NT}[c_1 M(NT) + c_2] \quad (N = 1, 2, \dots), \tag{9.5}$$

where c_1 = cost of replacement at each failure, and c_2 = cost of planned replacement at time NT. From the inequality $C_3(N+1) \geq C_3(N)$,
$$NM((N+1)T) - (N+1)M(NT) \geq \frac{c_2}{c_1} \quad (N = 1, 2, \dots). \tag{9.6}$$

Suppose that a density function of $F(t)$ is $f(t) = \lambda^2 t e^{-\lambda t}$ and $M(t) = (\lambda t/2) - (1/4) + (1/4)e^{-2\lambda t}$. Denoting the left-hand side of (9.6) by $L_3(N)$,
$$L_3(N) = \frac{1}{4}[1 + Ne^{-2\lambda(N+1)T} - (N+1)e^{-2\lambda NT}], \quad \lim_{N \to \infty} L_3(N) = \frac{1}{4}$$
$$L_3(N) - L_3(N-1) = \frac{N}{4}e^{-2\lambda(N-1)T}(1 - e^{-2\lambda T})^2 > 0.$$

Therefore, the optimum policy is as follows:

(i) If $c_2/c_1 < 1/4$ then there exists a finite and unique minimum N^* $(1 \leq N^* < \infty)$ that satisfies
$$1 - (N+1)e^{-2\lambda NT} + Ne^{-2\lambda(N+1)T} \geq \frac{4c_2}{c_1}.$$

(ii) If $c_2/c_1 \geq 1/4$ then $N^* = \infty$; i.e., a unit is replaced only at failure.

Next, suppose that a unit is replaced only at time NT and remains failed until the next replacement time. Then, from (5.10) in Section 5.2,

$$C_4(N) = \frac{1}{NT}\left[c_1 \int_0^{NT} F(t)\,\mathrm{d}t + c_2\right] \qquad (N = 1, 2, \ldots), \qquad (9.7)$$

where $c_1 =$ downtime cost per unit of time for the time elapsed between a failure and its replacement and $c_2 =$ cost of planned replacement at time NT. From the inequality $C_4(N+1) - C_4(N) \geq 0$,

$$N\int_0^{(N+1)T} F(t)\,\mathrm{d}t - (N+1)\int_0^{NT} F(t)\,\mathrm{d}t \geq \frac{c_2}{c_1} \qquad (N = 1, 2, \ldots). \qquad (9.8)$$

Denoting the left-hand side of (9.8) by $L_4(N)$, it is evident that $L_4(N)$ is increasing and

$$L_4(N) \geq TF(NT) - \int_0^T F(t)\,\mathrm{d}t.$$

Thus, if $\int_0^T \overline{F}(t)\mathrm{d}t > c_2/c_1$ then there exists a finite and unique minimum N^* $(1 \leq N^* < \infty)$ that satisfies (9.8).

(4) Inspection

The inspection is planned at times kT $(k = 1, 2, \ldots)$ for a fixed $T > 0$ and a failed unit is detected only by inspection. Then, a unit is replaced at time NT or at failure detection, whichever occurs first. It is assumed that both inspection and replacement times are negligible. The other assumptions are the same ones as age replacement.

From (8.3) in Chapter 8, the expected cost rate is

$$C_5(N) = \frac{c_1\sum_{j=0}^{N-1}\int_{jT}^{(j+1)T}[(j+1)T - t]\,\mathrm{d}F(t) + c_2}{T\sum_{j=0}^{N-1}\overline{F}(jT)} \qquad (N = 1, 2, \ldots), \qquad (9.9)$$

where $c_1 =$ downtime cost per unit of time for the time elapsed between a failure and its detection, and $c_2 =$ cost of planned replacement at time NT.

Suppose that the failure rate $h(t)$ is continuous and strictly increasing. Then, from the inequality $C_5(N+1) \geq C_5(N)$,

$$\int_0^{NT} \overline{F}(t)\,\mathrm{d}t - \frac{\int_{NT}^{(N+1)T}\overline{F}(t)\,\mathrm{d}t}{\overline{F}(NT)}\sum_{j=0}^{N-1}\overline{F}(jT) \geq \frac{c_2}{c_1} \qquad (N = 1, 2, \ldots). \qquad (9.10)$$

Denoting the left-hand side of (9.10) by $L_5(N)$, we have

$$L_5(N) - L_5(N-1) =$$
$$\sum_{j=0}^{N-1} \overline{F}(jT) \int_0^T \left[\frac{F(t+NT) - F(NT)}{\overline{F}(NT)} - \frac{F(t+(N-1)T) - F((N-1)T)}{\overline{F}((N-1)T)} \right] dt > 0.$$

If $\lim_{N\to\infty} [F(t+NT) - F(NT)]/\overline{F}(NT) = 1$ for any $t > 0$ then $\lim_{N\to\infty} L_5(N) = \mu$. Therefore, the optimum policy is

(i) If $\mu > c_2/c_1$ then there exists a finite and unique minimum N^* ($1 \le N^* < \infty$) that satisfies (9.10).
(ii) If $\mu \le c_2/c_1$ then $N^* = \infty$.

In particular, the failure time is uniformly distributed on $(0, nT)$; i.e., $f(t) = 1/(nT)$ for $0 < t \le nT$. Then, the expected cost rate is

$$C_5(N) = \frac{c_1(NT/2n) + c_2}{NT[1 - (N-1)/2n]} \qquad (N = 1, 2, \ldots, n).$$

The optimum number N^* is given by a unique minimum such that

$$\frac{N(N+1)T}{n(n-N)} \ge \frac{4c_2}{c_1} \qquad (N = 1, 2, \ldots, n-1).$$

The optimum policy is as follows.

(i) If $(n-1)T \ge 4c_2/c_1$ then $1 \le N^* \le n-1$.
(ii) If $(n-1)T < 4c_2/c_1$ then $N^* = n$; i.e., a unit should be replaced after failure.

For example, when $n = 10$ and $T = 10$, $N^* = 1, 2, 4, 5, 7, 8, 9, 9, 10$, respectively, for $c_2/c_1 = 0.05, 0.1, 0.5, 1, 3, 5, 10, 20, 30$.

9.2 Maintenance Policies for a Finite Interval

It is important to consider practical maintenance policies for a finite interval, because the working times of most units are finite in the actual field. This section converts the standard replacement models to the models for a finite interval, and derives optimum policies for each model, using the partition method derived in Section 8.6. Very few papers treated with replacements for a finite interval. In this section, we have considered the inspection model for a finite working time and given the optimum policies, by partitioning the working time into equal parts.

This section proposes modified replacement policies that convert three standard models of periodic replacement in Chapter 4, block replacement, and no replacement at failure in Chapter 5 to replacement models for a finite interval. The optimum policies for three replacements are analytically derived,

242   9 Modified Maintenance Models

using the partition method. Furthermore, it is shown that all equations for the three replacements can be written on general forms.

A unit has to be operating for a finite interval $[0, S]$; i.e., its working time is given by a specified value S ($0 < S < \infty$). To maintain a unit, an interval S is partitioned equally into N parts in which it is replaced at periodic times kT ($k = 1, 2, \ldots, N$) as shown in Figure 8.5, where $NT = S$. Then, we consider the replacement with minimal repair, the block replacement, and no replacement at failure.

(1) Periodic Replacement

A unit is replaced at periodic times kT ($k = 1, 2, \ldots, N$) and any unit becomes as good as new at each replacement. When a unit fails between replacements, only minimal repair is made. It is assumed that the repair and replacement times are negligible.

Suppose that the failure times of each unit are independent, and have the failure rate $h(t)$ and the cumulative hazard function $H(t)$. Then, from (4.16) in Chapter 4, the expected cost of one interval $[0, T]$ is

$$\widetilde{C}_1(1) \equiv c_1 H(T) + c_2 = c_1 H\left(\frac{S}{N}\right) + c_2,$$

where $c_1 =$ cost of minimal repair at failure, and $c_2 =$ cost of planned replacement at time kT. Thus, the total expected cost until time S is

$$C_1(N) \equiv N\widetilde{C}_1(1) = N\left[c_1 H\left(\frac{S}{N}\right) + c_2\right] \qquad (N = 1, 2, \ldots). \qquad (9.11)$$

We find an optimum partition number N^* that minimizes $C_1(N)$ in (9.11). Evidently,

$$C_1(1) = c_1 H(S) + c_2, \qquad C_1(\infty) \equiv \lim_{N \to \infty} C_1(N) = \infty.$$

Thus, there exists a finite N^* ($1 \le N^* < \infty$) that minimizes $C_1(N)$. Forming the inequality $C_1(N+1) - C_1(N) \ge 0$, we have

$$\frac{1}{NH(\frac{S}{N}) - (N+1)H(\frac{S}{N+1})} \ge \frac{c_1}{c_2} \qquad (N = 1, 2, \ldots). \qquad (9.12)$$

When the failure time has a Weibull distribution, i.e., $H(t) = \lambda t^m$ ($m > 1$), Equation (9.12) becomes

$$\frac{1}{\frac{1}{N^{m-1}} - \frac{1}{(N+1)^{m-1}}} \ge \frac{\lambda c_1}{c_2} S^m \qquad (N = 1, 2, \ldots). \qquad (9.13)$$

Because it is easy to prove that $[1/x]^\alpha - [1/(x+1)]^\alpha$ is strictly decreasing in x for $1 \le x < \infty$ and $\alpha > 0$, the left-hand side of (9.13) is strictly increasing in

9.2 Maintenance Policies for a Finite Interval

N to ∞. Thus, there exists a finite and unique minimum N^* $(1 \le N^* < \infty)$ that satisfies (9.13).

To obtain simply an optimum N^* in another method, putting that $T = S/N$ in (9.11), we have

$$C_1(T) = S\left[\frac{c_1 H(T) + c_2}{T}\right]. \tag{9.14}$$

Thus, the problem of minimizing $C_1(T)$ corresponds to the problem of the standard replacement with minimal repair given in Section 4.2. Let $\widetilde{T}$ be a solution to (4.18) in Chapter 4. Then, using the partition method in Section 8.6, we have the following optimum policy.

(i) If $\widetilde{T} < S$ then we put $[S/\widetilde{T}] \equiv N$ and calculate $C_1(N)$ and $C_1(N+1)$ from (9.11). If $C_1(N) \le C_1(N+1)$ then $N^* = N$, and conversely, if $C_1(N) > C_1(N+1)$ then $N^* = N+1$.
(ii) If $\widetilde{T} \ge S$ then $N^* = 1$.

(2) Block Replacement

A unit is replaced at periodic times kT ($k = 1, 2, \ldots, N$) and is always replaced at any failure between replacements. This is called block replacement and was already discussed in Chapter 5.

Let $M(t)$ be the renewal function of $F(t)$; i.e., $M(t) \equiv \sum_{j=1}^{\infty} F^{(j)}(t)$. Then, from (5.1), the expected cost of one interval $(0, T]$ is

$$\widetilde{C}_2(1) \equiv c_1 M(T) + c_2 = c_1 M\left(\frac{S}{N}\right) + c_2,$$

where c_1 = cost of replacement at each failure, and c_2 = cost of planned replacement at time kT. Thus, the total expected cost until time S is

$$C_2(N) \equiv N\widetilde{C}_2(1) = N\left[c_1 M\left(\frac{S}{N}\right) + c_2\right] \quad (N = 1, 2, \ldots). \tag{9.15}$$

From the inequality $C_2(N+1) - C_2(N) \ge 0$,

$$\frac{1}{NM\left(\frac{S}{N}\right) - (N+1)M\left(\frac{S}{N+1}\right)} \ge \frac{c_1}{c_2} \quad (N = 1, 2, \ldots) \tag{9.16}$$

and putting that $T = S/N$ in (9.15),

$$C_2(T) = S\left[\frac{c_1 M(T) + c_2}{T}\right] \tag{9.17}$$

which corresponds to the standard block replacement in Section 5.1.

Therefore, by obtaining $\widetilde{T}$ which satisfies (5.2) and applying it to the optimum policy (i) or (ii), we can get an optimum replacement number N^* that minimizes $C_2(N)$.

(3) No Replacement at Failure

A unit is replaced only at times kT ($k = 1, 2, \ldots$) as described in Section 5.2. When the failure distribution $F(t)$ is given, the expected cost of one interval $(0, T]$ is, from (5.9),

$$\widetilde{C}_3(1) \equiv c_1 \int_0^T F(t)\,dt + c_2 = c_1 \int_0^{S/N} F(t)\,dt + c_2,$$

where $c_1 = $ downtime cost per unit of time for the time elapsed between a failure and its replacement. Thus, the total expected cost until time S is

$$C_3(N) \equiv N\widetilde{C}_3(1) = N\left[c_1 \int_0^{S/N} F(t)\,dt + c_2 \right] \qquad (N = 1, 2, \ldots). \qquad (9.18)$$

Because

$$C_3(1) = c_1 \int_0^S F(t)\,dt + c_2, \qquad C_3(\infty) \equiv \lim_{N \to \infty} C_3(N) = \infty$$

there exists a finite N^* ($1 \leq N^* < \infty$) that minimizes $C_3(N)$. Forming the inequality $C_3(N+1) - C_3(N)$ implies

$$\frac{1}{N \int_0^{S/N} F(t)\,dt - (N+1) \int_0^{S/(N+1)} F(t)\,dt} \geq \frac{c_1}{c_2} \qquad (N = 1, 2, \ldots). \qquad (9.19)$$

Putting $T = S/N$ in (9.18),

$$C_3(T) = S\left[\frac{c_1 \int_0^T F(t)\,dt + c_2}{T} \right]. \qquad (9.20)$$

Therefore, by obtaining $\widetilde{T}$ which satisfies (5.11) and applying it to the optimum policy, we can get an optimum replacement number N^* that minimizes $C_3(N)$.

In general, the above results of three replacements are summarized as follows: The total expected cost until time S is

$$C(N) = N\left[c_1 \Phi\!\left(\frac{S}{N}\right) + c_2 \right] \qquad (N = 1, 2, \ldots), \qquad (9.21)$$

where $\Phi(t)$ is $H(t)$, $M(t)$, and $\int_0^t F(u)\,du$ for the respective models. Forming the inequality $C(N+1) - C(N) \geq 0$ yields

$$\frac{1}{N\Phi(\frac{S}{N}) - (N+1)\Phi(\frac{S}{N+1})} \geq \frac{c_1}{c_2} \qquad (N = 1, 2, \ldots). \qquad (9.22)$$

Putting $T = S/N$,

$$C(T) = S\left[\frac{c_1\Phi(T) + c_2}{T}\right] \qquad (9.23)$$

and differentiating $C(T)$ with respect to T and setting it equal to zero,

$$T\Phi'(T) - \Phi(T) = \frac{c_2}{c_1}. \qquad (9.24)$$

If there exists a solution $\widetilde{T}$ to (9.24) then we can get an optimum number N^* for each replacement, using the optimum partition method.

9.3 Random Maintenance Policies

Most systems in offices and industry successively execute jobs and computer processes. For such systems, it would be impossible or impractical to maintain them in a strictly periodic fashion. For example, when a job has a variable working cycle and processing time, it would be better to do some maintenance after it has completed its work and process. The reliability quantities of the random age replacement policy were obtained analytically [9], using a renewal theory. Furthermore, when a unit is replaced only at random times, the properties of replacement times between two successive failed units were investigated in [10]. The various schedules of jobs that have random processing times were summarized in [11].

This section proposes random replacement policies in which a unit is replaced at the same random times as its working times. However, it would be necessary to replace a working unit at planned times in the case where its working time becomes large. Thus, we suggest the extended models of age replacement, periodic replacement, and block replacement in Chapters 3, 4, and 5: a unit is replaced at either planned time T or at a random time that is statistically distributed according to a general distribution $G(x)$. Then, the expected cost rates of each model are obtained and optimum replacement times that minimize them are analytically derived by similar methods to those of Chapters 3, 4, and 5. Also, we consider the random inspection policy in which a unit is checked at the same random times as its working times. At first, we obtain the total expected cost of a unit with random checking times until failure detection. Next, we consider the extended inspection model where a unit is checked at both random and periodic times. Then, the total expected cost is derived, and optimum inspection policies that minimize it are analytically derived. Of course, we may consider the replacement as preventive maintenance (PM) in Chapter 6, where a unit becomes like new after PM. The replacement models with the Nth random times and the inspection model with random and successive checking times are introduced. Finally, numerical examples are given.

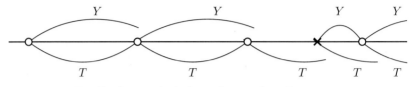

Fig. 9.1. Process of random and age replacement

9.3.1 Random Replacement

Suppose that the failure time X of each unit is independent and has an identical distribution $F(t)$ with finite mean μ and the failure rate $h(t)$, where, in general, $\overline{\Phi} \equiv 1 - \Phi$. A unit is replaced at planned time T or at random time Y which has a general distribution $G(x)$ and is independent of X. Then, we consider the random and periodic policies of age replacement, periodic replacement and block replacement, and obtain the expected cost rates of each model. Furthermore, we derive optimum replacement policies that minimize these cost rates.

(1) Age Replacement

A unit is replaced at time T, Y, or at failure, whichever occurs first, where T $(0 < T \leq \infty)$ is constant and Y is a random variable with distribution $G(x)$ in Figure 9.1.

The probability that a unit is replaced at time T is

$$\Pr\{T < X, T < Y\} = \overline{F}(T)\overline{G}(T), \qquad (9.25)$$

the probability that it is replaced at random time Y is

$$\Pr\{Y \leq T, Y \leq X\} = \int_0^T \overline{F}(t)\,dG(t), \qquad (9.26)$$

and the probability that it is replaced at failure is

$$\Pr\{X \leq T, X \leq Y\} = \int_0^T \overline{G}(t)\,dF(t). \qquad (9.27)$$

Note that the summation of (9.25), (9.26), and (9.27) is equal to 1. Thus, the mean time to replacement is

$$T\overline{G}(T)\overline{F}(T) + \int_0^T t\,\overline{F}(t)\,dG(t) + \int_0^T t\,\overline{G}(t)\,dF(t) = \int_0^T \overline{G}(t)\overline{F}(t)\,dt.$$

9.3 Random Maintenance Policies

From (3.3) in Chapter 3, the expected cost rate is

$$C_1(T) = \frac{(c_1 - c_2) \int_0^T \overline{G}(t)\,\mathrm{d}F(t) + c_2}{\int_0^T \overline{G}(t)\overline{F}(t)\,\mathrm{d}t}, \qquad (9.28)$$

where $c_1 =$ cost of replacement at failure, and $c_2 =$ cost of replacement at a planned or random time with $c_2 < c_1$. When $\overline{G}(x) \equiv 1$ for any $x \geq 0$, $C_1(T)$ agrees with the expected cost rate in (3.4), and when $T = \infty$, this represents only the random age replacement [9, p. 94; 12], whose cost rate is given by

$$C_1(\infty) = \frac{(c_1 - c_2) \int_0^\infty \overline{G}(t)\,\mathrm{d}F(t) + c_2}{\int_0^\infty \overline{G}(t)\overline{F}(t)\,\mathrm{d}t}. \qquad (9.29)$$

In addition, the mean time that a unit is replaced at failure for the first time is given by a renewal function

$$l(T) = \int_0^T t\,\overline{G}(t)\,\mathrm{d}F(t) + [T + l(T)]\overline{G}(T)\overline{F}(T) + \int_0^T [t + l(T)]\overline{F}(t)\,\mathrm{d}G(t);$$

i.e.,

$$l(T) = \frac{\int_0^T \overline{G}(t)\overline{F}(t)\,\mathrm{d}t}{\int_0^T \overline{G}(t)\,\mathrm{d}F(t)}$$

which agrees with (1.6) of Chapter 1 when $\overline{G}(x) \equiv 1$ for any $x \geq 0$.

Suppose that the failure rate $h(t)$ is continuous and strictly increasing with $h(\infty) \equiv \lim_{t \to \infty} h(t)$. Then, we seek an optimum T^* that minimizes $C_1(T)$ in (9.28). It is first noted that there exists an optimum T^* $(0 < T^* \leq \infty)$ because $\lim_{T \to 0} C_1(T) = \infty$. Differentiating $C_1(T)$ with respect to T and putting it equal to zero, we have

$$h(T) \int_0^T \overline{G}(t)\overline{F}(t)\,\mathrm{d}t - \int_0^T \overline{G}(t)\,\mathrm{d}F(t) = \frac{c_2}{c_1 - c_2}. \qquad (9.30)$$

Letting $Q_1(T)$ be the left-hand side of (9.30), we see that $\lim_{T \to 0} Q_1(T) = 0$,

$$Q_1(\infty) \equiv \lim_{T \to \infty} Q_1(T) = h(\infty) \int_0^\infty \overline{G}(t)\overline{F}(t)\,\mathrm{d}t - \int_0^\infty \overline{G}(t)\,\mathrm{d}F(t)$$

and for any $\Delta T > 0$,

$$Q_1(T+\Delta T) - Q_1(T) = h(T+\Delta T) \int_0^{T+\Delta T} \overline{G}(t)\overline{F}(t)\,\mathrm{d}t - \int_0^{T+\Delta T} \overline{G}(t)\,\mathrm{d}F(t)$$

$$- h(T) \int_0^T \overline{G}(t)\overline{F}(t)\,\mathrm{d}t + \int_0^T \overline{G}(t)\,\mathrm{d}F(t)$$

$$\geq h(T+\Delta T)\int_0^{T+\Delta T}\overline{G}(t)\overline{F}(t)\,\mathrm{d}t - h(T+\Delta T)$$

$$\times \int_T^{T+\Delta T}\overline{G}(t)\overline{F}(t)\,\mathrm{d}t - h(T)\int_0^T\overline{G}(t)\overline{F}(t)\,\mathrm{d}t$$

$$= [h(T+\Delta T) - h(T)]\int_0^T\overline{G}(t)\overline{F}(t)\,\mathrm{d}t > 0$$

because $h(T+\Delta T)\geq \int_T^{T+\Delta T}\overline{G}(t)\,\mathrm{d}F(t)/\int_T^{T+\Delta T}\overline{G}(t)\overline{F}(t)\,\mathrm{d}t$. Thus, $Q_1(T)$ is strictly increasing from 0 to $Q_1(\infty)$. Therefore, if $Q_1(\infty) > c_2/(c_1 - c_2)$ then there exists an optimum T_1^* $(0 < T_1^* < \infty)$ that satisfies (9.30), and its resulting cost rate is

$$C_1(T^*) = (c_1 - c_2)h(T^*). \tag{9.31}$$

Conversely, if $Q_1(\infty) \leq c_2/(c_1 - c_2)$ then $T^* = \infty$, and the expected cost rate is given in (9.29).

In particular, when $G(x) = 1 - \mathrm{e}^{-\theta x}$, the expected cost rates in (9.28) and (9.29) are, respectively,

$$C_1(T) = \frac{(c_1 - c_2)\int_0^T \mathrm{e}^{-\theta t}\,\mathrm{d}F(t) + c_2}{\int_0^T \mathrm{e}^{-\theta t}\overline{F}(t)\,\mathrm{d}t} \tag{9.32}$$

$$C_1(\infty) = \frac{(c_1 - c_2)F^*(\theta) + c_2}{[1 - F^*(\theta)]/\theta}, \tag{9.33}$$

where $F^*(\theta)$ is the Laplace–Stieltjes transform of $F(t)$; i.e., $F^*(\theta) \equiv \int_0^\infty \mathrm{e}^{-\theta t}\,\mathrm{d}F(t)$ for $\theta > 0$. Furthermore, Equation (9.30) can be rewritten as

$$h(T)\int_0^T \mathrm{e}^{-\theta t}\overline{F}(t)\,\mathrm{d}t - \int_0^T \mathrm{e}^{-\theta t}\,\mathrm{d}F(t) = \frac{c_2}{c_1 - c_2} \tag{9.34}$$

and

$$Q_1(\infty) = h(\infty)\frac{1 - F^*(\theta)}{\theta} - F^*(\theta).$$

Therefore, if $h(\infty)/\theta > [c_1/(c_1 - c_2)]/[1 - F^*(\theta)] - 1$ then there exists a finite and unique T^* $(0 < T^* < \infty)$ that satisfies (9.34), and it minimizes $C_1(T)$. It is easy to see that if θ increases then T^* increases, and tends to ∞ as $\theta \to \infty$, because the left-hand side of (9.34) is a decreasing function of θ. That is, the smaller the mean random time is, the larger the planned replacement time is.

Finally, suppose that the replacement cost at planned time T is different from that at a random time. In this case,

$$C_1(T) = \frac{(c_1 - c_2)\int_0^T \overline{G}(t)\,\mathrm{d}F(t) + (c_3 - c_2)\int_0^T \overline{F}(t)\,\mathrm{d}G(t) + c_2}{\int_0^T \overline{G}(t)\overline{F}(t)\,\mathrm{d}t}, \tag{9.35}$$

Table 9.2. Optimum replacement time T^* when $1/\lambda = 100$ and $c_1 = 5$, $c_2 = 1$

$1/\theta$	T^*		
	$m=1$	$m=2$	$m=3$
1	∞	13.704	3.177
5	∞	6.148	2.476
10	∞	5.584	2.403
20	∞	5.335	2.367
50	∞	5.196	2.347
∞	∞	5.107	2.333

where $c_1 =$ cost of replacement at failure, $c_2 =$ cost of replacement at planned time, and $c_3 =$ cost of replacement at a random time.

We seek an optimum T_1^* that minimizes $C_1(T)$ in (9.35). Differentiating $C_1(T)$ with respect to T and setting it equal to zero,

$$(c_1 - c_2)\left[h(T)\int_0^T \overline{G}(t)\overline{F}(t)\,\mathrm{d}t - \int_0^T \overline{G}(t)\,\mathrm{d}F(t)\right]$$
$$+ (c_3 - c_2)\left[r(T)\int_0^T \overline{G}(t)\overline{F}(t)\,\mathrm{d}t - \int_0^T \overline{F}(t)\,\mathrm{d}G(t)\right] = c_2 \quad (9.36)$$

and the minimum expected cost rate is

$$C_1(T_1^*) = (c_1 - c_2)h(T_1^*) + (c_3 - c_2)r(T_1^*), \quad (9.37)$$

where $r(t) \equiv g(t)/\overline{G}(t)$ and $g(t)$ is a density function of $G(t)$. It can be easily seen from (9.36) that when the random time is exponential, i.e., $G(x) = 1 - \mathrm{e}^{-\theta x}$, T_1^* is equal to T^* given in (9.30). Furthermore, when $r(t)$ is increasing, if $c_2 \geq c_3$ then $T_1^* \geq T^*$ and vice versa. This justifies a natural conclusion that if the periodic replacement cost is higher than the random one, then the planned replacement should be done later than the optimum T^*.

Example 9.2. Suppose that the failure time has a Weibull distribution and the random replacement is exponential; i.e., $F(t) = 1 - \exp(-\lambda t^m)$ and $G(x) = 1 - \mathrm{e}^{-\theta x}$. Table 9.2 shows the optimum replacement time T^* for $m = 1, 2, 3$ and $1/\theta = 1, 5, 10, 20, 50, \infty$ when $1/\lambda = 100$, $c_1 = 5$, and $c_2 = 1$. This indicates that the optimum times are decreasing with parameters $1/\theta$ and m. However, if the mean time $1/\theta$ exceeds some level, they do not vary remarkably for given m. Thus, it would be useful to replace a system at least at the smallest time T^* for large $1/\theta$. In particular, when $m = 1$, i.e., the failure time is exponential, T^* is infinity for any $1/\theta$. ∎

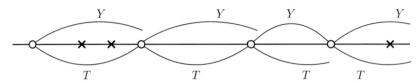

Fig. 9.2. Process of periodic replacement

(2) Periodic Replacement

A unit is replaced at planned time T or at random time Y, whichever occurs first, and undergoes only minimal repair at failures between replacements as described in Chapter 4.

Let $H(t)$ be the cumulative hazard function of a unit; i.e., $H(t) \equiv \int_0^t h(u)\,du$. By a similar method to that of Section 4.2, the expected cost until replacement is

$$\int_0^T [c_1 H(t) + c_2]\,dG(t) + [c_1 H(T) + c_2]\overline{G}(T) = c_1 \int_0^T \overline{G}(t)\,dH(t) + c_2$$

and the mean time to replacement is

$$\int_0^T t\,dG(t) + T\overline{G}(T) = \int_0^T \overline{G}(t)\,dt.$$

Thus, the expected cost rate is

$$C_2(T) = \frac{c_1 \int_0^T \overline{G}(t)\,dH(t) + c_2}{\int_0^T \overline{G}(t)\,dt}, \qquad (9.38)$$

where $c_1 =$ cost of minimal repair at failure, and $c_2 =$ cost of replacement at a planned or random time. When $\overline{G}(x) \equiv 1$ for any $x \geq 0$, $C_2(T)$ agrees with the expected cost rate given in (4.16) in Chapter 4.

Suppose that $h(t)$ is continuous and strictly increasing. Then, differentiating $C_2(T)$ with respect to T and setting it equal to zero,

$$h(T) \int_0^T \overline{G}(t)\,dt - \int_0^T \overline{G}(t)\,dH(t) = \frac{c_2}{c_1}. \qquad (9.39)$$

Letting $Q_2(T)$ be the left-hand side of (9.39), we have $\lim_{T \to 0} Q_2(T) = 0$,

$$Q_2(\infty) \equiv \lim_{T \to \infty} Q_2(T) = h(\infty) \int_0^\infty \overline{G}(t)\,dt - \int_0^\infty \overline{G}(t)\,dH(t)$$

9.3 Random Maintenance Policies

and for any $\Delta T > 0$,

$$Q_2(T + \Delta T) - Q_2(T)$$
$$= h(T + \Delta T) \int_0^{T+\Delta T} \overline{G}(t)\,\mathrm{d}t - h(T) \int_0^T \overline{G}(t)\,\mathrm{d}t - \int_T^{T+\Delta T} \overline{G}(t)\,\mathrm{d}H(t)$$
$$\geq [h(T + \Delta T) - h(T)] \int_T^{T+\Delta T} \overline{G}(t)\,\mathrm{d}t + \int_T^{T+\Delta T} [h(T) - h(t)]\overline{G}(t)\,\mathrm{d}t > 0.$$

Therefore, if $Q_2(\infty) > c_2/c_1$ then there exists an optimum T^* ($0 < T^* < \infty$) that satisfies (9.39), and its resulting cost rate is

$$C_2(T^*) = c_1 h(T^*). \tag{9.40}$$

If the replacement costs at planned time T and at a random time are different from each other, then the expected cost rate is

$$C_2(T) = \frac{c_1 \int_0^T \overline{G}(t)\,\mathrm{d}H(t) + c_2 \overline{G}(T) + c_3 G(T)}{\int_0^T \overline{G}(t)\,\mathrm{d}t}, \tag{9.41}$$

where c_2 and c_3 are given in (9.35). We seek an optimum T_2^* that minimizes $C_2(T)$ in (9.41). Differentiating $C_2(T)$ with respect to T and setting it equal to zero,

$$c_1 \left[h(T) \int_0^T \overline{G}(t)\,\mathrm{d}t - \int_0^T \overline{G}(t)\,\mathrm{d}H(t) \right] + (c_3 - c_2)\left[r(T) \int_0^T \overline{G}(t)\,\mathrm{d}t - G(T) \right] = c_2 \tag{9.42}$$

and the resulting cost rate is

$$C_2(T_2^*) = c_1 h(T_2^*) + (c_3 - c_2) r(T_2^*), \tag{9.43}$$

where $r(t)$ is given in (9.36) and (9.37). From these equations, we have $T_2^* = T^*$ in (9.39) when $G(x) = 1 - e^{-\theta x}$. Also, when $r(t)$ is increasing, if $c_2 \geq c_3$ then $T_2^* \geq T^*$ and *vice versa*. In particular, when the failure time is exponential, i.e., $H(t) = t/\mu$, Equation (9.42) takes the same form as (3.9) in Chapter 3. In this case, if $c_2 \geq c_3$ then $T_2^* = \infty$.

(3) Block Replacement

A unit is replaced at planned time T or at a random time and also at each failure. Let $M(t)$ be the renewal function of $F(t)$; i.e., $M(t) \equiv \sum_{j=1}^{\infty} F^{(j)}(t)$. Then, by a similar method to that of Section 5.1, the expected cost until replacement is

$$\int_0^T [c_1 M(t) + c_2]\,\mathrm{d}G(t) + [c_1 M(T) + c_2]\overline{G}(T) = c_1 \int_0^T \overline{G}(t)\,\mathrm{d}M(t) + c_2$$

and the mean time to replacement is given by $\int_0^T \overline{G}(t)\,\mathrm{d}t$ which is the same as the periodic replacement. Thus, the expected cost rate is

$$C_3(T) = \frac{c_1 \int_0^T \overline{G}(t)\,\mathrm{d}M(t) + c_2}{\int_0^T \overline{G}(t)\,\mathrm{d}t}, \qquad (9.44)$$

where c_1 = cost of replacement at each failure, and c_2 = cost of replacement at a planned or random time.

If the replacement costs at planned time T and at a random time are different from each other, then the expected cost rate is

$$C_3(T) = \frac{c_1 \int_0^T \overline{G}(t)\,\mathrm{d}M(t) + c_2 \overline{G}(T) + c_3 G(T)}{\int_0^T \overline{G}(t)\,\mathrm{d}t}, \qquad (9.45)$$

where c_2 and c_3 are given in (9.35).

Next, if a unit is not replaced at failure, and hence, it remains failed for the time interval from a failure to its replacement as described in Section 5.2, then, because the expected cost until replacement is

$$\int_0^T \left[c_1 \int_0^x (x-t)\,\mathrm{d}F(t) + c_2 \right] \mathrm{d}G(x) + \overline{G}(T) \left[c_1 \int_0^T (T-t)\,\mathrm{d}F(t) + c_2 \right]$$

$$= c_1 \int_0^T \overline{G}(t) F(t)\,\mathrm{d}t + c_2$$

the expected cost rate is

$$C_4(T) = \frac{c_1 \int_0^T \overline{G}(t) F(t)\,\mathrm{d}t + c_2}{\int_0^T \overline{G}(t)\,\mathrm{d}t}, \qquad (9.46)$$

where c_1 = downtime cost per unit of time for the time elapsed between a failure and its replacement, and c_2 = cost of replacement at a planned or random time.

Furthermore, the mean time that a unit is replaced after failure for the first time is given by a renewal function

$$l(T) = T\overline{G}(T)F(T) + \int_0^T t F(t)\,\mathrm{d}G(t)$$

$$+ [T + l(T)]\overline{G}(T)\overline{F}(T) + \int_0^T [t + l(T)]\overline{F}(t)\,\mathrm{d}G(t);$$

i.e.,

$$l(T) = \frac{1}{\int_0^T \overline{G}(t)\,\mathrm{d}F(t)} \int_0^T \overline{G}(t)\,\mathrm{d}t$$

9.3 Random Maintenance Policies   253

which agrees with the result of [10] when $T = \infty$.

Differentiating $C_4(T)$ with respect to T and setting it equal to zero,

$$F(T) \int_0^T \overline{G}(t)\,\mathrm{d}t - \int_0^T \overline{G}(t) F(t)\,\mathrm{d}t = \frac{c_2}{c_1}. \tag{9.47}$$

It is easy to prove that the left-hand side of (9.47) is strictly increasing from 0 to $\int_0^\infty \overline{G}(t)\overline{F}(t)\,\mathrm{d}t$. Therefore, if $\int_0^\infty \overline{G}(t)\overline{F}(t)\,\mathrm{d}t > c_2/c_1$ then there exists a finite and unique T^* that satisfies (9.47), and its resulting cost rate is

$$C_4(T^*) = c_1 F(T^*). \tag{9.48}$$

In particular, when $\overline{G}(x) \equiv 1$ for any $x \geq 0$, the above results correspond to those of Section 5.2.

Until now, it has been assumed that a unit is replaced at one random time. Next, we suppose that a unit is replaced at either planned time T $(0 < T \leq \infty)$ or at the Nth random time $(N = 1, 2, \dots)$. Then, the expected cost rates of each model can be rewritten as

$$C_1(T, N) = \frac{(c_1 - c_2) \int_0^T [1 - G^{(N)}(t)]\,\mathrm{d}F(t) + c_2}{\int_0^T [1 - G^{(N)}(t)] \overline{F}(t)\,\mathrm{d}t} \tag{9.49}$$

$$C_2(T, N) = \frac{c_1 \int_0^T [1 - G^{(N)}(t)]\,\mathrm{d}H(t) + c_2}{\int_0^T [1 - G^{(N)}(t)]\,\mathrm{d}t} \tag{9.50}$$

$$C_3(T, N) = \frac{c_1 \int_0^T [1 - G^{(N)}(t)]\,\mathrm{d}M(t) + c_2}{\int_0^T [1 - G^{(N)}(t)]\,\mathrm{d}t} \tag{9.51}$$

$$C_4(T, N) = \frac{c_1 \int_0^T [1 - G^{(N)}(t)] F(t)\,\mathrm{d}t + c_2}{\int_0^T [1 - G^{(N)}(t)]\,\mathrm{d}t}. \tag{9.52}$$

9.3.2 Random Inspection

Suppose that a unit works for an infinite time span and is checked at successive times Y_j $(j = 1, 2, \dots)$, where $Y_0 \equiv 0$ and $Z_j \equiv Y_j - Y_{j-1}$ $(j = 1, 2, \dots)$ are independently and identically distributed random variables, and also, independent of its failure time. It is assumed that each Z_j has an identical distribution $G(x)$ with finite mean; i.e., $\{Z_j\}_{j=1}^\infty$ form a renewal process in Section 1.3, and the distribution of Y_j is represented by the j-fold convolution $G^{(j)}$ of G with itself.

Furthermore, a unit has a failure distribution $F(t)$ with finite mean μ, and its failure is detected only by some check. It is assumed that the failure rate of a unit is not changed by any check, and times needed for checks are negligible. A unit is checked at successive times Y_j $(j = 1, 2, \dots)$ and also at periodic times kT $(k = 1, 2, \dots)$ for a specified $T > 0$ (see Figure 9.3). The failure is detected by either random or periodic inspection, whichever occurs first.

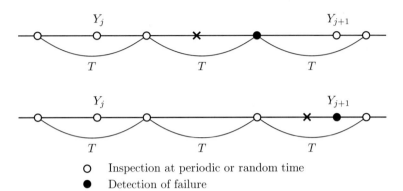

Fig. 9.3. Process of random and periodic inspections

The probability that the failure is detected by periodic check is

$$\sum_{k=0}^{\infty} \int_{kT}^{(k+1)T} \left[\sum_{j=0}^{\infty} \int_0^t \overline{G}[(k+1)T - x] \, dG^{(j)}(x) \right] dF(t) \qquad (9.53)$$

and the probability that it is detected by random check is

$$\sum_{k=0}^{\infty} \int_{kT}^{(k+1)T} \left[\sum_{j=0}^{\infty} \int_0^t \{G[(k+1)T - x] - G(t - x)\} \, dG^{(j)}(x) \right] dF(t), \qquad (9.54)$$

where note that the summation of (9.53) and (9.54) is equal to 1.

Let c_{pi} be the cost of the periodic check, c_{ri} be the cost of the random check and c_2 be the downtime cost per unit of time for the time elapsed between a failure and its detection at the next check. Then, the total expected cost until failure detection is

$$C(T) = \sum_{k=0}^{\infty} \int_{kT}^{(k+1)T} \left[\sum_{j=0}^{\infty} \{(k+1)c_{pi} + jc_{ri} + c_2[(k+1)T - t]\} \right] dF(t)$$

$$\times \int_0^t \overline{G}[(k+1)T - x] \, dG^{(j)}(x) + \sum_{k=0}^{\infty} \int_{kT}^{(k+1)T} dF(t)$$

$$\times \sum_{j=0}^{\infty} \int_0^t \left\{ \int_{t-x}^{(k+1)T - x} [kc_{pi} + (j+1)c_{ri} + c_2(x+y-t)] \, dG(y) \right\} dG^{(j)}(x)$$

9.3 Random Maintenance Policies

$$= c_{pi} \sum_{k=0}^{\infty} \overline{F}(kT) + c_{ri} \sum_{j=0}^{\infty} j \int_0^{\infty} [G^{(j)}(t) - G^{(j+1)}(t)] \, dF(t)$$

$$- (c_{pi} - c_{ri}) \sum_{k=0}^{\infty} \int_{kT}^{(k+1)T} \left\{ G[(k+1)T] - G(t) \right.$$

$$+ \left[\int_0^t \{G[(k+1)T - x] - G(t-x)\} \, dM(x) \right] \right\} dF(t)$$

$$+ c_2 \sum_{k=0}^{\infty} \int_{kT}^{(k+1)T} \left\{ \int_t^{(k+1)T} \overline{G}(y) \, dy + \left[\int_0^t \left(\int_{t-x}^{(k+1)T-x} \overline{G}(y) \, dy \right) dM(x) \right] \right\} dF(t), \tag{9.55}$$

where $M(x) \equiv \sum_{j=1}^{\infty} G^{(j)}(x)$ represents the expected number of checks during $(0, x]$.

We consider the following two particular cases.

(i) *Random inspection.* If $T = \infty$, i.e., a unit is checked only by random inspection, then the total expected cost is

$$\lim_{T \to \infty} C(T) = c_{ri} \sum_{j=0}^{\infty} (j+1) \int_0^{\infty} [G^{(j)}(t) - G^{(j+1)}(t)] \, dF(t)$$

$$+ c_2 \left\{ \int_0^{\infty} F(t) \overline{G}(t) \, dt + \int_0^{\infty} \left[\int_0^{\infty} [F(x+t) - F(x)] \overline{G}(t) \, dt \right] dM(x) \right\}. \tag{9.56}$$

(ii) *Periodic and random inspections.* When $G(x) = 1 - e^{-\theta x}$, the total expected cost $C(T)$ in (9.55) can be rewritten as

$$C(T) = c_{pi} \sum_{k=0}^{\infty} \overline{F}(kT) + c_{ri} \theta \mu - \left(c_{pi} - c_{ri} - \frac{c_2}{\theta} \right)$$

$$\times \sum_{k=0}^{\infty} \int_{kT}^{(k+1)T} \left\{ 1 - e^{-\theta[(k+1)T-t]} \right\} dF(t). \tag{9.57}$$

We find an optimum checking time T^* that minimizes $C(T)$. Differentiating $C(T)$ with respect to T and setting it equal to zero,

$$\frac{\sum_{k=0}^{\infty}(k+1)\int_{kT}^{(k+1)T} \theta e^{-\theta[(k+1)T-t]} \, dF(t)}{\sum_{k=0}^{\infty} kf(kT)} - (1 - e^{-\theta T}) = \frac{c_{pi}}{c_{ri} - c_{pi} + c_2/\theta} \tag{9.58}$$

for $c_{ri} + c_2/\theta > c_{pi}$. This is a necessary condition that an optimum T^* minimizes $C(T)$.

In particular, when $F(t) = 1 - e^{-\lambda t}$ for $\lambda < \theta$, the expected cost $C(T)$ in (9.57) becomes

$$C(T) = \frac{c_{pi}}{1-e^{-\lambda T}} + c_{ri}\frac{\theta}{\lambda} - \left(c_{pi} - c_{ri} - \frac{c_2}{\theta}\right)\left[1 - \frac{\lambda}{\theta - \lambda}\frac{e^{-\lambda T} - e^{-\theta T}}{1 - e^{-\lambda T}}\right]. \tag{9.59}$$

Clearly, we have $\lim_{T \to 0} C(T) = \infty$,

$$C(\infty) \equiv \lim_{T \to \infty} C(T) = c_{ri}\left(\frac{\theta}{\lambda} + 1\right) + \frac{c_2}{\theta}. \tag{9.60}$$

Equation (9.58) can be simplified as

$$\frac{\theta}{\theta - \lambda}[1 - e^{-(\theta - \lambda)T}] - (1 - e^{-\theta T}) = \frac{c_{pi}}{c_{ri} - c_{pi} + c_2/\theta} \tag{9.61}$$

whose left-hand side is strictly increasing from 0 to $\lambda/(\theta - \lambda)$.

Therefore, if $\lambda/(\theta - \lambda) > c_{pi}/(c_{ri} - c_{pi} + c_2/\theta)$, i.e., $c_{ri} + c_2/\theta > (\theta/\lambda)c_{pi}$, then there exists a finite and unique T^* ($0 < T^* < \infty$) that satisfies (9.61), and it minimizes $C(T)$. The physical meaning of the condition $c_{ri} + c_2/\theta > [(1/\lambda)/(1/\theta)]c_{pi}$ is that the total of the checking cost and the downtime cost of the mean interval between random checks is greater than the periodic cost for the expected number of random checks until failure detection. Conversely, if $c_{ri} + c_2/\theta \le (\theta/\lambda)c_{pi}$ then periodic inspection is not needed.

Furthermore, using the approximation of $e^{-at} \approx 1 - at + (at)^2/2$ for small $a > 0$, we have, from (9.61),

$$\tilde{T} = \sqrt{\frac{2}{\lambda\theta}\frac{c_{pi}}{c_{ri} - c_{pi} + c_2/\theta}} \tag{9.62}$$

which gives the approximate time of optimum T^*.

Example 9.3. Suppose that the failure time has a Weibull distribution and the random inspection is exponential; i.e., $F(t) = 1 - \exp(-\lambda t^m)$ and $G(x) = 1 - e^{-\theta x}$. Then, from (9.58), an optimum checking time T^* satisfies

$$\frac{\sum_{k=0}^{\infty}(k+1)\int_{kT}^{(k+1)T} \theta e^{-\theta[(k+1)T-t]}\lambda m t^{m-1}e^{-\lambda t^m}\,dt}{\sum_{k=0}^{\infty} k\lambda m(kT)^{m-1}e^{-\lambda(kT)^m}} - (1 - e^{-\theta T})$$

$$= \frac{c_{pi}}{c_{ri} - c_{pi} + c_2/\theta}. \tag{9.63}$$

In particular, when $m = 1$, i.e., the failure time is exponential, Equation (9.63) is identical to (9.61). Also, when $1/\theta$ tends to infinity, Equation (9.63) reduces to

$$\frac{\sum_{k=0}^{\infty} e^{-\lambda(kT)^m}}{\sum_{k=0}^{\infty} k\lambda m(kT)^{m-1}e^{-\lambda(kT)^m}} - T = \frac{c_{pi}}{c_2} \tag{9.64}$$

which corresponds to the periodic inspection with Weibull failure time in Section 8.1.

Table 9.3. Optimum checking time T^* when $1/\lambda = 100$ and $c_{pi}/c_2 = 2$, $c_{ri}/c_2 = 1$

$1/\theta$	$\widetilde{T}$	T^* $m=1$	$m=2$	$m=3$
1	∞	∞	∞	∞
5	22.361	∞	12.264	6.187
10	21.082	∞	8.081	5.969
20	20.520	32.240	6.819	5.861
50	20.203	22.568	6.266	5.794
∞	20.000	19.355	5.954	5.748

Table 9.4. Value of $\widehat{T} = 1/\widehat{\theta}$ in Equation (9.63)

$1/\widehat{\theta}$		
$m=1$	$m=2$	$m=3$
26.889	11.712	6.687

Table 9.3 shows the optimum checking time T^* for $m = 1, 2, 3$ and $1/\theta = 1, 5, 10, 20, 50, \infty$, and approximate time $\widetilde{T}$ in (9.62) when $1/\lambda = 100$, $c_{pi}/c_2 = 2$, and $c_{ri}/c_2 = 1$. This indicates that the optimum times are decreasing with parameters $1/\theta$ and m. However, if the mean time $1/\theta$ exceeds some level, they do not vary remarkably for given m. Thus, it would be useful to check a unit at least at the smallest time T^* for large $1/\theta$, which satisfies (9.58). Approximate times $\widetilde{T}$ give a good approximation for large $1/\theta$ when $m = 1$.

Furthermore, it is noticed from Table 9.3 that values of T^* are larger than $1/\theta$ for some $\widehat{\theta} < \theta$, and *vice versa*. Hence, there would exist numerically a unique $\widehat{T}$ that satisfies $T = 1/\theta$ in (9.63), and it is given by a solution of the following equation:

$$\left\{ \left[\left(\frac{c_{ri}}{c_2} - \frac{c_{pi}}{c_2} \right) \frac{1}{T} + 1 \right] \times \frac{\sum_{k=0}^{\infty}(k+1) \int_{kT}^{(k+1)T} e^{-[(k+1)-t/T]} \lambda m t^{m-1} e^{-\lambda t^m} \, dt}{\sum_{k=0}^{\infty} k \lambda m (kT)^{m-1} e^{-\lambda (kT)^m}} \right\} - (1 - e^{-1}) = \frac{c_{pi}}{c_2}. \tag{9.65}$$

The values of $\widehat{T} = 1/\widehat{\theta}$ for $m = 1, 2, 3$ are shown in Table 9.4 when $c_{pi}/c_2 = 2$ and $c_{ri}/c_2 = 1$. If the mean working time $1/\theta$ is previously estimated and is smaller than $1/\widehat{\theta}$, then we may check a unit at a larger interval than $1/\widehat{\theta}$, and *vice versa*. ∎

Until now, we have considered the random inspection policy and discussed the optimum checking time that minimizes the expected cost. If a working unit is checked at successive times T_k ($k = 1, 2, \ldots$), where $T_0 \equiv 0$ and at random times, the expected cost in (9.55) can be easily rewritten as

$$C(T_1, T_2, \ldots) = c_{pi} \sum_{k=0}^{\infty} \overline{F}(T_k) + c_{ri} \sum_{j=0}^{\infty} j \int_0^{\infty} [G^{(j)}(t) - G^{(j+1)}(t)] \, dF(t)$$

$$- (c_{pi} - c_{ri}) \sum_{k=0}^{\infty} \int_{T_k}^{T_{k+1}} \left\{ G(T_{k+1}) - G(t) \right.$$

$$\left. + \left[\int_0^t [G(T_{k+1} - x) - G(t - x)] \, dM(x) \right] \right\} dF(t)$$

$$+ c_2 \sum_{k=0}^{\infty} \int_{T_k}^{T_{k+1}} \left\{ \int_t^{T_{k+1}} \overline{G}(y) \, dy + \left[\int_0^t \left(\int_{t-x}^{T_{k+1}-x} \overline{G}(y) \, dy \right) dM(x) \right] \right\} dF(t). \tag{9.66}$$

In particular, when $G(x) = 1 - e^{-\theta x}$,

$$C(T_1, T_2, \ldots) = c_{pi} \sum_{k=0}^{\infty} \overline{F}(T_k) + c_{ri} \theta \mu$$

$$- \left(c_{pi} - c_{ri} - \frac{c_2}{\theta} \right) \sum_{k=0}^{\infty} \int_{T_k}^{T_{k+1}} [1 - e^{-\theta(T_{k+1} - t)}] \, dF(t). \tag{9.67}$$

9.4 Replacement Maximizing MTTF

System reliability can be improved by providing spare units. When failures of units during actual operation are costly or dangerous, it is important to know when to replace or to do preventive maintenance before failure.

This section suggests the following replacement policy for a system with n spares: If a unit fails then it is replaced immediately with one of the spares. Furthermore, to prevent failures in operation, a unit may be replaced before failure at time T_k when there are k spares ($k = 1, 2, \ldots, n$). The mean time to failure (MTTF) is obtained and the optimum replacement time T_k^* that maximizes it is derived. It is of interest that T_k^* is decreasing in k; i.e., a unit should be replaced earlier as many times as the system has spares, and MTTF is approximately given by $1/h(T_k^*)$, where $h(t)$ is the failure rate of each unit.

A unit begins to operate at time 0 and there are n spares, which are statistically independent and have the same function as the operating unit. Suppose that each unit has an identical distribution $F(t)$ with finite mean μ and the failure rate $h(t)$, where $\overline{F} \equiv 1 - F$. An operating unit with k spares ($k = 1, 2, \ldots, n$) is replaced at failure or at time T_k from its installation, whichever occurs first. When there is no spare, the last unit has to operate until failure.

When there are unlimited spares and each unit is replaced at failure or at periodic time T, from Example 1.2 in Chapter 1,

$$MTTF = \frac{1}{F(T)} \int_0^T \overline{F}(t)\,dt. \qquad (9.68)$$

Similarly, when there is only one spare, MTTF is

$$M_1(T_1) = \int_0^{T_1} \overline{F}(t)\,dt + \overline{F}(T_1)\mu \qquad (9.69)$$

and when there are k spares,

$$M_k(T_1, T_2, \ldots, T_k) = \int_0^{T_k} \overline{F}(t)\,dt + \overline{F}(T_k) M_{k-1}(T_1, T_2, \ldots, T_{k-1})$$

$$(k = 2, 3, \ldots, n). \qquad (9.70)$$

It is trivial that M_k is increasing in k because $M_k(T_1, T_2, \ldots, T_{k-1}, 0) = M_{k-1}(T_1, T_2, \ldots, T_{k-1})$.

When the failure rate $h(t)$ is continuous and strictly increasing, we seek an optimum replacement time T_k^* that maximizes $M_k(T_1, T_2, \ldots, T_k)$ by induction.

When $n = 1$, *i.e.*, there is one spare, we have, from (9.69),

$$M_1(\infty) = M_1(0) = \mu$$

$$\frac{dM_1(T_1)}{dT_1} = \overline{F}(T_1)[1 - \mu h(T_1)].$$

Because $h(t)$ is strictly increasing and $h(0) < 1/\mu < h(\infty)$, in Example 1.2 of Section 1.1, there exists a finite and unique T_1^* that satisfies $h(T_1) = 1/\mu$.

Next, suppose that T_1^*, T_2^*, ..., and T_{k-1}^* are already determined. Then, differentiating $M_k(T_1^*, \ldots, T_{k-1}^*, T_k)$ in (9.70) with respect to T_k implies

$$\frac{dM_k(T_1^*, \ldots, T_{k-1}^*, T_k)}{dT_k} = \overline{F}(T_k)[1 - h(T_k) M_{k-1}(T_1^*, \ldots, T_{k-1}^*)]. \qquad (9.71)$$

First, we prove the inequalities $h(0) < 1/M_{k-1}(T_1^*, \ldots, T_{k-1}^*) \leq 1/\mu < h(\infty)$. Because $1/\mu < h(\infty)$, we need to show only the inequalities $h(0) < 1/M_{k-1}(T_1^*, \ldots, T_{k-1}^*) \leq 1/\mu$. Also, because M_k is increasing in k from (9.70),

$$M_{k-1}(T_1^*, \ldots, T_{k-1}^*) \geq M_1(T_1^*) \geq M_1(\infty) = M_1(0) = \mu.$$

Moreover, we prove that $M_{k-1}(T_1^*, \ldots, T_{k-1}^*) < 1/h(0)$ for $h(0) > 0$ by induction. It is trivial that $h(0) < 1/M_{k-1}(T_1^*, \ldots, T_{k-1}^*)$ when $h(0) = 0$. From the assumption that $h(t)$ is strictly increasing, we have

$$M_1(T_1^*) = \int_0^{T_1^*} \overline{F}(t)\,dt + \frac{\overline{F}(T_1^*)}{h(T_1^*)}$$

$$< \int_0^{T_1^*} \overline{F}(t)\,dt + \frac{\overline{F}(T_1^*)}{h(0)} < \frac{1}{h(0)}.$$

Suppose that $M_{k-2}(T_1^*, \ldots, T_{k-2}^*) < 1/h(0)$. From (9.70),

$$M_{k-1}(T_1^*, \ldots, T_{k-1}^*) = \int_0^{T_{k-1}^*} \overline{F}(t)\,\mathrm{d}t + \overline{F}(T_{k-1}^*) M_{k-2}(T_1^*, \ldots, T_{k-2}^*)$$

$$< \int_0^{T_{k-1}^*} \overline{F}(t)\,\mathrm{d}t + \frac{\overline{F}(T_{k-1}^*)}{h(0)} < \frac{1}{h(0)}$$

which completes the proof that $h(0) < 1/M_{k-1}(T_1^*, \ldots, T_{k-1}^*) < 1/h(\infty)$.

Using the above results, there exists a finite and unique T_k^* that satisfies $\mathrm{d}M_k/\mathrm{d}T_k = 0$ in (9.71), i.e.,

$$h(T_k) = \frac{1}{M_{k-1}(T_1^*, \ldots, T_{k-1}^*)} \qquad (k = 2, 3, \ldots, n), \qquad (9.72)$$

and the resulting maximum MTTF is

$$M_k(T_k^*) = \int_0^{T_k^*} \overline{F}(t)\,\mathrm{d}t + \frac{\overline{F}(T_k^*)}{h(T_k^*)} \qquad (k = 1, 2, \ldots, n). \qquad (9.73)$$

Note that optimum T_k^* is decreasing in k.

Furthermore, when $h(t)$ is strictly increasing, it can be easily proved that for any $T > 0$,

$$\int_0^T \overline{F}(t)\,\mathrm{d}t + \frac{\overline{F}(T)}{h(T)} > \frac{1}{h(T)}$$

$$\int_0^T \overline{F}(t)\,\mathrm{d}t + \frac{\overline{F}(T)}{h(T)} < \int_0^T \overline{F}(t)\,\mathrm{d}t + \frac{\overline{F}(T)}{F(T)} \int_0^T \overline{F}(t)\,\mathrm{d}t = \frac{1}{F(T)} \int_0^T \overline{F}(t)\,\mathrm{d}t$$

which is given in (9.68), and hence,

$$\frac{1}{h(T_k^*)} < M_k(T_k^*) < \frac{1}{F(T_k^*)} \int_0^{T_k^*} \overline{F}(t)\,\mathrm{d}t. \qquad (9.74)$$

From the above discussions, we can specify the computing procedure for obtaining the optimum replacement schedule:

(i) Solve $h(T_1^*) = 1/\mu$ and compute $M_1(T_1^*) = \int_0^{T_1^*} \overline{F}(t)\,\mathrm{d}t + \mu \overline{F}(T_1^*)$.
(ii) Solve $h(T_k^*) = 1/M_{k-1}(T_{k-1}^*)$ and compute $M_k(T_k^*) = \int_0^{T_k^*} \overline{F}(t)\,\mathrm{d}t + \overline{F}(T_k^*)/h(T_k^*)$ $(k = 2, 3, \ldots, n)$.
(iii) Continue until $k = n$.

Example 9.4. Suppose that $F(t) = 1 - \exp(-t^2)$. Table 9.5 shows the optimum replacement time T_n^*, MTTF $M_n(T_n^*)$, the lower bound $1/h(T_n^*)$ for n $(1 \le n \le 15)$ spares, and MTTF $\int_0^{T_n^*} \overline{F}(t)\,\mathrm{d}t/F(T_n^*)$ for unlimited spares.

Table 9.5. Optimum T_n^*, lower bound $1/h(T_n^*)$, and MTTF $M_n(T_n^*)$ for n spares, and MTTF $\int_0^{T_n^*} \overline{F}(t)dt/F(T_n^*)$ for unlimited spares

n	T_n^*	$1/h(T_n^*)$	$M_n(T_n^*)$	$\int_0^{T_n^*} \overline{F}(t)dt/F(T_n^*)$
1	0.564	0.886	1.154	1.869
2	0.433	1.154	1.364	2.382
3	0.367	1.364	1.543	2.790
4	0.324	1.543	1.702	3.141
5	0.294	1.702	1.847	3.454
6	0.271	1.847	1.981	3.740
7	0.252	1.981	2.106	4.004
8	0.237	2.106	2.223	4.252
9	0.225	2.223	2.334	4.484
10	0.214	2.334	2.440	4.704
11	0.205	2.440	2.542	4.915
12	0.197	2.542	2.640	5.117
13	0.189	2.640	2.734	5.312
14	0.183	2.734	2.825	5.498
15	0.177	2.825	2.913	5.679

For example, when $n = 5$, a unit should be replaced before failure at intervals 0.294, 0.324, 0.367, 0.433, 0.564, and MTTF is 1.847 and is twice as long as the mean $\mu = 1/h(T_1^*) = 0.886$ of each unit. It is of interest that the lower bound $1/h(T_n^*)$ equals $M_{n-1}(T_{n-1}^*)$ and is a fairly good approximation of MTTF, and $\int_0^{T_n^*} \overline{F}(t)dt/F(T_n^*)$ is about twice as long as the lower bound $1/h(T_n^*)$. ∎

9.5 Discrete Replacement Maximizing MTTF

Consider the modified discrete age replacement policy for an operating unit with n spares where the replacement is planned only at times kT ($k = 1, 2, \ldots$) for a specified T defined in Section 9.1: An operating unit with n spares is replaced at time $N_n T$ for constant $T > 0$. By a similar method to that of Section 9.4, when there is one spare,

$$M_1(N_1) = \int_0^{N_1 T} \overline{F}(t)\,dt + \overline{F}(N_1 T)\mu \qquad (9.75)$$

and when there are k spares,

$$M_k(N_1, N_2, \ldots, N_k) = \int_0^{N_k T} \overline{F}(t)\,dt + \overline{F}(N_k T) M_{k-1}(N_1, N_2, \ldots, N_{k-1})$$
$$(k = 2, 3, \ldots, n) \qquad (9.76)$$

which is increasing in k because $M_k(N_1, \ldots, N_{k-1}, 0) = M_{k-1}(N_1, \ldots, N_{k-1})$.

When the failure rate $h(t)$ is strictly increasing, we seek an optimum number N_k^* that maximizes $M_k(N_1, N_2, \ldots, N_k)$ by induction. When $n = 1$, we have that $M_1(\infty) = M_1(0) = \mu$ from (9.75).

The inequality $M_1(N_1) \geq M_1(N_1 + 1)$ implies

$$\frac{F((N_1+1)T) - F(N_1 T)}{\int_{N_1 T}^{(N_1+1)T} \overline{F}(t)\, dt} \geq \frac{1}{\mu}. \tag{9.77}$$

Because $h(t)$ is strictly increasing, we have

$$h((N+1)T) > \frac{F((N+1)T) - F(NT)}{\int_{NT}^{(N+1)T} \overline{F}(t)\, dt} > h(NT) > \frac{F(NT) - F((N-1)T)}{\int_{(N-1)T}^{NT} \overline{F}(t)\, dt}$$

$$\frac{F(T)}{\int_0^T \overline{F}(t)\, dt} < \frac{1}{\mu} < h(\infty).$$

Therefore, the left-hand side of (9.77) is strictly increasing in N_1 from $F(T)/\int_0^T \overline{F}(t)dt$ to $h(\infty)$, and hence, N_1^* $(1 \leq N_1^* < \infty)$ is given by a unique minimum that satisfies (9.77).

Next, suppose that $N_1^*, N_2^*, \ldots,$ and N_{k-1}^* are determined. Then, the inequality $M_k(N_1^*, \ldots, N_{k-1}^*, N_k) \geq M_k(N_1^*, \ldots, N_{k-1}^*, N_k + 1)$ implies

$$\frac{F((N_k+1)T) - F(N_k T)}{\int_{N_k T}^{(N_k+1)T} \overline{F}(t)\, dt} \geq \frac{1}{M_{k-1}(N_1^*, \ldots, N_{k-1}^*)}. \tag{9.78}$$

Because M_{k-1} is increasing in k and $1/M_{k-1}(N_1^*, \ldots, N_{k-1}^*) \leq 1/\mu < h(\infty)$, a finite and unique minimum that satisfies (9.78) exists, and is decreasing in k.

Therefore, we can specify the computing procedure as follows.

(i) Obtain a minimum N_1^* such that

$$\frac{F((N_1+1)T) - F(N_1 T)}{\int_{N_1 T}^{(N_1+1)T} \overline{F}(t)\, dt} \geq \frac{1}{\mu}$$

and compute $M_1(N_1^*)$ in (9.75).
(ii) Obtain a minimum N_k^* that satisfies (9.78), and compute $M_k(N_1^*, \ldots, N_k^*)$ in (9.76).
(iii) Continue until $k = n$.

Example 9.5. Suppose that the failure time of each unit has a gamma distribution with order 2; i.e., $F(t) = 1 - (1+t)e^{-t}$ and $\mu = 2$. Table 9.6 gives the optimum replacement time T_k^*, MTTF $M_k(T_k^*)$ derived in Section 9.4, and number N_k^*, MTTF $M_k(N_k^*)$ $(k = 1, 2, \ldots, 10)$ for $T = 0.1$. MTTF $M_k(T_k^*)$ are a little longer than $M_k(N_k^*)$. When $k = 9$, both MTTFs are twice as long as μ. Conversely speaking, we should provide 9 spares to assure that MTTF is twice as long as that of the unit. ∎

Table 9.6. Optimum time T_k^*, MTTF $M_k(T_k^*)$, and number N_k^*, MTTF $M_k(N_k^*)$ for $T = 0.1$

k	T_k^*	$M_k(T_k^*)$	N_k^*	$M_k(N_k^*)$
1	1.000	2.368	10	2.368
2	0.731	2.659	7	2.658
3	0.603	2.908	6	2.907
4	0.524	3.129	5	3.129
5	0.470	3.331	5	3.330
6	0.429	3.518	4	3.517
7	0.397	3.693	4	3.691
8	0.371	3.857	4	3.855
9	0.350	4.014	4	4.009
10	0.332	4.163	3	4.157

9.6 Other Maintenance Policies

Units are assumed to have only two possible states: operating or failed. However, some units such as power systems and plants may deteriorate with time and be in one of multiple states that can be observed through planned inspections. This is called a *Markovian deteriorating system*. The maintenance policies for such systems have been studied by many authors [13–15]. Using these results, the inspection policies for a multistage production system were discussed in [16, 17], and the reliability of systems with multistate units was summarized in [18]. Furthermore, multipleunits may fail simultaneously due to a single underlying cause. This is called *common-cause failure*. An extensive reference list of such failures that are classified into four categories was provided in [19]. Most products are sold with a *warranty* that offers protection to buyers against early failures over the warranty period. The literature that links and deals with warranty and maintenance was reviewed in [20, 21].

The notions of maintenance, techniques, and methods discussed in this book could spread to other fields. Fundamental reliability theory has already been widely applied to *fault-tolerant design and techniques* [22–24]. Some viewpoints from inspection policies have been applied to recovery techniques and checkpoint generations of computer systems [25–28]. Recently, various schemes of *self-checking* and *self-testing* [29, 30] for digital systems, and *fault diagnosis* [31] for control systems, which are one modification of inspection policies, have been proposed. Furthermore, data transmission schemes in a communication system were discussed in [32], using the technique of Markov renewal processes. Analytical tools of risk analysis such as *risk-based inspection* and *risk-based maintenance* have been rapidly developed and applied generally to the maintenance of big plants [33]. After this, maintenance with due regard to risk evaluation would be a main policy for large-scale and complex systems [34, 35].

This book might be difficult for those learning reliability for the first time. We recommend three recently published books [36–38] for such readers.

References

1. Nakagawa T (1987) Modified, discrete replacement models. IEEE Trans R-36:243–245.
2. Nakagawa S, Okuda Y, Yamada S (2003) Optimal checking interval for task duplication with spare processing. In: Ninth ISSAT International Conference on Reliability and Quality in Design:215–219.
3. Mizutani S, Teramoto K, Nakagawa T (2004) A survey of finite inspection models. In: Tenth ISSAT International Conference on Reliability and Quality in Design:104–108.
4. Sugiura T, Mizutani S, Nakagawa T (2003) Optimal random and periodic inspection policies. In: Ninth ISSAT International Conference on Reliability and Quality in Design:42–45.
5. Sugiura T, Mizutani S, Nakagawa T (2004) Optimal random replacement policies. In: Tenth ISSAT International Conference on Reliability and Quality in Design:99–103.
6. Nakagawa T (1989) A replacement policy maximizing MTTF of a system with several spare units. IEEE Trans Reliab 38:210–211.
7. Nakagawa T, Goel AL, Osaki S (1975) Stochastic behavior of an intermittently used system. RAIRO Oper Res 2:101–112.
8. Mine H, Kawai H, Fukushima Y (1981) Preventive replacement of an intermittently-used system. IEEE Trans Reliab R-30:391–392.
9. Barlow RE, Proschan F (1965) Mathematical Theory of Reliability. J Wiley & Sons, New York.
10. Stadje W (2003) Renewal analysis of a replacement process. Oper Res Letters 31:1–6.
11. Pinedo M (2002) Scheduling Theory, Algorithms, and Systems. Prentice-Hall, Upper Saddle River, NJ.
12. Gertsbakh I (2000) Reliability Theory with Applications to Preventive Maintenance. Springer, New York.
13. Yeh RH (1996) Optimal inspection and replacement policies for multi-state deterioration systems. Eur J Oper Res 96:248–259.
14. Stadje W, Zuckerman D (1996) A generalized maintenance model for stochastically deteriorating equipment. Eur J Oper Res 89:285–301.
15. Kawai H, Koyanagi J, Ohnishi M (2002) Optimal maintenance problems for Markovian deteriorating systems. In: Osaki S (ed) Stochastic Models in Reliability and Maintenance. Springer, New York:193–218.
16. Hurst EG (1973) Imperfect inspection in multistage production process. Manage Sci 20:378–384.
17. Gupta A, Gupta H (1981) Optimal inspection policy for multistage production process with alternate inspection plans. IEEE Trans Reliab R-30:161–162.
18. Lisnianski A, Levitin G (2003) Multi-State Reliability. World Scientific, Singapore.
19. Dhillon BS, Anude OC (1994) Common-cause failures in engineering systems: A review. Inter J Reliab Qual Saf Eng 1:103–129.
20. Blischke WR, Murthy DNP (1996) Product Warranty Handbook. Marcel Dekker, New York.
21. Murthy DNP, Jack N (2003) Warranty and maintenance. In: Pham H (ed) Handbook of Reliability Engineering. Springer, London:305–316.

22. Trivedi K (1982) Probability and Statistics with Reliability, Queueing and Computer Science Applications. Prentice-Hall, Englewood Cliffs, NJ.
23. Lala PK (1985) Fault Tolerant and Fault Testable Hardware Design. Prentice-Hall, London.
24. Gelenbe E (2000) System Performance Evaluation. CRC, Boca Raton FL.
25. Reuter A (1984) Performance analysis of recovery techniques. ACM Trans Database Syst 9:526–559.
26. Fukumoto S, Kaio N, Osaki S (1992) A study of checkpoint generations for a database recovery mechanism. Comput Math Appl 24:63–70.
27. Vaidya N (1998) A case for two-level recovery schemes. IEEE Trans Comput 47:656–666.
28. Nakagawa S, Fukumoto S, Ishii N (2003) Optimal checkpointing intervals of three error detection schemes by a double modular redundancy. Math Comput Model 38:1357–1363.
29. Lala PK (2001) Self-Checking and Fault-Tolerant Digital Design. Academic, San Francisco.
30. O'Connor PDT (ed) (2001) Test Engineering. J Wiley & Sons, Chichester England.
31. Korbicz J, Kościelny JM, Kowalczuk Z, Cholewa W (eds) (2004) Fault Diagnosis. Springer, New York.
32. Yasui K, Nakagawa T, Sandoh H (2002) Reliability models in data communication systems. In: Osaki S (ed) Stochastic Models in Reliability and Maintenance. Springer, New York:281–306.
33. Modarres M, Martz M, Kaminskiy (1996) The accident sequence precursor analysis: Review of the methods and new insights. Nuclear Sci Eng 123:238–258.
34. Aven T (1992) Reliability and Risk Analysis. Elsevier Applied Science, London.
35. Bari RA (2003) Probabilistic risk assessment. In: Pham H (ed) Handbook of Reliability Engineering. Springer, London:543–557.
36. Dhillon BS (2002) Engineering Maintenance. CRC, Boca Raton FL.
37. O'Connor PDT (2002) Practical Reliability Engineering. J Wiley & Sons, Chichester England.
38. Rausand M, Høyland A (2004) System Reliability Theory. J Wiley & Sons, Hoboken NJ.

Index

age replacement 2, 69–92, 117, 118,125, 127–131, 136, 224, 235–237, 245–249
aging 5
allowed time 25, 26, 46, 47
alternating renewal process 19, 24–26, 34, 40, 135
availability 2–5, 9–11, 39, 47–51, 70, 102, 135, 136, 139, 145, 150–154, 171, 172, 188–192, 201, 204

bathtub curve 6
binomial distribution 13
block replacement 2, 70, 117–132, 235, 236, 239, 241–243, 246, 251–253

calendar time 3
catastrophic failure 3, 69
characteristic life 6
common-cause failure 263
corrective maintenance, replacement 2, 39, 69, 135
Cox's proportional hazard model 5
cumulative hazard function 6, 23, 75, 76, 96–104, 217–219, 238, 239, 242, 243, 250, 251
cumulative process 23
current age 22, 23

decreasing failure rate (DFR) 6–9, 13
degenerate distribution 12, 137, 147, 212, 213
degraded failure 3, 69
delay time 202

discounting 70, 78–80, 107, 108, 119, 120, 125, 126
discrete distribution 9, 13, 14
discrete time 3, 13, 16, 70, 76, 80–92, 95, 107, 108
downtime 11, 24, 25, 39, 45–47, 120, 122, 135, 201, 240, 254

earning 8, 55, 139
Erlang distribution 15
excess time 45
expected cost 3, 39, 51–56, 59–62, 69–92, 101–114, 117–132, 152, 157–160, 166, 167, 171–183, 187, 192–196, 201–229, 236–258
expected number of failures 2, 6, 39, 40, 45, 46, 56, 58, 59, 64, 102, 104, 118, 135, 136, 156, 157
exponential distribution 6–8, 12–17, 22, 43, 46, 49, 50, 54, 62, 63, 85, 90, 92, 140, 153, 203, 212, 214, 217, 218, 221, 222, 225, 248, 249, 251, 255, 256
extreme distribution 13, 15–18

failure rate 4–9, 14–17, 23, 42, 60–62, 70, 73–75, 79–91, 96, 98–103, 107–114, 126–132, 141–144, 150–153, 176–180, 183–184, 193–196, 202, 209, 215, 236–240, 242, 246–262
fault 110, 160–164, 202, 220–223
finite interval, time 4, 9, 69, 224–228, 241–245

267

Index

first-passage time 20, 27, 29–34, 39, 56, 57, 64, 65, 148, 149

gamma distribution 13, 15, 44, 62, 76, 80, 124, 143, 153, 175, 215, 237, 239, 262
geometric distribution 13, 14, 17, 88, 181

hazard rate 5–7
hidden fault 188, 201, 202
human error 172, 187

imperfect maintenance 2, 135, 171–197
imperfect repair 39, 172
increasing failure rate (IFR) 6–9, 13
inspection 3, 4, 171, 172, 183–187, 201–229, 235, 236, 240, 241, 245, 253–258
inspection intensity 201, 207–210, 224, 227–229
intensity function 23, 155–167
intermittent failure, fault 3, 155, 172, 202, 220–224
intermittently used system 3, 236
interval reliability 11, 48–50, 135, 140–144

job scheduling 11–13

k-out-of-n system 66, 83, 190

log normal distribution 14

mass function 28–34, 41, 146–149
Markov chain 19, 20, 26–28
Markov process 19, 20, 26–34
Markov renewal process 19, 26, 28–34, 39–42, 136, 146
Markovian deteriorating 263
mean time to failure (MTTF) 2, 3, 5, 8, 9, 18, 39, 40, 48, 56, 58, 63–66, 69, 92, 111, 135, 144, 145, 148, 149–154, 171, 172, 183, 186, 188, 191, 235, 258–263
mean time to repair (MTTR) 40, 48
mean value function 6, 23, 98, 155–166
minimal repair 23, 75, 95–110, 126–132, 156–160, 172, 175–182, 192, 238, 239, 242, 243, 250

negative binomial distribution 13, 14, 17, 81, 92
nonhomogeneous Poisson process 6, 23, 98, 155–166
normal distribution 12, 14, 46, 51

one-unit system 19, 24, 31, 39–55, 135–144, 176, 183, 192
opportunistic replacement 70, 135, 145

parallel system 2, 17, 32, 39, 65, 66, 70, 76, 82, 83, 136, 145, 166
partition method 4, 202, 225, 235, 241–244
percentile point 72, 75, 76
periodic replacement 2, 95–114, 117, 125-131, 235, 236, 238, 239, 241–243, 246, 250, 251
Poisson distribution 13, 14, 23
Poisson process 15, 23, 156
preventive maintenance 2, 4, 8, 11, 31, 51, 56, 60, 62, 95, 135–167, 171–197, 202, 205, 245
preventive replacement 2, 171
protective unit 202

random replacement 3, 235, 245–253
regeneration point 30–34, 136, 146–149
reliability function 5, 12, 217
renewal density 21, 118–120, 123
renewal function 20–22, 29–34, 40–44, 58, 118–123, 135–138, 239, 243, 251, 252
renewal process 19–24, 28, 71, 83, 123, 253
renewal reward 23, 24
repair limit 2, 39, 40, 51–55, 135
repair rate 42, 53, 54
repairman problem 39
residual lifetime 9, 20, 22, 23, 98, 121
reversed hazard rate 6

semi-Markov process 19, 26, 28–30, 39
sequential maintenance 191–197
series system 9, 135
shock 17, 18, 23, 136, 166
spare unit, part 2, 3, 8, 9, 24, 39, 56–63, 117, 135, 235, 236, 258–263

standby unit, system   2, 3, 24, 39, 55–65, 144–154, 182, 201, 202, 212–216
stochastic process   4, 19–34, 39, 45
storage unit, system   3, 113, 202, 216–220

transition probability   20, 26–34, 39–44, 63–66, 135–139, 149, 150, 221
two types of failure   96, 110–112
two types of units   96, 112–114
two-unit system   31, 34, 39, 117, 135, 144–154

uniform distribution   12, 208, 241
uptime   11, 48
used unit   74, 95, 107–109, 121

warranty policy   263
wearout failure   107, 109
Weibull distribution   6, 13, 15–18, 54, 70, 76, 92, 103, 107, 111, 181, 182, 185, 192, 194–196, 207, 210, 211, 217, 219, 220, 227, 228, 242, 249, 256, 260